T0245255

CAMBRIDGE LIBRARY COLLECTION

Books of enduring scholarly value

Mathematical Sciences

From its pre-historic roots in simple counting to the algorithms powering modern desktop computers, from the genius of Archimedes to the genius of Einstein, advances in mathematical understanding and numerical techniques have been directly responsible for creating the modern world as we know it. This series will provide a library of the most influential publications and writers on mathematics in its broadest sense. As such, it will show not only the deep roots from which modern science and technology have grown, but also the astonishing breadth of application of mathematical techniques in the humanities and social sciences, and in everyday life.

Principles of Geometry

Henry Frederick Baker (1866–1956) was a renowned British mathematician specialising in algebraic geometry. He was elected a Fellow of the Royal Society in 1898 and appointed the Lowndean Professor of Astronomy and Geometry in the University of Cambridge in 1914. First published between 1922 and 1925, the six-volume *Principles of Geometry* was a synthesis of Baker's lecture series on geometry and was the first British work on geometry to use axiomatic methods without the use of co-ordinates. The first four volumes describe the projective geometry of space of between two and five dimensions, with the last two volumes reflecting Baker's later research interests in the birational theory of surfaces. The work as a whole provides a detailed insight into the geometry which was developing at the time of publication. This, the fifth volume, describes the birational geometry of curves.

Cambridge University Press has long been a pioneer in the reissuing of out-of-print titles from its own backlist, producing digital reprints of books that are still sought after by scholars and students but could not be reprinted economically using traditional technology. The Cambridge Library Collection extends this activity to a wider range of books which are still of importance to researchers and professionals, either for the source material they contain, or as landmarks in the history of their academic discipline.

Drawing from the world-renowned collections in the Cambridge University Library, and guided by the advice of experts in each subject area, Cambridge University Press is using state-of-the-art scanning machines in its own Printing House to capture the content of each book selected for inclusion. The files are processed to give a consistently clear, crisp image, and the books finished to the high quality standard for which the Press is recognised around the world. The latest print-on-demand technology ensures that the books will remain available indefinitely, and that orders for single or multiple copies can quickly be supplied.

The Cambridge Library Collection will bring back to life books of enduring scholarly value (including out-of-copyright works originally issued by other publishers) across a wide range of disciplines in the humanities and social sciences and in science and technology.

Principles
of Geometry

VOLUME 5:
ANALYTICAL PRINCIPLES OF
THE THEORY OF CURVES

H.F. BAKER

CAMBRIDGE UNIVERSITY PRESS

Cambridge, New York, Melbourne, Madrid, Cape Town, Singapore,
São Paolo, Delhi, Dubai, Tokyo, Mexico City

Published in the United States of America by Cambridge University Press, New York

www.cambridge.org
Information on this title: www.cambridge.org/9781108017817

© in this compilation Cambridge University Press 2010

This edition first published 1933
This digitally printed version 2010

ISBN 978-1-108-01781-7 Paperback

This book reproduces the text of the original edition. The content and language reflect
the beliefs, practices and terminology of their time, and have not been updated.

Cambridge University Press wishes to make clear that the book, unless originally published
by Cambridge, is not being republished by, in association or collaboration with, or
with the endorsement or approval of, the original publisher or its successors in title.

PRINCIPLES OF GEOMETRY

LONDON
Cambridge University Press
FETTER LANE

NEW YORK · TORONTO
BOMBAY · CALCUTTA · MADRAS
Macmillan

TOKYO
Maruzen Company Ltd

All rights reserved

PRINCIPLES OF GEOMETRY

BY

H. F. BAKER, Sc.D., LL.D., F.R.S.,

LOWNDEAN PROFESSOR, AND FELLOW OF ST JOHN'S COLLEGE,
IN THE UNIVERSITY

VOLUME V

ANALYTICAL PRINCIPLES OF
THE THEORY OF CURVES

CAMBRIDGE

AT THE UNIVERSITY PRESS

1933

PRINTED IN GREAT BRITAIN

PREFACE

THE present volume is an account of the analytic principles of the theory of curves, of the rational functions belonging thereto and of the integrals of these functions, with a brief account of the methods, by loops and by Riemann surfaces, for dealing with the periods of these integrals. But the theory of correspondence, and some necessary references to involutions in a plane, find themselves in the succeeding volume, which is mainly devoted to the theory of surfaces.

It is perhaps desirable to explain the origin of these volumes. In the last fifty years a remarkable advance has been made in the theory of surfaces, and of algebraic loci in general; the English reader may find a description of the nature of this in a Presidential Address to the London Mathematical Society given in November 1912 (*Proceedings*, Vol. XII). But attempts, since the War, to expound these new results have continually shewn the necessity for a precise appreciation of the ideas out of which this advance has developed; in mathematics it is not sufficient to know the enunciation of a result; it is necessary to understand the proof. These two volumes have grown up in the attempt to meet this need. The further need of a volume explaining the applications of topological theory, especially to the periods of the integrals belonging to the higher loci, may, I hope, appeal to another. The volumes are necessarily very incomplete in their inclusion of detail, as the specialist in any branch will easily find; their object is to lay the foundations for a more detailed study.

The pursuit of the analytical principles has a fascination in itself; but since, for reasons of space, these volumes are so largely devoted to this, I may be allowed to add another remark. The study of the fundamental notions of geometry is not itself geometry; this is more an Art than a Science, and requires the constant play of an agile imagination, and a delight in exploring the relations of geometrical figures; only so do the exact ideas find their value. As when, upon a landscape of rugged hill and ruffled water, there breaks the morning sun, scattering the clouds, and anon bathing the whole in a glory of contrasting colour. If these volumes should help to increase the number of those to whom the comparison does not seem an exaggeration, they will have been worth the making.

To the University Press very special acknowledgments are due, for the care, and speed, with which the volumes have been printed.

H. F. B.

29 *August*, 1933.

TABLE OF CONTENTS

CHAPTER IV. THE GENUS OF A CURVE. FUNDA-
MENTALS OF THE THEORY OF LINEAR SERIES

CHAPTER V. THE PERIODS OF ALGEBRAIC INTEGRALS.
LOOPS IN A PLANE. RIEMANN SURFACES

Contents

CHAPTER VI. THE VARIOUS KINDS OF ALGEBRAIC INTEGRALS. RELATIONS AMONG PERIODS

CHAPTER VII. THE MODULAR EXPRESSION OF RATIONAL FUNCTIONS AND INTEGRALS

CHAPTER VIII. ENUMERATIVE PROPERTIES OF CURVES

PAGES

CHAPTER I

INTRODUCTORY ACCOUNT OF RATIONAL AND ELLIPTIC CURVES

THE present volumes V, VI are an introduction to the more important of the algebraical and functional relations which are necessary for a clear and precise understanding of the principles of algebraic geometry. These relations are as the bones of the structure, to be clothed finally with a body of purely geometrical doctrine.

For the expression of these relations we make free use of coordinates, of which the justification has been examined in Vols. I and II. With their use we can define an algebraic construct (curve, surface, manifold, etc.) as the aggregate of points whose coordinates satisfy a set of algebraic equations, taken with points (limiting points, and other) which it may be necessary conventionally to add thereto. And, it is to be understood that all coordinates, and parameters, that enter in the equations employed, are capable of complex values; in particular, an aggregate will be said to be of dimension, or freedom, r, or simply to be ∞^r, when it depends on the values of r parameters not restricted to real values. In the elementary geometry two figures, or two manifolds, are regarded as essentially identical when they are projectively related to one another, that is (as we have seen) transformable into one another by equations which are linear in the (homogeneous) coordinates; in general algebraic geometry, two manifolds are regarded as essentially identical when the coordinates of the points of either are expressible as rational algebraic functions of the coordinates of the points of the other, whether linear functions or not.

The simplest algebraic construct from this point of view is then a line, which may be regarded as the locus of a point identified by one parameter fixing the position of the point upon the line. But the points of a conic, of which one point is known, are equally expressible in terms of a parameter, by taking the intersection of the conic with a variable line drawn through the known point. Or again, a plane cubic curve with a double point of known coordinates, is likewise the locus of a point whose coordinates are rational in one parameter, the intersection of the curve with a variable line through the double point. More generally, there is an unlimited family of curves with the property that the coordinates of any point of the curve are expressible as rational functions of a parameter; and this parameter can be chosen so that, conversely, it is a rational function of the coordinates of the point of the curve with which it is associated.

It will be seen, moreover, though this is a subsidiary property at the present stage, that, in the rational functions of the parameter which express the coordinates of a point of the curve, the coefficients which enter may be taken to be rational in the coefficients of the equations by which the curve is given and in the coordinates of one point of the curve; this is illustrated by the case of a conic. Regarding the variable parameter as representing a point of a line, such a curve as we have spoken of is therefore in (1, 1) birational correspondence with the line. Such a curve is called a *rational* curve. In the present chapter we give some fundamental properties of such curves, partly to illustrate general ideas, and partly because these properties should be known. And we treat also, for the same reasons, of curves which, in order of formal difficulty, next follow, those called *Elliptic Curves*.

Linear series on a line. One fundamental notion, which can be stated for a line, and has application not only to rational curves but to algebraic curves in general, is that of a *series of sets of points*, and, in particular, of a *linear* series. If a general point of a line be given by a parameter θ, a *set* of n points will be given by an equation $\theta^n + a_1\theta^{n-1} + \dots + a_n = 0$, with given coefficients $a_1, \dots, a_n$. These coefficients in turn may depend on other parameters $\xi, \eta, \dots$, so that when these vary the coefficients vary, and consequently the original set of n points of the line also varies, and gives rise to a *series* of sets, of each n points. The coefficients $a_1, \dots, a_n$ may be *rational* functions of $\xi, \eta, \dots$, all of these being capable of independent variation; then we have a *rational* series of sets of n points on the line. As a particular case of this, the coefficients $a_1, \dots, a_n$ may be linear (fractional) functions of $\xi, \eta, \dots$, all with the same denominator, of the forms $a_i = u_i/u$, where $u = p\xi + q\eta + \dots$, $u_i = p_i\xi + q_i\eta + \dots$, $(i = 1, \dots, n)$, where $p, q, \dots, p_i, q_i, \dots$ are constants. Then the sets on the line are given by $u\theta^n + u_1\theta^{n-1} + \dots + u_n = 0$, and hence by $\xi U + \eta V + \dots = 0$, where $U = p\theta^n + p_1\theta^{n-1} + \dots$, $V = q\theta^n + q_1\theta^{n-1} + \dots$; these $U, V, \dots$ are then definite polynomials of order n in θ. Such a series is called a *linear* series of sets of points on the line. But in regard to the polynomials $U, V, \dots$ two facts must be clear. It may happen that they all vanish for a certain number, say k, of definite values of θ; then, as $\xi, \eta, \dots$ vary, the sets of the series consist of $n - k$ variable points, and of k points common to all the sets, these being fixed. Further, it may happen that $U, V, \dots$ are not linearly independent, but connected by one or more linear equations, with constant coefficients, satisfied identically for all values of θ; and this will always happen if the number, say $r + 1$, of the polynomials $U, V, \dots$ is greater than $n + 1$. When the $r + 1$ polynomials $U, V, \dots$ are linearly independent, we can obviously determine a set of the linear series, of which r points have arbitrarily assigned positions, the other points of the set being thereby determined; in this case we speak of the series as being of *freedom* (or dimension) r, or as being ∞^r. And we may speak of $n-k$ as the *grade* of the series, this being the number of points which vary when ξ, η, ... vary. It is obviously necessary that the freedom be not greater than the grade. In particular cases it is convenient to consider linear series of which all the sets have a certain number of fixed common points, and it may be convenient to modify the definition of grade accordingly.

A series of sets of points on the line which is not linear, nor rational, may be *algebraical* in a more general sense. This will be so if the coefficients a_1, ..., a_n, in the equation which determines the points of a set, be algebraic functions, of one or more parameters ξ, η, ..., of such character as not to be capable of being expressed as rational functions of other parameters. The simplest case of this is when they are algebraic functions of one parameter, ξ. This means, in the first instance, that each coefficient, a_i, satisfies an (irreducible) algebraic equation with coefficients which are rational in ξ; in this case, however, it can be shewn that a single algebraic function of ξ can be chosen, say σ, such that all the coefficients a_1, ..., a_n are expressible rationally in terms of ξ and σ, say $a_i = \psi_i(\xi, \sigma)$; this can be done so that the aggregate of values of all these coefficients, each determined by its own equation in terms of ξ, is obtained by taking $a_i = \psi_i(\xi, \sigma)$, and allowing ξ, σ to take all possible simultaneous values consistent with the algebraic equation by which σ is determined from ξ. We may thus *define* an algebraic series of sets of points, upon the line, as that given by an equation $\theta^n + \Sigma \psi_i(\xi, \sigma) \theta^{n-i} = 0$, wherein σ, ξ have all values which satisfy a definite (irreducible) polynomial equation $f(\sigma, \xi) = 0$. And we may similarly have an algebraic series of sets depending on r parameters ξ_1, ..., ξ_r, wherein each coefficient a_i, of the determining equation, is rationally expressible in terms of ξ_1, ..., ξ_r and a further variable σ, satisfying a rational polynomial equation, say $f(\sigma, \xi_1, ..., \xi_r) = 0$, all values of σ and ξ_1, ..., ξ_r which satisfy this equation being taken; it can be shewn that this covers all cases of an algebraic series.

A general linear series of sets of n points, in which the freedom is also n, is that expressed by the original equation $\theta^n + \Sigma a_i \theta^{n-i} = 0$, in which all the coefficients vary independently of one another. Every algebraic series of sets of n points evidently consists of sets selected from this general series; it may belong to a linear series, of freedom less than n, arising by imposing linear restriction of the values allowed to a_1, ..., a_n in the general linear series spoken of.

In an algebraic series of freedom 1, there will generally be more than one set which contains a particular point of the line; for an equation $\theta_0{}^n + \Sigma \psi_i(\xi, \sigma) \theta_0{}^{n-i} = 0$ will generally be satisfied, with the same θ_0, by several pairs (ξ_1, σ_1), (ξ_2, σ_2), ... satisfying the fundamental equation $f(\sigma, \xi) = 0$. The number of such sets is called the *index* of the series. This is evidently equal to the number of zeros of the rational function, for the curve $f(\sigma, \xi) = 0$, which is expressed by $\theta_0{}^n + \Sigma \psi_i(\xi, \sigma) \theta_0{}^{n-i}$. From this it can be proved that the index cannot be 1 unless the curve $f(\sigma, \xi) = 0$ is rational. And then, for the index to be 1, each of the rational functions $\psi_i(\xi, \sigma)$ must be expressible as a fractional linear function of the parameter by which the curve $f(\sigma, \xi) = 0$ is expressed, with a denominator the same for all. In other words, an algebraic series on the line, of freedom 1 and of index 1, must be given by an equation of the form $\lambda \phi + \mu \psi = 0$, where ϕ, ψ are definite polynomials in the parameter θ of the line, and λ, μ are independently variable. More generally, in an algebraic series of freedom r, on a line, there is generally more than one set of the series of which r points are assigned; when the series is linear there is only one such set.

A familiar application of these ideas arises in the definition of a rational curve. Suppose that the coordinates of any point of a given curve are expressible rationally in terms of a parameter, θ, whose variation gives all the points of the curve. For example, for the curve whose equation is $x^2 + y^2 = z^2$, we may take $x = 1 - \theta^4$, $y = 2\theta^2$, $z = 1 + \theta^4$; then $\theta^2 = y/(x+z)$, and to each point of the curve correspond two points of the line on which θ is represented. We thus have, on the line, an algebraic series of sets of two points, of index 1; but these sets belong to a linear series, expressed by $\lambda + \mu \theta^2 = 0$, where λ, μ are variable. The coordinates of a point of the curve are then expressible rationally in terms of the single parameter λ/μ; and this is a *representative* parameter, as having only one value for a point of the curve, $-y/(x+z)$. Such a parameter can always be chosen to express a rational curve, as we have indicated. (Cf. Vol. II, p. 136.)

Rational curves. Consider now a plane curve, of which the ratios of the coordinates x, y, z of a point, are expressible rationally in terms of a parameter, θ, so chosen that only one value of this belongs to any general point of the curve. Thus, if ρ denote a factor of proportionality, we may say that each of ρx, ρy, ρz is equal to a polynomial in θ; let one at least of these polynomials be of order n (in general, all of this order); then n is the number of values of θ for which an arbitrary linear form $ax + by + cz$ vanishes, namely the number of points of the curve on an arbitrary line, or the *order* of the curve. For illustration, suppose $n = 3$; and imagine the plane (x, y, z) to be in space, wherein the coordinates are X, Y, Z, T. To compare now with the equations $\rho x = f_1$, $\rho y = f_2$, $\rho z = f_3$ for the plane cubic curve, take in the space (X, Y, Z, T) the curve expressed by $\sigma X = f_1$, $\sigma Y = f_2$, $\sigma Z = f_3$, $\sigma T = f_4$, where f_4 is an arbitrary cubic polynomial, such that no identity $Af_1 + Bf_2 + Cf_3 + Df_4 = 0$ holds, for values of A, B, C, D independent of θ; this is possible since evidently no identity $af_1 + bf_2 + cf_3 = 0$ holds, or the points of the plane cubic would be in a line. The curve in space is likewise a cubic curve, having three points in a plane. Since now any point θ, or (x, y, z), of the plane cubic curve is associated with the point θ, or (X, Y, Z, T), of the cubic curve in space by the equations $X/x = Y/y = Z/z$, we have the conclusion that any rational cubic curve in a plane may be regarded as the projection, from a point, of a rational cubic curve in space. Similarly, suppose we have a rational quartic curve in a plane, given by equations $\rho x = \phi_1$, $\rho y = \phi_2$, $\rho z = \phi_3$, where ϕ_1, ϕ_2, ϕ_3 are linearly independent quartic polynomials in a parameter θ; and suppose the plane to lie in a fourfold space of (homogeneous) coordinates X, Y, Z, T, U. Take two other, arbitrary, quartic polynomials in θ, namely ϕ_4, ϕ_5, and consider the rational curve, in this fourfold space, which is given by $\sigma X = \phi_1$, $\sigma Y = \phi_2$, $\sigma Z = \phi_3$, $\sigma T = \phi_4$, $\sigma U = \phi_5$, it being understood that no identity

$$A\phi_1 + B\phi_2 + C\phi_3 + D\phi_4 + E\phi_5 = 0$$

holds, for constant values of A, B, C, D, E. This new curve is related to the given quartic by the equations $X/x = Y/y = Z/z$; and, in space of four dimensions, the three equations $X = 0$, $Y = 0$, $Z = 0$ represent a line. Hence we say, a rational plane quartic curve may be looked upon as arising by projection *from a line*, from a rational curve in space of four dimensions; this curve is also of order 4, since, with a, b, c, d, e arbitrary, the equation

$$aX + bY + cZ + dT + eU = 0$$

is satisfied by 4 values of θ. A rational quartic curve in space of three dimensions is similarly derivable from a rational quartic curve in four dimensions, by taking only one additional coordinate, the derivation in this case being by projection from a point.

In general, a rational curve in space of r dimensions, where the homogeneous coordinates are $x_0, x_1, \ldots, x_r$, is given by equations $\rho x_i = f_i$, in which the functions f_i are polynomials in a parameter, θ. One at least of these polynomials must be of order as great as r, since otherwise constants $a_0, \ldots, a_r$ can be found to render the equation $\Sigma a_i f_i = 0$ identically true; in which case there exist one or more linear equations $\Sigma a_i x_i = 0$ connecting the coordinates of a point of the curve, which then lies in space of less than r dimensions. On the other hand, if one (or all) of the polynomials f_i is of order greater than r, the curve may be regarded as derived by projection from a rational curve lying in space of more than r dimensions, in the manner we have illustrated. We may then suppose that these polynomials are of order r, and are linearly independent. The order of the curve represented by the equations, being, by definition, the

number of points lying on a general prime of the space, whose
equation is of the form $\Sigma c_i x_i = 0$, is then r. Conversely, any algebraic curve of order r, lying in space of r dimensions, is rational; for
a variable prime, drawn through $(r-1)$ fixed points of the curve,
meets the curve in one further point; and this point is identified by
the single parameter which fixes the particular prime; moreover, by
supposing the $(r-1)$ points to coincide at one point of the curve, we
see that the only irrationality entering into the rational expression
of the points of the curve, beyond those which determine the curve,
is that of the coordinates of some particular point of the curve.

We see then that all rational curves are reducible to curves of
order r, in space of r dimensions. Such a curve is called a rational
normal curve; it is given, so far, by equations of the form $\rho x_i = f_i$,
where the $(r+1)$ polynomials f_i are linearly independent; by taking
suitable linear functions of $x_0, ..., x_r$, say $\xi_0, ..., \xi_r$, it is thus
capable of being expressed by $\xi_0/\theta^r = \xi_1/\theta^{r-1} = ... = \xi_r/1$. Moreover,
we see that we can pass from any rational curve of order r, in space
of r dimensions, to any other such curve, by linear transformation
of the coordinates. Upon such a curve there can be no point such
that the value of the parameter θ appropriate to this point occurs as
a multiple root in the equation, of order r, which gives the intersections of the curve with a general prime of the space; for such a
prime can be put through $r-1$ arbitrary other points of the curve,
beside this one. In other words, the curve has no multiple point.

But a rational plane curve, of order greater than 2, must needs
have multiple points. We prove in fact that a rational plane curve
of order n has $\frac{1}{2}(n-1)(n-2)$ double points, or multiple points
equivalent to as many double points; and, conversely, that an
algebraic plane curve of order n with $\frac{1}{2}(n-1)(n-2)$ double points,
is necessarily rational. To prove the former, assume the curve
given by an equation $f(x, y, z) = 0$, and to have δ nodes, and κ cusps.
We prove $\delta + \kappa = \frac{1}{2}(n-1)(n-2)$. The equation of the tangent of the
curve, at an ordinary point (x, y, z), being known, the t tangent
lines of the curve, from an arbitrary point (ξ, η, ζ), touch the curve
at the ordinary intersections of the curve with the curve of order
$(n-1)$ given by $\xi \partial f/\partial x + \eta \partial f/\partial y + \zeta \partial f/\partial z = 0$; but it is easy to prove
that this curve has 2 intersections with the original curve at a node,
and has 3 intersections at a cusp. Thus we have $t + 2\delta + 3\kappa = n(n-1)$.
On the other hand, assuming the expression of the coordinates of a
point of the curve in terms of a parameter, θ, if we join (ξ, η, ζ) to
an arbitrary point (x_1, y_1, z_1) of the curve, for which the value of the
parameter is θ_1, the parameters of the remaining $(n-1)$ intersections of the joining line, with the curve, are given by the equa-
tion $\xi(yz_1 - y_1z) + \eta(zx_1 - z_1x) + \zeta(xy_1 - x_1y) = 0$; if herein, x, y, z and
x_1, y_1, z_1 are replaced by their values in terms of θ and θ_1, respec-

tively, and the result divided by $\theta - \theta_1$, there remains an equation $(\theta, \theta_1) = 0$, of order $n-1$ in regard both to θ and θ_1, and symmetrical in regard to these. There are therefore $2(n-1)$ values of θ_1 for which the points (θ), (θ_1) coincide, given by the equation $(\theta_1, \theta_1) = 0$, which, for a general position of (ξ, η, ζ), will be of aggregate order $2(n-1)$ in θ_1. Such coincidences arise, however, only in two ways, if we assume that the curve has no multiple points beside nodes and cusps: (i) when (θ_1) is a point of the curve at which the tangent line passes through (ξ, η, ζ); (ii) when (θ_1) is a cusp. Thus we infer that $t + \kappa = 2(n-1)$. From the former equation obtained we therefore have $\delta + \kappa = \frac{1}{2}(n-1)(n-2)$. The corresponding equation when the curve has higher singularities requires an appropriate definition of δ and κ, into which we do not enter now.

To prove the converse result, that a plane curve of order n, whose only multiple points are nodes and cusps, whose aggregate number is $\frac{1}{2}(n-1)(n-2)$, can have the coordinates of its points expressed rationally by a reversible parameter, we shew that such a curve can be changed, by transformations which are rational in the coordinates, and also rational in the coefficients in the given equation of the curve, either to a straight line, or to a conic. For this, consider, in conjunction with the given curve f, of order n, the most general plane curve ψ, of order $n-2$, which passes through each of the $\frac{1}{2}(n-1)(n-2)$ double points of f. The number of terms in the equation of the general plane curve of order $n-2$ is $\frac{1}{2}n(n-1)$; for this to contain a double point of f one linear condition must be imposed upon the coefficients in ψ. The form of ψ under consideration will thus contain $\frac{1}{2}n(n-1) - \frac{1}{2}(n-1)(n-2)$, or $n-1$, homogeneously entering arbitrary coefficients, or *more* if the $\frac{1}{2}(n-1)(n-2)$ conditions for the double points are not independent; thus the curve ψ will have an equation of the form

$$\lambda_0 \psi_0 + \ldots + \lambda_{n-2} \psi_{n-2} + \mu_1 V_1 + \mu_2 V_2 + \ldots = 0,$$

where $\psi_0, \ldots, \psi_{n-2}, V_1, V_2, \ldots$ are definite polynomials of order $n-2$ in the coordinates, linearly independent of one another upon the curve f, and $\lambda_0, \lambda_1, \ldots, \lambda_{n-2}, \mu_1, \mu_2, \ldots$ are arbitrary. As the double points of f are the common solutions of three rational equations $\partial f/\partial x = 0$, $\partial f/\partial y = 0$, $\partial f/\partial z = 0$, symmetrical functions of the coordinates of all these points are expressible rationally by the coefficients in the equation of f, and therefore the coefficients in the polynomials $\psi_0, \ldots, \psi_{n-2}, V_1, \ldots$ are rational in the coefficients in f. The curve ψ will have intersections with f not at the double points, of number $n(n-2) - (n-1)(n-2)$, or $n-2$; the number of coefficients left arbitrary in ψ cannot therefore be enough to enable us to prescribe a particular ψ having, other than at the double points, more than $n-2$ intersections with the given curve f; thus the terms

$\mu_1 V_1 + \mu_2 V_2 + \dots$ are unnecessary; and the double points of f do furnish independent conditions for ψ. Moreover, the curves $\psi_0 = 0, \dots, \psi_{n-2} = 0$ cannot have common zeros on the curve f, other than at the double points of f, because the number of intersections remaining would then be less than the number which can be prescribed arbitrarily by proper choice of $\lambda_0, \dots, \lambda_{n-2}$. In particular $\psi_0, \dots, \psi_{n-2}$ have no common factor.

We remark in passing that we can now at once see that the coordinates of a point of the curve f are expressible rationally by a parameter, if the coordinates of some arbitrarily taken point of this curve be assumed known. For, if, first, $n-3$ arbitrary points be taken on f, the curves ψ through these points and through the double points will, by what we have seen, have an equation of the form $\theta u - v = 0$, where u, v are definite polynomials in x, y, z, involving the coordinates of the $n-3$ points taken, and θ is variable. The combination of this equation with the equation of f will lead to the coordinates of the only remaining intersection of this curve with f, expressed rationally in terms of θ. If we now suppose the $n-3$ arbitrarily taken fixed points of f to be made to coincide at one point of f, we thus have an expression rational in θ, and in the coordinates of one point of f; and θ is conversely rational in the coordinates of the point which it represents, being equal to v/u.

But we may proceed by a succession of steps, from the general equation of ψ involving $n-1$ homogeneous parameters $\lambda_0, \lambda_1, \dots, \lambda_{n-2}$. Take $(n-4)$ arbitrary points of the plane, whose coordinates may then be reckoned rational. The curves ψ passing through these will then have an equation of the form $c_0 \Psi_0 + c_1 \Psi_1 + c_2 \Psi_2 = 0$, where Ψ_0, Ψ_1, Ψ_2 are definite linear functions of $\psi_0, \dots, \psi_{n-2}$, likewise, therefore, rational in the coefficients in f; while c_0, c_1, c_2 are arbitrary. Take now ξ, η, ζ, so that $\xi/\Psi_0 = \eta/\Psi_1 = \zeta/\Psi_2$; then, as (x, y, z) describes the curve f, the point of which ξ, η, ζ are the coordinates will describe another curve, ϕ. The order of this curve ϕ, equal to the number of zeros of a general form $u\xi + v\eta + w\zeta = 0$, on ϕ, is equal to the number of variable zeros of $u\Psi_0 + v\Psi_1 + w\Psi_2$ on f, other than the common zeros of Ψ_0, Ψ_1, Ψ_2 at the double points of f, namely is $n-2$ in general, that is, when the double points are distinct. To any ordinary point of f will correspond a single point of ϕ; to a node of f will in general correspond two points of ϕ, each obtained by one of the modes of approach on f to this double point; but these will be fixed points on ϕ. To a general point of ϕ, say (ξ, η, ζ), there will, by the construction, correspond a point (x, y, z) of the curve f; but there will not correspond *two* points. For equations of the form

$$\Psi_0(x', y', z')/\Psi_0(x, y, z) = \Psi_1(x', y', z')/\Psi_1(x, y, z)$$
$$= \Psi_2(x', y', z')/\Psi_2(x, y, z)$$

would involve that every curve $c_0\Psi_0 + c_1\Psi_1 + c_2\Psi_2 = 0$ which passes through the point (x, y, z) of f passes likewise through (x', y', z'); it is not true that every curve ψ of order $n-2$, through the double points of f, which is drawn through a general point of f likewise passes through another point of f determined thereby, since, as we have seen, such a curve ψ involves a number $n-2$ of arbitrary parameters equal to the number of its unassigned intersections with f; we may therefore assume, if the $n-4$ fixed points of the plane are taken with sufficient generality, that the same is true of the system $c_0\Psi_0 + c_1\Psi_1 + c_2\Psi_2 = 0$. Wherefore, the two equations

$$\xi/\Psi_0 = \eta/\Psi_1 = \zeta/\Psi_2,$$

taken with $f = 0$, lead, for a general point (ξ, η, ζ) of the curve ϕ, to a single point (x, y, z) of f. This point may be determined by rational processes of elimination; so that the ratios of x, y, z are not only rational in (ξ, η, ζ), but equally rational in the coefficients in Ψ_0, Ψ_1, Ψ_2, and hence also in the coefficients in f.

We have thus found a new curve ϕ, of points (ξ, η, ζ), of order m say, where m is in general $n-2$, which is wholly in birational correspondence with f. We have seen above, in passing, that the coordinates of the points of f are expressible rationally in terms of a reversible parameter (the expressions involving the coordinates of a point of f); the points of ϕ are therefore expressible rationally by a parameter. From this, by what was proved above, it follows that ϕ has the equivalent of $\frac{1}{2}(m-1)(m-2)$, in general $\frac{1}{2}(n-3)(n-4)$ double points. If we assume that these are distinct, the same process can be applied to ϕ as was applied to f; it can be placed in $(1, 1)$ birational correspondence with a curve of order $n-4$; and the argument can be repeated, until finally we reach either a line (when n is odd), or a conic. The coordinates of the points of a conic can be expressed rationally in terms of a parameter and the coordinates of one particular point arbitrarily taken on the conic; the coordinates of the points of a line are wholly rational in terms of a parameter. The conclusion which has been stated thus follows in general. But the reasoning assumes that at each stage the curve obtained has distinct nodes or cusps. This condition is evidently unnecessary when, for instance, there arises, instead, a k-ple point with separated tangents; for by prescribing the reducing curve to have there a $(k-1)$-ple point, equivalent to $\frac{1}{2}k(k-1)$ conditions, we thereby prescribe $k(k-1)$ intersections, or twice the number of conditions; the multiple point thus has just the effect of $\frac{1}{2}k(k-1)$ separated double points. The examination of the corresponding necessary modification of the reasoning in more complicated cases must be omitted at this stage. Another possibility may be illustrated by considering the simple case where the process is applied to reduce a curve of order k which has a general $(k-1)$-ple point O. The reducing curves would then be taken to be curves of order $k-2$ with a $(k-2)$-ple point at the given multiple point O; these reducing curves would then consist of $k-2$ lines through this point, and the system $u\Psi_0 + v\Psi_1 + w\Psi_2 = 0$ would consist of two variable lines through O, together with $(k-4)$ fixed lines through O. The process thus reduces the given curve to the conic $\xi\zeta - \eta^2 = 0$, which is rational without the assignment of any point thereon. The original curve is in fact obviously rational, being met in one variable point by a variable line through O, even when k is even. The proof that essentially no other case needs remark must be omitted here.

Greatest possible number of double points of a plane curve. It is natural to suppose, since the condition for a curve of order n to be rational has been shewn to be the possession of $\frac{1}{2}(n-1)(n-2)$ double points, in general, that this is the maximum number possible. We prove now definitely that this is so. More generally, for an irreducible curve of order n with multiple points, of which the general one is of multiplicity denoted by k (varying from point to point), we prove that $\frac{1}{2}\Sigma k(k-1) \leqslant \frac{1}{2}(n-1)(n-2)$. This inequality we prove in the equivalent form

$$n(n-1) \geqslant \Sigma k(k-1) + \tfrac{1}{2}(n-1)(n+2) - \tfrac{1}{2}\Sigma k(k-1).$$

For the curve $f = 0$, of order n, it can easily be proved that the so-called first polar curve, whose equation is

$$\xi \partial f/\partial x + \eta \partial f/\partial y + \zeta \partial f/\partial z = 0,$$

has $k(k-1)$ intersections with $f = 0$ at an ordinary k-ple point of f; it may have more when the tangents at the multiple point are not distinct (for instance, at a cusp there are 3 intersections). As the total number of intersections is $n(n-1)$, we infer that

$$\Sigma k(k-1) \leqslant n(n-1),$$

so that we have $\frac{1}{2}(n-1)(n+2) - \frac{1}{2}\Sigma k(k-1) \geqslant n-1$, and, for $n > 1$, the number, say ξ, occurring on the left is positive. Now consider the most general curve of order $n-1$, prescribed to have a $(k-1)$-ple point at every k-ple point of the curve f. If these conditions at the multiple points are independent, the curve will have $\xi + 1$ homogeneously entering arbitrary coefficients; for a curve of order m has $\frac{1}{2}m(m+3) + 1$ homogeneously entering coefficients, and, for a curve to have an h-ple point at a given point requires $\frac{1}{2}h(h+1)$ conditions.

Thus such curve of order $n-1$ has always at least $\xi + 1$ homogeneous coefficients, and, by choice of these, can be made to pass through at least ξ other points of the curve f. By being made to have a $(k-1)$-ple point at a k-ple point of f, a number of intersections at least $k(k-1)$ is secured; but it may happen, for special kinds of the multiple point, that there are more than this. Thus the curve of order $n-1$ has in all at least $\xi + \Sigma k(k-1)$ intersections; its total number is however $n(n-1)$, which is thus equal to or greater than $\xi + \Sigma k(k-1)$. And this is the fact we set out to prove.

The application of the result is most generally to the case when every one of the k-ple points has distinct tangents; but it may be noticed that the argument does not require this. And further, when $\frac{1}{2}\Sigma k(k-1) = \frac{1}{2}(n-1)(n-2)$, the argument shews (a) that, for curves of order $n-1$, the conditions of having a $(k-1)$-ple point at every k-ple point are independent, (b) that the number of other intersections of such curve with f is precisely equal to the number of disposable coefficients in such curve, so that all these other intersections can be taken arbitrarily, (c) that all the k-ple points of f are such that the curve of order $n-1$ prescribed to have thereat a $(k-1)$-ple point has just $k(k-1)$ intersections with f at this point, and no more. The whole argument, however, as in previous cases, assumes that f is an irreducible curve; otherwise the curve of order $n-1$, by containing a part of this, may have infinity, and not $n(n-1)$, as the greatest possible number of its intersections with f.

Elliptic curves. Having proved that $\frac{1}{2}(n-1)(n-2)$ is the greatest possible number of double points for an irreducible curve of order n, and considered the case when this is the actual number, we consider now the case of a curve with $\frac{1}{2}(n-1)(n-2)-1$ double points $(n>2)$. We prove that such a curve is birationally equivalent with a plane cubic curve.

Let a general curve, ψ, of order $n-2$ be described through the $\frac{1}{2}(n-1)(n-2)-1$ double points of the given curve f; this will have $n(n-2)-(n-1)(n-2)+2$, or n further intersections, and its equation, containing at least $\frac{1}{2}n(n-1)-\frac{1}{2}(n-1)(n-2)+1$, or n arbitrary coefficients, will be of the form

$$\lambda_0\psi_0 + \ldots + \lambda_{n-1}\psi_{n-1} + \mu_1 U_1 + \mu_2 U_2 + \ldots = 0,$$

where $\psi_0, \ldots, \psi_{n-1}, U_1, U_2, \ldots$ are definite polynomials of order $n-2$, linearly independent on the given curve, and the terms in $U_1, U_2, \ldots$ may be omitted if the double points of f furnish independent conditions for the curve of order $n-2$ described through them. In fact there are not more than n terms in ψ; for, if so, take a form of ψ involving $n+1$ terms (say $\mu_2 = \mu_3 = \ldots = 0$); and then, that particular ψ which further passes through $n-1$ prescribed general points of f; this ψ will have an equation of the form $u + v\theta = 0$, where u, v are definite, and θ is variable, and this ψ will have one further intersection with f. There is then a correspondence between this point, and the value of θ, either determining the other; by elimination it then follows that the coordinates of this point are rational functions of θ. But this is not the case, or the curve f would have $\frac{1}{2}(n-1)(n-2)$ double points. We may thus suppose, in the equation of ψ, that $\mu_1 = \mu_2 = \ldots = 0$. Further, it is not the case that if ψ be made to pass through an arbitrary general point of f, it will necessarily pass through a further point of f determined by the first point; for, by suitable choice of the ratios of $\lambda_0, \ldots, \lambda_{n-1}$, the curve ψ can be made to pass through $n-1$ arbitrary points of f, and if each of these determined another point of f lying on ψ, we should thus have $2(n-1)$ intersections, which is greater than the number n, of intersections of ψ with f which are not at the double points of f, if $n > 2$.

Now, of such curves ψ, consider those which have $n-3$ prescribed general points of intersection with the curve f; such curves will have 3 further intersections with f, and will have an equation of the form $c_0 \Psi_0 + c_1 \Psi_1 + c_2 \Psi_2 = 0$, where c_0, c_1, c_2 are variable. Such a curve can, by proper choice of the ratios of c_0, c_1, c_2, be made to pass through two arbitrary points of f; thus, the assignment of one arbitrary point of such a curve, upon f, does not involve another intersection determined by the former. Hence, if we put

$$\xi / \Psi_0 = \eta / \Psi_1 = \zeta / \Psi_2,$$

the point (ξ, η, ζ), as (x, y, z) describes f, describes another curve, to any point of which corresponds conversely only a single point of f; this curve, in other words, is birationally equivalent with f. The order of this curve is 3, unless Ψ_0, Ψ_1, Ψ_2 have a common intersection with f in addition to the $n-3$ prescribed intersections of ψ; as this curve cannot be a conic or a line, not being rational, the order is therefore really 3. This shews that the curve f is birationally equivalent with a cubic curve, as stated. The equations of transformation, however, involve the coordinates of the $n-3$ points through which ψ was made to go; by taking these points to coincide at one point, the equations of transformation become rational in the original coefficients of f, and the coordinates of one arbitrary point thereon.

The simplest example of this argument is that of a quartic curve with two double points. Conics through these, and one further arbitrary point of the quartic curve, suffice to transform this curve into a cubic curve. If the two double points, and the third given point, be the vertices of the triangle of reference for (x, y, z), the transformation is effectively $\xi x = \eta y = \zeta z$, as is easy to see. But we may take another point of view, which is suggestive of generalisation. With a, b as constants and u_1, v_1 as homogeneous of order 1 in y, z only, and u_2 homogeneously quadratic in y, z, the equation of the quartic curve may be taken to be

$$ax^4 + x^3 u_1 + x^2 u_2 + xyzv_1 + by^2 z^2 = 0,$$

(the double points being at the vertices (y), (z)). The curve is therefore the projection, on to the original plane, of the curve of intersection of two quadric surfaces; namely the projection, from $(0, 0, 0, 1)$, of the intersection of the two quadric surfaces whose equations are $xt - yz = 0$ and $ax^2 + xu_1 + u_2 + tv_1 + bt^2 = 0$. The cubic curve, to which the plane quartic curve is transformable, may be taken to be the projection of the quartic curve in the space (x, y, z, t), from any point of itself. The plane quartic curve is not transformable into a cubic curve with equations rational only in the coefficients in the equation of the quartic curve.

Returning to the plane curve of order n, with $\frac{1}{2}(n-1)(n-2) - 1$ double points, if the equation of the general curve of order $n - 2$ through these double points be written as before in the form $\lambda_0 \psi_0 + \ldots + \lambda_{n-1} \psi_{n-1} = 0$, and if we take

$$\frac{\xi_0}{\psi_0} = \frac{\xi_1}{\psi_1} = \ldots = \frac{\xi_{n-1}}{\psi_{n-1}},$$

and regard $(\xi_0, \ldots, \xi_{n-1})$ as coordinates of a point in space $[n-1]$, of $n - 1$ dimensions, we may shew that as (x, y, z) describes the curve f, the point $(\xi_0, \ldots, \xi_{n-1})$ describes a curve in the space $[n-1]$ which is in birational $(1, 1)$ correspondence with f. The order of this curve, defined as the number of its intersections with a general *prime* locus of this space, given by an equation of the form

$$\lambda_0 \xi_0 + \ldots + \lambda_{n-1} \xi_{n-1} = 0,$$

is n. Conversely, a curve of order n, in space $[n-1]$, is necessarily either rational, or in $(1, 1)$ birational correspondence with a plane cubic curve (without double point). This we see by projecting the curve, in the space $[n-1]$, on to a plane, by means of variable spaces $[n-3]$ drawn through $n-3$ fixed arbitrary points of the curve. The projection is a cubic curve.

Curves which are in $(1, 1)$ birational correspondence with a non-singular plane cubic curve are generally called *Elliptic*. They are evidently distinct from rational curves. But whereas any two rational curves, being in $(1, 1)$ correspondence with a line, are in $(1, 1)$ correspondence with one another, and, in that sense, essentially identical, it is not the case that any two elliptic curves, though they have the same character, are in such $(1, 1)$ correspondence. There belongs in fact to an elliptic curve a single numerical constant, called its modulus; and only when the moduli of two elliptic curves are the same are they essentially identical in the sense explained. To make these statements quite clear, it is desirable to have a detailed theory of a plane cubic curve; and to this end it is proper to reduce its equation to as simple form as possible. The equation of a general plane cubic curve contains 10 homogeneous coefficients, and so depends on 9 constants. The general linear homogeneous transformation in a plane contains 8 constants; it is to be expected therefore that the equation of a plane cubic curve may be reduced, by linear transformation only, to contain only one constant. It may in fact be reduced to the form $\xi^3 + \eta^3 + \zeta^3 + 6\mu\,\xi\eta\zeta = 0$, involving the constant μ. But the nature of the irrationalities, in the original coefficients, involved in this reduction, is not thereby made clear. In fact this form is easily deduced when the points of inflexion of the curve are known. Such an inflexion would be obtainable by substituting, in the original general equation, for z, an expression $mx + ny$, and expressing the conditions, for m and n, that the resulting cubic equation for x/y should have three equal roots. The examination shews that we are thus led to require the solution of an equation of order 9. The theory of the solution of this equation is indeed interesting. But we shall not enter into it here. Instead, we shew how, by a birational transformation which is not linear in the original coordinates, we can reduce the equation of the cubic curve to the form $Y^2Z = 4X^3 - g_2XZ^2 - g_3Z^3$, by equations rational in the original coefficients and the coordinates of one point assumed to be given (arbitrarily) on the curve. It is easy to see that this equation essentially depends only on the one modulus g_2^3/g_3^2. A knowledge of the curve given by this equation then enables us to determine all the properties of the original curve.

If the equation of the cubic curve be given in general form, and the coordinates of a point O thereon, we can determine, rationally, the coordinates of the point Q where the tangent line at O meets the curve again. By a linear transformation we can then take O for the point (y) of a triangle of reference, and Q for the point (x). Take now an arbitrary line through O for $x = 0$, and an arbitrary line through Q for $y = 0$. The equation of the cubic curve thus takes a form $pz^3 + ax^2y + bx^2z + cy^2z + dz^2x + ez^2y + fxyz = 0$, $z = 0$ being the tangent at (y). It is assumed here that O is not a point of inflexion (the case when this is so is dealt with below); also it is assumed that the curve is irreducible and has no double point; in particular there is no double point at O or Q and the coefficients c and a do not vanish. The general conic through the point Q, or (x), and through the two points, other than O, in which $x = 0$ meets the curve, has then an equation $l(pz^2 + cy^2 + eyz) + hxy + kxz = 0$, where l, h, k are arbitrary; and such conic has three other intersections with the curve, of which two may be arbitrarily assigned by choice of l, h, k. As in the preceding cases, we define new (non-homogeneous) coordinates by the ratios of the coefficients of l, h, k, putting $\xi = y/z$, $\eta = (pz^2 + cy^2 + eyz)/xz$. Considering these as rational functions on the curve, ξ becomes infinite to the second order at O, and has for zeros the two points, other than Q, where $y = 0$ meets the curve, while η becomes infinite to the third order at O, and vanishes at the three points constituted by Q, and the two points, not on

$x=0$, where the lines joining Q to the intersections of the curve with $x=0$ meet the curve again. For the reverse transformation we then have $x/z=(p+e\xi+c\xi^2)/\eta$, $y/z=\xi$. If now we make the linear transformation to coordinates ξ_1, η_1 which is expressed by $-\xi_1=ac\xi+\frac{1}{3}(ae+bc-\frac{1}{4}f^2)$, $\eta_1/ac=f\xi+2\eta+d$, it is easy to verify that the equation of the curve is the same as $\eta_1^2=4\xi_1^3-g_2\xi_1-g_3$, where, in terms of the three constants given by $u=ae+bc-\frac{1}{4}f^2$, $v=be+ap-\frac{1}{2}df$, $w=bp-\frac{1}{4}d^2$, the values of g_2, g_3 are given by $\frac{1}{4}g_2=\frac{1}{3}u^2-acv$, $\frac{1}{4}g_3=\frac{2}{27}u^3-\frac{1}{3}acuv+a^2c^2w$. As the coefficient ac is not zero, ξ and η are reversely expressible in terms of ξ_1 and η_1.

The construction for the line $y=0$ fails if the point O be a point of inflexion. In this case, it is easy to see that if a variable line through O meet the curve again in H and K, the locus of the harmonic conjugate of O in regard to H and K is a line. If this line be taken for $y=0$, and a suitable line through O be taken for $x=0$, it is at once verified that the equation is reducible, by linear transformation only, to the form obtained in general.

Combining the result now obtained, with what was proved above, we can assert that a plane curve of order n, with $\frac{1}{2}(n-1)(n-2)-1$ double points, can be birationally reduced to have its equation in the form $\eta^2=4\xi^3-g_2\xi-g_3$, the necessary equations being rational in the coefficients of the given curve and the coordinates of an arbitrary point thereon. This enables us to specify in detail the difference between a rational curve and the curves now under consideration. A rational curve has its coordinates reversibly expressible by a parameter θ; this statement may be modified by saying that its coordinates are reversibly expressible by singly periodic functions (trigonometrical functions) of an argument u; for we have only to take u so that $\tan\frac{1}{2}u=\theta$, or $\cos u=(1-\theta^2)/(1+\theta^2)$, $\sin u=2\theta/(1+\theta^2)$, leading to $\theta=\sin u/(1+\cos u)$; then to any point of the curve corresponds a definite value of $\sin u$ and $\cos u$, and hence a definite value of u, save for multiples of 2π; while conversely any value of u gives a definite point of the curve. For the curve of order n with $\frac{1}{2}(n-1)(n-2)-1$ double points, there is an analogous reversible expression of the coordinates of a point of the curve by single-valued functions of an argument u, the functions namely which are called elliptic functions. These differ however from the trigonometrical functions in having *two* periods (whose ratio is not real), say Ω and Ω'; to any value of u belongs a definite point of the curve; to any point of the curve belongs a definite value of u save for an additive term $k\Omega+k'\Omega'$, in which k and k' are integers. We refer to books dealing with these elliptic functions for the establishment of their properties; but the theory, in its simplest form, is directly associated with the equation $\eta^2=4\xi^3-g_2\xi-g_3$ which we have obtained, it being assumed that this curve has no double point. (The condition for this curve to have a double point is easily verified to be $g_3{}^2 = (\tfrac{1}{3}g_2)^3$.) So much of the elements of the theory as will indicate the connexion with the equation to which we have reduced the cubic curve may be curtly stated. Let two values of the argument u whose difference is of the form $k\Omega + k'\Omega'$, in which k and k' are integers, be called equivalent. Any of the elliptic functions considered is expressible, in the neighbourhood of any finite value, u_0, of the argument, in the form

$$A_r(u-u_0)^{-m_r} + \ldots + A_1(u-u_0)^{-m_1} + B + B_1(u-u_0) + B_2(u-u_0)^2 + \ldots,$$

in which m_r, m_{r-1}, ..., m_1 are positive integers with

$$m_r > m_{r-1} > \ldots > m_1 > 0, \text{ and } A_r, \ldots, A_1, B, B_1, \ldots$$

are constants, in which there is a converging infinite series of positive integer powers of $u-u_0$, and a finite number of negative integer powers. It is only for a finite number of non-equivalent values of the particular argument u_0 that these negative powers of $u-u_0$ are actually present in the expansion; when they are present, the argument u_0 is called a *pole* of the function, and the integer m_r is called the order of the pole. The sum of the orders of the poles actually existing, at non-equivalent values of u_0, is called the *order of the function*. For such a function there is also a number of non-equivalent values of the argument for which the function vanishes, the *zeros* of the function, and the sum of the orders of vanishing at all the non-equivalent zeros is equal to the order of the function. More generally, if A be any arbitrary value, and ϕ denote the function, the sum of the orders of vanishing of $\phi - A$ at all non-equivalent values of the argument is equal to the order of the function ϕ. We may agree to say that if the difference $\phi - A$ vanishes to order m for a particular value of the argument, then ϕ takes the value A for this argument m times. With this phraseology there is the further property, that the sum of the non-equivalent values of the argument u at which the function takes any specified value, A, is independent of A, being the same as when A is infinite. There exists no such function whose order is less than 2; there exists, however, a function with simple poles for two arbitrary non-equivalent values of the argument, and no other non-equivalent poles; and there exists a function with a pole of the second order for an arbitrary value of the argument, whose only other poles are for equivalent values of the argument. This latter function, suitably specialised, is the simplest doubly periodic, or elliptic, function. The pole being taken to be at $u=0$, the function can in fact be chosen to have an expression, in the neighbourhood of this value, of the form $u^{-2} + A_1u^2 + A_2u^4 + \ldots$, where only even powers of u arise, and there is no term independent of u. This function being denoted by $\wp(u)$, it can be shewn that its differential coefficient, $\wp'(u)$, is connected with it by an equation of the form

$$[\wp'(u)]^2 = 4\wp^3(u) - g_2\wp(u) - g_3,$$

where g_2, g_3 are definite functions of the periods Ω, Ω'. It can further be shewn that every elliptic function, having the character we have described, which has the periods Ω, Ω', can be expressed as a rational function of $\wp(u)$ and $\wp'(u)$. Thus the points of the cubic curve above obtained can be expressed, in the form $\xi = \wp(u)$, $\eta = -\wp'(u)$, in terms of u, the value of u which belongs to any point of the curve being given by the equation

$$u = \int_{(\xi, \eta)}^{\infty} \frac{d\xi}{\eta},$$

wherein the general value of the integral, for all possible paths of integration, is of the form $u + k\Omega + k'\Omega'$, in which k, k' are integers.

Many of the geometrical properties of a cubic curve, and, more generally, of any curve which is birationally represented by this cubic curve, become very simple with the help of the ideas which we have sketched. For instance, consider the intersections of the cubic curve with a general plane curve of order m, whose equation we may represent by $f(\xi, \eta) = 0$. These intersections correspond to the non-equivalent values of the argument u for which the doubly periodic function $f(\wp u, -\wp' u)$ vanishes. If the term in η^m is present in $f(\xi, \eta)$, this function has a pole of order $3m$ at $u = 0$, and no other non-equivalent pole, since $\wp'(u)$ has a pole of order 3 at $u = 0$, and no other non-equivalent pole, while $\wp(u)$ has a pole of order 2 at $u = 0$, and no other non-equivalent pole. Hence, by what we have said, the function $f(\wp u, -\wp' u)$ vanishes for $3m$ non-equivalent values of u, say $u_1, ..., u_{3m}$, and the sum $u_1 + ... + u_{3m}$, save for an additive term $k\Omega + k'\Omega'$, in which k, k' are integers, is equal to the sum of the values of u at the non-equivalent poles of the function, that is, equal to zero. Conversely, it can be shewn that, if $3m - 1$ arbitrary points be taken on the cubic curve, then a curve of order m can be put through these, and is thereby determined, so far as its intersections with the cubic curve are concerned, the argument of its remaining intersection being given by the equation remarked.

Of the intersections with a given *rational* curve, however, of another curve, *all* can be arbitrarily assigned. For a given curve which is neither rational, nor elliptic, it will be seen, in what follows, that *more than one* of its intersections with another curve are determined by the others; the number of such intersections determined by the others will be seen to be a distinguishing mark of the character of the given curve.

Examples of the properties of elliptic curves. 1. All plane cubic curves which pass through 8 of the nine intersections of two given cubic curves $U = 0$, $V = 0$, pass through their remaining intersection. In fact these cubics all have an equation of the form $U + \lambda V = 0$. We have remarked above on the general case of the intersection of a given cubic curve with curves of order m. A particular consequence is that, if three lines meet a given cubic curve respectively in P, Q, R, in P', Q', R', and in P'', Q'', R'', and $PP'P''$ be in line and $QQ'Q''$ be in line, then also $RR'R''$ are in line.

2. If the tangent line of the cubic curve $f(x, y, z) = 0$, at a point (x, y, z), meet the curve again in (ξ, η, ζ), we have $\xi \partial f/\partial x + \eta \partial f/\partial y + \zeta \partial f/\partial z = 0$. Regarding (ξ, η, ζ) as given, and (x, y, z) as current coordinates, this equation represents a conic, called the *polar conic* of (ξ, η, ζ); it can easily be shewn to touch the cubic curve at (ξ, η, ζ); thus it has 4 other intersections with the curve; so that 4 tangents can be drawn to the curve from any point of itself. We have seen that the coordinates of the points of the cubic curve can be represented by elliptic functions of an argument u, in such a way that the three values of u at the intersections of the curve with a line, have a sum which vanishes, or has a form $k\Omega + k'\Omega'$, where k, k' are integers, Ω, Ω' being the periods of the functions. In particular, if the tangent of the curve, at a point u, meet the curve again in v, we have $2u + v = k\Omega + k'\Omega'$. When v is given, this leads only to 4 non-equivalent values of u, namely $-\frac{1}{2}v$, $-\frac{1}{2}(v+\Omega)$, $-\frac{1}{2}(v+\Omega')$, $-\frac{1}{2}(v+\Omega+\Omega')$; these give the 4 points of contact of tangents to the curve from the point v. Denote these points in order by A, B, C, D; then AB meets the curve again in the point $v + \frac{1}{2}\Omega$, and CD meets the curve again in the point $v + \frac{1}{2}\Omega + \Omega'$, which is equivalent to $v + \frac{1}{2}\Omega$, and is the same as the former point. Similarly AC, BD meet on the curve, in the point $v + \frac{1}{2}\Omega'$, and AD, BC meet on the curve, in the point $v + \frac{1}{2}(\Omega + \Omega')$. And the tangents of the curve at the four points v, $v + \frac{1}{2}\Omega$, $v + \frac{1}{2}\Omega'$, $v + \frac{1}{2}(\Omega + \Omega')$ all meet the curve again in the same point, whose argument is $-2v$.

3. Let X, Y, T be three collinear points of the cubic curve, and Z be the point of contact of any one of the four tangents which can be drawn to the curve from T. From what is remarked (and proved algebraically) in Ex. 1, it follows that if an arbitrary line from X meet the curve again in P and P', and ZP, ZP' meet the curve again respectively in Q and Q', then Q, Q' are in line with Y. The line YQQ' is thus determined algebraically and unambiguously from the line XPP', and conversely. The pencil of all lines XPP' through X is thus projectively related to the pencil of corresponding lines YQQ'. The locus of the point of intersection of XPP' and YQQ' can in fact be shewn to be a conic which touches XZ, YZ respectively at X and Y, and passes through the points of contact of the four tangents of the curve which can be drawn from Z. A particular consequence is that the pencil of four tangents of the curve from any point X lying thereon is related to the pencil of four tangents from any other point Y. We may in particular suppose Y to be one of the inflexions of the curve, and then that the equation of the curve is reduced (by linear transformation), as is explained above, to the form $y^2z = 4x^3 - g_2xz^2 - g_3z^3$, ($y$) being the inflexion Y, and $z = 0$ the inflexional tangent. Then the pencil of the 4 tangents from Y is that given by the values of x/z which satisfy the quartic equation $4x^3z - g_2xz^3 - g_3z^4 = 0$. Assuming algebraic properties, the quadratic and cubic invariants of this equation are g_2 and g_3 and the cross ratio of its roots satisfies a sextic equation whose coefficients are rational in $g_2{}^3/g_3{}^2$. This ratio is the absolute

invariant of the elliptic functions in terms of which the curve is expressible; and the necessary and sufficient condition for two cubic curves to be birationally transformable into one another is that this invariant have the same value for both curves.

4. The following fundamental so-called theorem of coresiduation for a cubic curve follows at once from the result stated above in regard to the sum of the elliptic arguments of the intersections of any curve with a given cubic curve: let (A), (P) be two sets of points of the cubic curve, forming together the complete intersection of this curve with another curve; let (A), (Q) be likewise, together, the complete intersection of the cubic curve with another curve of the same order as before. The sets (P), (Q) are each said to be *residual* to (A), and *coresidual* with one another. Now draw any other curve, of sufficiently high order, through the points (P), its residual set of intersections with the cubic curve being (B). Then the theorem is that the sets (Q), (B) form together the complete intersection of the cubic curve with another curve. Stated briefly, of two coresidual sets, on the cubic curve, either is the residual of an arbitrary residual of the other. Algebraically, there exists a function, rational in the coordinates by which the cubic curve is expressed, which, considered on the curve, has the set (P) for its zeros, and the set (Q) for its poles (or infinities).

5. By eliminating y between the equation of a general cubic curve, and the equation of a line $ly = \mu x + z$, and expressing that the resulting cubic equation in x/z has three equal roots, we may prove that the determination of the inflexions of the cubic curve depends upon an equation of order 9. There are thus 9 inflexions, and, when the coefficients in the equation of the cubic curve are real, it appears at once that at least one of these inflexions is real. Assuming one inflexion known, the equation of the curve is at once reducible, as we have seen, to the form

$$y^2 z = x^3 + axz^2 + bz^3.$$

Taking this form, prove that the conditions for $ly = \mu x + z$ to be an inflexional tangent are $a^2 \lambda^2 + \lambda(3b\mu^2 - 4a\mu) + \mu^2 = 0$, $3a\lambda^2 - 6\lambda\mu - \mu^4 = 0$, where $\lambda = l^2$. Putting c for $a^3/27b^2$, using instead of μ the value $\theta = \frac{1}{3}a/b\mu$, and putting $f(\theta) = 3\theta^4 - 2\theta^3 + 6c\theta^2 - 6c\theta + c - c^2$, prove that the conditions are equivalent to $f(\theta) = 0$, $\lambda = 2(\theta^2 + c)^2/b\theta^2 f'(\theta)$, where $f'(\theta) = \partial f/\partial \theta$. The form $f(\theta)$ is one for which the quadratic invariant vanishes. When c is real the equation $f(\theta) = 0$ has two real, and two imaginary roots, as we may verify, using $(\theta - \frac{1}{6})f'(\theta) = f(\theta) + (3\theta^2 - \theta + c)^2$. A corresponding inflexion is then given by $x/z = a^2 f'(\theta)/54b(\theta^2 + c)^2$, $y\lambda^{\frac{1}{2}} = \mu x + z$. Thus, through any inflexion of the curve there can be drawn four straight lines each of which contains two further inflexions; and if the original inflexion be real, and a, b be real, two of these lines are real, and two are conjugate imaginaries; of the real lines, one contains two real inflexions beside the original, and the other contains two conjugate imaginary inflexions (the signs of $f'(\theta)$ being opposite for the two real roots of $f(\theta)$, as is easy to see). Let A, B, C be the real inflexions on the first line, and APP' the second real line containing the two conjugate imaginary inflexions P and P'; through B there will similarly be a real line BQQ', containing the two conjugate imaginary inflexions Q, Q', which will therefore be, one on each of the two conjugate imaginary lines through A; through C there will similarly pass a real line CRR', with the imaginary inflexion R, say, on AQ, and the imaginary inflexion R' on AQ'. The three real lines APP', BQQ', CRR' thus contain all the nine inflexions, each line containing one real, and two conjugate imaginary inflexions. But the line joining any two inflexions whatever contains a third inflexion. This follows from the remark made in Ex. 1. Here, to use the equations, we may prove this anew. Let $z = 0$ be the line joining two inflexions, $P = 0$, $Q = 0$ being the inflexional tangents at these; the equation of the cubic curve is expressible by the vanishing of a homogeneous cubic polynomial in z, P, Q, and here this reduces to the single term z^3 both when $P = 0$ and when $Q = 0$. The equation of the curve is thus of the form $z^3 + PQ(aP + bQ + cz) = 0$, where a, b, c are constants. There is thus a third inflexion on $z = 0$, given by $aP + bQ = 0$. Now write this equation more symmetrically in the form $z^3 + PQR = 0$, and take the constants l, m, n, k, and the linear functions of P, Q, z denoted by X and Y, in such a way that $lP + mQ + nR = 3kz$, and also $lP = X + Y + kz$, $mQ = \omega X + \omega^2 Y + kz$, $nR = \omega^2 X + \omega Y + kz$, where $\omega = \exp(2\pi i/3)$; and then take $Z = (lmn + k^3)^{\frac{1}{3}} z$, $\mu = k/(lmn + k^3)^{\frac{1}{3}}$. Then the equation of the cubic curve takes the form

$$X^3 + Y^3 + Z^3 - 3\mu XYZ = 0.$$

Each of the lines $X = 0$, $Y = 0$, $Z = 0$ contains three inflexions, and, in particular, the inflexional tangents at the three inflexions on $Z = 0$ are $X + Y + \mu Z = 0$, $\omega X + \omega^2 Y + \mu Z = 0$, $\omega^2 X + \omega Y + \mu Z = 0$. We thus see that any two inflexions determine a triangle whose sides contain all the nine inflexions, and that there are four lines through any inflexion each containing two others. The inflexions thus lie, by threes, on twelve lines; and there are four triangles for each of which the sides contain all the inflexions. Beside $XYZ = 0$, found above, it is easy to see that the other three triangles are $(X + Y + Z)(\omega X + \omega^2 Y + Z)(\omega^2 X + \omega Y + Z) = 0$, and the two whose equations are obtainable from this last by replacing X by ωX and by $\omega^2 X$ respectively, where $\omega = \exp(2\pi i/3)$.

The number and configuration of the inflexions may also be obtained by using the representation in terms of elliptic functions. The argument, u, of such a point, is such that $3u = k\Omega + k'\Omega'$, where k and k' are integers. The inflexions have thus the nine non-equivalent arguments $\frac{1}{3}k\Omega + \frac{1}{3}k'\Omega'$, where $k = 0$, 1, 2 and $k' = 0$, 1, 2.

The inflexions are the intersections of the cubic curve $f = 0$ with another cubic curve, commonly called the Hessian of the original, whose equation is obtained by equating to zero the three-rowed determinant whose general element is f_{ij}, in which $f_{11} = \partial^2 f/\partial x^2$, $f_{12} = \partial^2 f/\partial x \partial y$, etc. The Hessian is a covariant of the cubic form, giving the same curve if formed, by the same rule, from any linear transformation of the original equation of the cubic.

Of the three methods of considering the inflexions here suggested, the first, by direct algebra, requires modification when the cubic curve has a double point; the second, by elliptic functions, is inapplicable; but the third may be used. The Hessian of a cubic curve which has a double point, is a cubic curve also with a double point whereat the tangents are the same as for the original curve—the other three intersections give the only existing inflexions, three in number, which are in line. The inflexions of $xyz + u(x, y) = 0$ lie on $2z + \partial^2 u/\partial x \partial y = 0$.

In the direct algebraic consideration, we have assumed one of the inflexions to be known. It is, however, the case that the equation of the ninth order on which the determination of the inflexions depends is one which is capable of algebraic solution. This can be seen from what has been proved if we add thereto the facts (a) that the Hessian of a given general cubic curve has its inflexions at the inflexions of the given cubic and (b) that every cubic curve through the inflexions of a given general cubic curve has likewise its inflexions at these points; these follow by remarking that the Hessian of a cubic curve given by $y^2 z + (x, z)_3 = 0$ is given by $y^2 L + [x, z]_3 = 0$, where L is of the form $px + qz$. From this it follows, if $f = 0$, $H = 0$ be a general cubic curve and its Hessian respectively, that the Hessian of a curve $\lambda f + \mu H = 0$, passing through the intersections of $f = 0$, $H = 0$ has an equation $\lambda' f + \mu' H = 0$, where then λ', μ' are polynomials in λ, μ, of the third order, as we see by considering the determinantal formation of the Hessian. A triangle containing the nine inflexions of $f = 0$, as passing through the intersections with $H = 0$, has an equation $\lambda f + \mu H = 0$; and the Hessian of the cubic curve constituted by three lines consists of these lines, as we see at once; thus, when $\lambda f + \mu H = 0$ is a triangle, $\lambda \mu' - \lambda' \mu = 0$. This leads to four values of λ/μ. We have seen, however, that there are just four triangles containing the inflexions. Whence the condition that the Hessian of $\lambda f + \mu H = 0$ coincides with this is a sufficient as well as necessary condition that $\lambda f + \mu H = 0$ should be a triangle. The cubic polynomials λ', μ' in λ, μ can be computed by using a particular form of f, for example the form $y^2 z + (x, z)_3$. Whence the quartic equation $\lambda \mu' - \lambda' \mu = 0$ is known in terms of g_2 and g_3, which are the invariants of the form f, as may be proved. Whence, given f in its general form, and solving a quartic equation whose coefficients are functions of the invariants of f, we can determine the triangles each containing the nine inflexions. Moreover, the formation of the Hessian of a general cubic, the formation of the Hessian of $\lambda f + \mu H$, and the expression of the conditions that this Hessian agrees with $\lambda f + \mu H$, are direct algebraic problems, from which, by elimination, the quartic equation $\lambda \mu' - \lambda' \mu = 0$ can be formed without knowledge of the invariants of f. When finally we have a cubic form which is known to be a triangle, the vertices of this triangle are obtainable by solving the algebraic problem of finding the three common points of three conics, known to have three points in common (these conics being the first polars of the cubic). Thence the sides of two such triangles are known, and the inflexions of the original cubic are the intersections of these sides. Full consideration of the cubic form may be found in Clebsch-Lindemann-Benoist, *Leçons sur la Géométrie*, II, Paris, 1880. That the equation of the ninth degree on which the inflexions depend can be solved algebraically is a consequence, however, of the relation of three roots of this equation which follows from the fact that the line joining two of the inflexions contains a third. For this see Weber, *Algebra*, II, 1896, p. 322. Also, Dickson, *Linear Groups*, 1901, p. 77; Steiner, *Werke*, II, p. 435; Netto, *Combinatorik*, 1901, p. 202.

6. A problem for the cubic curve, suggested by the use of the elliptic functions, is that of finding a point on the curve such that the conic described to have there 5 intersections may have its sixth intersection also there. Of these points, called *sextactic* points, there are 27, lying in threes upon 9 lines, a set of 3 being the points of contact of the three tangents to the curve from an inflexion (other than the inflexion itself). The elliptic argument of a sextactic point is in fact clearly given by

$$u = \tfrac{1}{6}k\Omega + \tfrac{1}{6}k'\Omega', \quad k = 0, 1, \ldots, 5; \; k' = 0, 1, \ldots, 5,$$

where simultaneously even values of k, k' are to be excluded. The remaining $36 - 9$ or 27 possibilities are in 9 such sets of three as

$$\tfrac{1}{6}(2h+3)\Omega + \tfrac{1}{3}h'\Omega', \quad \tfrac{1}{3}h\Omega + \tfrac{1}{6}(2h'+3)\Omega', \quad \tfrac{1}{6}(2h+3)\Omega + \tfrac{1}{6}(2h'+3)\Omega',$$
$$h, h' = 0, 1, 2.$$

More generally, it may easily be proved that if the tangent at a point A of the cubic curve meet the curve again in B, and the tangent at B meet the curve again in C, and the chord AC meet the curve again in P, then the conic with five intersections at A contains P.

Cf. Cayley, *Papers*, IV, p. 227; V, p. 221. For the application of general methods to determine the sextactic points of a general curve, see Segre, *Ann. d. Mat.* XXII, 1894, p. 90 and Enriques-Chisini, *Teoria geometrica*, I, p. 277; II, p. 289.

7. A fundamental theorem is that the triads of points in which an arbitrary line meets a cubic curve, and its Hessian, are apolar to one another. It can be proved at once by verifying that the two curves $x^3 + y^3 + z^3 = 0$, $xyz = 0$ are met in apolar triads by an arbitrary line, and then noticing that the Hessian of the general cubic curve expressed by $x^3 + y^3 + z^3 - 3\mu\,xyz = 0$ has an equation of the form $x^3 + y^3 + z^3 - 3\mu_1 xyz = 0$.

If two binary cubics be symbolically expressed by a_x^3, b_x^3, the condition that they should be apolar is symbolically expressed by $(ab)^3 = 0$; when this is so, if p, q, r be the roots of $a_x^3 = 0$, we not only have $a_p a_q a_r = 0$, but also $b_p b_q b_r = 0$. From this it follows that, if f_x^3, h_x^3 be the symbolical forms of a ternary cubic and its Hessian, and (u_1, u_2, u_3) be arbitrary, we have $(fhu)^3 = 0$; and if (x), (y), (z) be the three points of intersection of the line $u_x = 0$ with the cubic curve $f_x^3 = 0$, we have $f_x f_y f_z = 0$ and $h_x h_y h_z = 0$. Another form of the theorem is that, if (x), (y) be any two points of the cubic curve, we have $f_x^2 f_y / h_x^2 h_y = f_x f_y^2 / h_x h_y^2$.

8. It may be proved, if (c) be an arbitrary point, and $(cx\,dx)$ denote the three-rowed determinant whose rows are (c_1, c_2, c_3), (x_1, x_2, x_3), (dx_1, dx_2, dx_3), that the integral

$$w = \int_{(y)}^{(x)} (cx\,dx)/f_x^2 f_c,$$

which is independent of (c), is finite for all positions of (x), (y) upon the cubic curve $f_x^3 = 0$, along which it is taken. And then that the elliptic function $\wp(w)$, defined to have the periods obtained from this integral, is given by

$$3(cxy)^2 \wp(w) = (f_x f_y f_c)^2 + 2f_x^2 f_c \cdot f_y^2 f_c - f_x f_c^2 \cdot f_x f_y^2 - f_y f_c^2 \cdot f_y f_x^2.$$

In particular, taking

$$f_x^3 = \tfrac{3}{2} [x_2^2 x_3 - 4x_1^3 + g_2 x_1 x_3^2 + g_3 x_3^3],$$

and also

$$(c) = (0, 1, 0), \quad \xi = x_1/x_3, \quad \eta = -x_2/x_3, \quad \xi' = y_1/y_3, \quad \eta' = -y_2/y_3,$$

with also

$$u = \int_{(\xi)}^{\infty} \frac{d\xi}{\eta}, \quad u' = \int_{(\xi')}^{\infty} \frac{d\xi}{\eta},$$

$(\xi, \eta, 1)$ and $(\xi', \eta', 1)$ being points of $f_x^3 = 0$, the integral w is $u - u'$, and the formula for $\wp(w)$ becomes $\wp(u - u') + \xi + \xi' = \tfrac{1}{4}(\eta + \eta')^2(\xi - \xi')^{-2}$, leading when ξ', η' are infinite to $\wp(u) = \xi$.

The general formula for $\wp(w)$ was given by G. Pick, *Math. Annal.* XXVIII, 1887, p. 309. It can, however, also be shewn that

$$\wp(u - u') = h_y^2 h_x / f_y^2 f_x, \quad = h_y h_x^2 / f_y f_x^2.$$

In the quotient $h_y^2 h_x / f_y^2 f_x$, if we regard (y) as a fixed point of the cubic curve, the denominator, $f_y^2 f_x$, regarded as a function of (x), vanishes to the second order at (y); and the numerator $h_y^2 h_x$ vanishes at the tangential point of (y), where the tangent of the cubic at (y) meets the curve again. Thus this quotient, regarded as an algebraic function of (x) on the cubic curve, has only the pole (y), of the second order. Considering the value of the function when (x) is near to (y), it may be verified that its infinite part is $(u - u')^{-2}$, and that $(u - u')^2 h_y^2 h_x / f_y^2 f_x$ is of the form $1 + H$, where H vanishes to the fourth order when (x) approaches to (y). Hence the identity $\wp(u - u') = h_y^2 h_x / f_y^2 f_x$ follows from the theory of the function $\wp$. An algebraic proof of this, and other formulae of this character, will be found in Halphen, *Fonctions elliptiques*, 2me partie, Chap. XI, 1888, p. 433, etc.

9. If $\phi = 0$, $\psi = 0$ be two independent cubic curves through 8 arbitrary points, the sextic curves, with these 8 points as double points, whose number should be $\frac{1}{2}7.8 - 24$, or 4, are given by an equation

$$a\phi^2 + 2h\phi\psi + b\psi^2 + cU = 0,$$

where a, h, b, c are variable parameters, and $U = 0$ is a particular sextic curve with these double points. These sextic curves meet the cubic $\phi = 0$ in $18 - 16$ or 2 further points, determining thereon ∞^1 sets of pairs of points given by $b\psi^2 + cU = 0$. These pairs are then determinable by a variable line passing through a fixed point of the cubic curve $\phi = 0$, and there are 4 points of this curve at which the points of a pair coincide; one of these is however the ninth intersection of the curves ψ and ϕ, the pair given by $b\psi^2 + cU = 0$ coinciding when $c = 0$. Let $\psi^2 - eU = 0$ be one of the three other curves $b\psi^2 + cU = 0$ whose two remaining intersections with $\phi = 0$ coincide, say in A. Then there are ∞^1 sextic curves, given by the equation $\lambda\phi^2 + \psi^2 - eU = 0$, in which λ is variable, which have 9 double points, consisting of A and the original 8 points. By what is proved in this chapter, for curves of order n with $\frac{1}{2}(n-1)(n-2) - 1$ double points, all these curves are elliptic. More generally, it can be proved in a similar way, that it is possible to have ∞^1 curves, of order $3n$, all having 9 multiple points, not double points but multiple points of order n. And these will be elliptic curves, since $\frac{1}{2}(3n-1)(3n-2) - \frac{9}{2}n(n-1) = 1$. This circumstance was remarked by Halphen, *Bull. d. l. Soc. Math. d. France*, X, 1882, p. 162.

CHAPTER II

THE ELIMINATION OF THE MULTIPLE POINTS OF A PLANE CURVE

As a large part of the difficulty of the theory of curves depends upon the possible intricacy of their multiple points, it seems desirable to shew at once, with the most elementary considerations, that any plane curve is in (1, 1) birational correspondence with a curve, in space of three or more dimensions, which is without multiple points. Some of the ideas involved have already been met with in Chap. I; but the present chapter is designed to be complete in itself.

Consider a given plane curve $f(x, y, z) = 0$, of order μ, with any multiple points; the curve is supposed to be irreducible, that is, the polynomial $f(x, y, z)$ cannot be written as the product of other polynomials also rational in x, y, z. A curve whose equation is of the form $\lambda_0 \phi_0 + \lambda_1 \phi_1 + \ldots + \lambda_r \phi_r = 0$, wherein $\phi_0 = 0, \ldots, \phi_r = 0$ are definite curves of the same order, which may have common points, upon $f = 0$, or elsewhere in the plane, determines, by its intersections with $f = 0$, a *set* of points thereon. Of these points, some may be the same for all values of the parametric coefficients $\lambda_0, \ldots, \lambda_r$, being common intersections with $f = 0$ of all the curves $\phi_0 = 0, \ldots, \phi_r = 0$; the others will, in general, vary when these coefficients vary. It is naturally assumed that there exists no identical linear homogeneous relation, with constant coefficients, connecting the polynomials $\phi_0, \ldots, \phi_r$; it is also assumed that there is no such relation which is true in virtue of $f = 0$, as would be the case if there were an identity of the form $c_0 \phi_0 + \ldots + c_r \phi_r = \psi f$, in which $c_0, \ldots, c_r$ are constants, and ψ is a polynomial in x, y, z. Thus we say that $\phi_0, \ldots, \phi_r$ are assumed to be linearly independent upon $f = 0$. Then we say that, as $\lambda_0, \ldots, \lambda_r$ vary, the curve $\lambda_0 \phi_0 + \ldots + \lambda_r \phi_r = 0$ cuts upon $f = 0$ a *linear series of sets of points of freedom* r; the number of points of a set which vary when $\lambda_0, \ldots, \lambda_r$ vary is called the *grade* of the series, and will be denoted by n; as, by proper choice of $\lambda_0, \ldots, \lambda_r$, a set of the series can be found including r arbitrary points of $f = 0$, it is clear that $n \geqslant r$.

There are two properties of such a linear system in regard to which it is desirable to be clear. First, there may exist, on $f = 0$, a batch of, say, s points, of definite positions, such that the curves of the linear system $\lambda_0 \phi_0 + \ldots + \lambda_r \phi_r = 0$ which are made, by suitable limitation of $\lambda_0, \ldots, \lambda_r$, to pass through some number less than s of the points of this batch, necessarily pass through the other points;

a simple illustration is that of a double point of $f=0$, with distinct tangents, so that we can distinguish the two points of the curve which coincide thereat; the curves of the linear system passing through one of these points necessarily passes through the other. In general, to make a curve of the linear system pass through a point of $f=0$ not common to all of $\phi_0=0, \ldots, \phi_r=0$, we must prescribe one linear relation connecting $\lambda_0, \ldots, \lambda_r$; a batch of s points of the kind under consideration is one which imposes less than s conditions for the curves of the linear system which are to contain the points of this batch. Such a batch is sometimes spoken of as *neutral*. The other possible property spoken of, refers to a batch of points on $f=0$ of which one point is arbitrary thereon; it may be that the curves of the linear system which are made to pass through *any* arbitrary point of $f=0$ necessarily then pass through other points, which vary when the first point is varied. A simple illustration is that of the linear system of lines (of freedom 1) which pass through a fixed point; any such line, made to pass through an arbitrary point of the curve $f=0$, necessarily passes through the other intersections of this line with the curve. We say, of a linear system $\lambda_0\phi_0+ \ldots +\lambda_r\phi_r=0$, that it is *stiff*, when any curve of it which passes through a general point of $f=0$, then necessarily passes through other, say $(k-1)$, points of $f=0$, determined by the first point, and variable with it. As a curve of the system can be put through r points of $f=0$, each eligible arbitrarily, independently of the others, the value of k must be such that $kr \leqslant n$. A linear system which is not stiff may be called *pliable*; the series determined on $f=0$ by this system is most often said to be *simple*.

We now consider linear systems *contained* in the given system; such a system is given by $\mu_0\psi_0+ \ldots +\mu_\rho\psi_\rho=0$, with $\rho+1, <r+1$, homogeneously entering parameters $\mu_0, \ldots, \mu_\rho$, wherein $\psi_0, \ldots, \psi_\rho$ are definite linear functions of $\phi_0, \phi_1, \ldots, \phi_r$ with constant co-efficients, which are linearly independent of one another. For instance all the curves of the original system which pass through an assigned point form such a contained system, with $\rho=r-1$. What we are interested in here is the possibility of such contained systems for which there is a lessening of the grade, or number of variable intersections with $f=0$, which is *at least twice* the lessening of the freedom. This arises, for instance, if there be a double point of $f=0$ which is not a common zero of $\phi_0, \ldots, \phi_r$, when we form, from the original system, the system of freedom $r-1$, of which all curves pass through this double point; the new system has a grade which is less by two, at least, than that of the original. If we denote $n-r$ by s, and call it momentarily, the *sequence*, and also introduce $s-r$, or $n-2r$, and call it the *efficiency* of the system, what we are interested in here is whether the given system contains another system whose

efficiency is less than, or at most equal to its own. We may momentarily say that a system containing such another system is *capable of reduction.*

We can shew that a linear system of freedom $r \geqslant 2$ which is pliable, or cuts on $f = 0$ a linear series which is simple, and is not capable of reduction, enables us to put the curve $f = 0$ into (1, 1) birational correspondence with another curve, in space of r dimensions, which has no multiple point. For perfect clearness we shall make the assumption, fully justified independently in a following chapter, that in every possible method of approach along the algebraic curve $f = 0$, to a simple or multiple point of this curve, the co-ordinates of the approaching point may be given by converging series of a parameter vanishing at the multiple point. Let the linear system be given as before by $\lambda_0 \phi_0 + \ldots + \lambda_r \phi_r = 0$, and consider the point $(\xi_0, \ldots, \xi_r)$, in space $[r]$, of dimension r, which is given by $\xi_0/\phi_0 = \xi_1/\phi_1 = \ldots = \xi_r/\phi_r$; as (x, y, z) describes $f = 0$, this point describes a curve in the space $[r]$; for any general (simple) point of $f = 0$, this point (ξ) is definite; for any point of $f = 0$ at which all of $\phi_0, \ldots, \phi_r$ vanish, the values of $\xi_0 : \xi_1 : \ldots : \xi_r$ are to be found by a limiting process, from the expansions of x, y, z referred to; and when the point of $f = 0$ is a multiple point there will correspond in general several points (ξ), each found from an appropriate expansion. Conversely, to any point (ξ) of the new curve, corresponds the point (x, y, z) of $f = 0$ from which it arose, but there corresponds no other point (x', y', z') of $f = 0$, unless, when ϕ_i' denotes what ϕ_i becomes when x', y', z' are put for x, y, z, we have

$$\phi_0'/\phi_0 = \phi_1'/\phi_1 = \ldots = \phi_r'/\phi_r;$$

these equations would involve, however, that any curve

$$\lambda_0 \phi_0 + \ldots + \lambda_r \phi_r = 0$$

which passes through the point (x, y, z) necessarily passes also through the point (x', y', z'). This possibility is expressly excluded, for an arbitrary point (x, y, z), by the hypothesis that the linear system is pliable. It is not excluded thereby for definite fixed positions of (x, y, z), which might then give rise to points (ξ) through which several branches of the curve in the space $[r]$ would pass. We deal with this below. The algebraic equations $\xi_0/\phi_0 = \ldots = \xi_r/\phi_r$, $f(x, y, z) = 0$, thus lead only to one set of values for (x, y, z), in terms of $(\xi_0, \ldots, \xi_r)$, for general values of these latter; and these, being obtainable by rational processes, must be capable of rational (necessarily homogeneous) expression in terms of $\xi_0, \ldots, \xi_r$. The new curve is thus in (1, 1) birational correspondence with $f = 0$. The new curve, however, can have no multiple point, namely cannot have a point such that the general prime of the space $[r]$, given by an equation, $c_0 \xi_0 + \ldots + c_r \xi_r = 0$, which is made to pass through this point, has a number of other intersections, with the curve, less by 2, or more, than the number of intersections of a general prime (the order of the curve). This order is clearly the number of intersections, with $f = 0$, of the curve $\lambda_0\phi_0 + \ldots + \lambda_r\phi_r = 0$, which are not common to all of $\phi_0 = 0, \ldots, \phi_r = 0$, that is, to the grade n of the series on $f = 0$ determined by this system. If the new curve, in $[r]$, had a multiple point, this system would contain another, of one freedom less, and of grade at least two less—and this we have excluded by the hypothesis that the system is not capable of reduction.

Passing now to the consideration of the reduction of the given linear system, of grade n and freedom r, suppose it contains another system, of grade n_1 and freedom r_1, (with $r_1 < r$), with not greater efficiency than itself, namely $n_1 - 2r_1 \leqslant n - 2r$, the conditions $1 \leqslant r_1 \leqslant n_1$ being satisfied. Suppose this new system likewise contains a further system of grade n_2 and freedom r_2, with

$$r_2 < r_1, \quad n_2 - 2r_2 \leqslant n_1 - 2r_1, \quad 1 \leqslant r_2 \leqslant n_2;$$

and so on. Each system is contained in the preceding, and, therefore, in the original system. There must be a final system beyond which reduction cannot go, either because a further step would lead to increased efficiency (as would be the case, for example, if a lessening of one in the freedom led only to a lessening of one in the grade, $\nu - 1 - 2(\rho - 1)$ being equal to $\nu - 2\rho + 1$), or because the freedom, having sunk to 1, cannot be further lessened, and still leave a variable system. Let this final system be of freedom r' and grade n', with $1 \leqslant r' \leqslant n'$ and $n' - 2r' \leqslant \ldots \leqslant n_1 - 2r_1 \leqslant n - 2r$. When $r' \geqslant 2$, this final system is pliable, since else, taking the system contained therein of which all the curves pass through a definite point of $f = 0$, we should have a new system, of freedom at least 1, with grade at least two less than n'. Thus, when $r' \geqslant 2$, the final system can be used to put $f = 0$ into $(1, 1)$ birational correspondence with a curve, in space $[r']$, having no multiple points. But $r' \geqslant 2r - n + s'$, where $s' = n' - r'$ is the sequence of the final system; thus $r' \geqslant 2$ when $s' \geqslant 2$, provided $2r - n \geqslant 0$. We prove below that if $s' = 0$ or $s' = 1$, the curve $f = 0$ has a special character, being either rational (for $s' = 0$), or of a kind called *hyperelliptic*. But whatever be $s'(\geqslant 0)$, we have $r' \geqslant 2$ if $2r - n \geqslant 2$.

And we can always take the original system so that $2r - n \geqslant 2$. For let the system be given by the most general polynomial of order N, where $N > \mu - 3$, in which μ is the order of the curve $f = 0$. Such a polynomial U has $\frac{1}{2}(N + 1)(N + 2)$ variable coefficients, entering homogeneously; when $N \geqslant \mu$, this polynomial vanishes on $f = 0$ at the same points as a polynomial $U - uf$, where u is any homogeneous polynomial of order $N - \mu$, containing therefore

$$\tfrac{1}{2}(N - \mu + 1)(N - \mu + 2)$$

homogeneous coefficients; these can be used to reduce to zero so
many of the coefficients in $U - uf$, or a fewer number if the linear
equations necessary for this be not independent. Thus $U = 0$ cuts
on $f = 0$ a series whose freedom r is such that

$$r \geqslant \tfrac{1}{2}N(N+3) - \tfrac{1}{2}(N-\mu+1)(N-\mu+2),$$

and this, in terms of the grade $n(=N\mu)$, is the same as

$$r \geqslant n - \tfrac{1}{2}(\mu-1)(\mu-2).$$

Since $(N-\mu+1)(N-\mu+2) = 0$, when $N = \mu-1$, or $N = \mu-2$, this
is true also for $N > \mu-3$, even if $N < \mu$. For $N = \mu-3$, with
$n = \mu(\mu-3)$, the corresponding formula is $r = n - \tfrac{1}{2}(\mu-1)(\mu-2)+1$.
For $N = \mu-1$, the formula gives $2r-n \geqslant 2(\mu-1)$, which is $\geqslant 2$ if
$\mu \geqslant 2$; for greater values of N, the value of $2r-n$ is still greater.
For $N = \mu-2$, the formula gives $2r-n \geqslant \mu-2$, which is $\geqslant 2$ if
$\mu \geqslant 4$. Thus certainly if $\mu \geqslant 4$, the curve $f = 0$ can be put into $(1, 1)$
birational correspondence with a curve having no multiple points,
in space of sufficiently high dimensions, the initial system of curves
being of order $\geqslant \mu-2$. For $N = \mu-3$, the formula gives $2r-n = 0$,
and we must have, for the final system in the preceding process of
reduction, $s' \geqslant 2$. This is so, we have said, and now prove, unless
$f = 0$ be a rational or hyperelliptic curve.

When, for the final system, s', $= n'-r' = 0$, $r' \geqslant 1$, consider the
curves of the final system put through $r'-1$ points, arbitrarily
fixed on $f = 0$; as the process of possible reduction has terminated,
these particular curves will give, on $f = 0$, a series of grade 1 and
freedom 1; these curves will be given by an equation $u - v\theta = 0$,
where u, v are definite polynomials, and θ is variable, and each
curve has one variable intersection with $f = 0$. The coordinates of
this variable point are then obtainable, by rational processes, as
rational functions of θ; conversely any value of θ determines such
an intersection; and the curve is therefore rational. Similarly when,
in the final system, $n' - r' = 1$, we can obtain an ∞^1 system of curves
$u - v\theta = 0$ having *two* variable intersections with $f = 0$. The ratios of
the coordinates of these intersections are then obtainable as roots
of a quadratic equation whose coefficients are rational in θ; thus the
coordinates of a general point of $f = 0$ are expressible as rational
functions of two variables ϕ, θ, of which ϕ^2 may be taken to be a
polynomial in θ; while, conversely, θ, ϕ are expressible rationally
by the coordinates of the point of the curve. It is such a curve
which is known as hyperelliptic; in particular it may be elliptic
(though the other word was originally introduced to express the
distinction).

Remark. In passing we may refer to a question which naturally
arises, whether the contemplated reduction is obtained by utilising,
at any stage of the process, the fact that the linear system to be

reduced is stiff, determining on $f=0$ a series which is not simple. The answer is immediate in case, for the original system, $N > \mu - 3$, and therefore $2r - n \geqslant 2$. For this involves

$$2r' - n' \geqslant \ldots \geqslant 2r_1 - n_1 \geqslant 2r - n;$$

and we have remarked that, for a system of grade ν and freedom ρ to be stiff, we must have $2\rho - \nu \leqslant 0$. In this case, therefore, neither the original system, nor any contained system obtained in the method of reduction, is stiff. If for the original system $N = \mu - 3$ and $2r - n = 0$, the answer to the question appears implicitly in subsequent more detailed theory; but the case of a rational or hyperelliptic curve evidently stands apart from the general case.

Suppose then that we have placed the original curve $f=0$ in (1, 1) birational correspondence with a curve Ψ', having no multiple points, in space $[r']$, starting with polynomials of order $\geqslant \mu - 3$ in general, but with polynomials, of order $> \mu - 3$ when the curve is rational or hyperelliptic. We proceed to deduce that when $r' > 2$ we can obtain a (1, 1) birational representation by means of a curve without multiple points in ordinary space of three dimensions, and also such a representation by a plane curve having no other multiple points than double points with distinct tangents. The first statement is proven when $r' = 3$; suppose then $r' > 3$. The points of the chords and tangents of a curve Ψ' in space $[r']$ form a locus, M_3, of 3 dimensions, there being ∞^2 chords of the curve (including tangents), and ∞^1 points on each chord, at most. In a space of r' dimensions, a space of $\check{r}' - 4$ dimensions, say a $[r' - 4]$, the intersection of four primes of the space (cf. the general explanation in Vol. I of the writer's *Principles of Geometry*), can be chosen to have no point in common with M_3; in general, two spaces $[p]$, $[q]$, in $[r]$, meet in a point only if $p + q \geqslant r$. Take now an arbitrary threefold space $[3]$, say S, in the space $[r']$, and join the chosen $[r' - 4]$ to any point P of the curve Ψ', by means of a space $[r' - 3]$. This will meet the space S in a point, P', which we speak of as obtained from P by projection from the chosen $[r' - 4]$. As P describes the curve Ψ' we thus obtain a curve Ψ'', in the space S, described by the point P'. Conversely, there are not two points P, Q, of the curve Ψ', thus giving rise to the same point P' of Ψ''; for, if so, the chord PQ, which lies in a definite $[r' - 3]$ through the chosen $[r' - 4]$, would meet this $[r' - 4]$, while we have supposed this space to be so chosen as not to contain any point of any chord or tangent of Ψ'. There is thus a definite (1, 1) correspondence between the two curves Ψ', Ψ''; and as Ψ' is without multiple points, so also is Ψ''. From the nature of the process, this correspondence may be expressed by rational equations, the coordinates of a point of either curve being rational in the coordinates of the point of the other curve to which it corresponds. Proceeding now further to place the curve Ψ'' in correspondence with a plane curve, we assume that this curve, in space [3], has ∞^1 trisecants, that is chords meeting the curve in three points, which then form a ruled surface; further we assume that there are ∞^1 chords of Ψ'', likewise forming a ruled surface, each chord having the property that the tangents of the curve, at the two ends of the chord, intersect one another; finally there are ∞^1 tangents of the curve, also forming a surface. We can then find points of the space in which Ψ'' lies which do not lie on any one of these three surfaces. Let O be such a point, and ϖ be a general plane. Then the curve Ψ'' projects from O into a curve Ψ''' of the plane ϖ; this curve Ψ''' will be without cusps (which would arise if O were on a tangent of Ψ''), and will be without multiple points of the third or higher order (which would arise if O were on a trisecant or a line meeting Ψ'' in more than three points). There may, and there will in general be double points of the curve Ψ''', corresponding to the finite number of (bisecant) chords of Ψ'' which pass through O; but as O is not on a chord of Ψ'' whereat the tangents at the two ends intersect one another, the tangents at such a double point of the curve Ψ''' will be distinct. Thus, to any general point of Ψ'' will correspond a single point of Ψ''', and conversely; but a double point of Ψ''', which may be reached along either branch of the curve passing through it, will correspond to two points of Ψ''. Thus again the coordinates of a point of either curve are rational functions of the coordinates of the point of the other to which it corresponds; but the coordinates of a point of Ψ'' are (ratios of) four polynomials, in the coordinates for Ψ''', having the property that two at least of these polynomials vanish at any double point of Ψ''', so that their ratio may be capable of taking two values as we approach the double point of Ψ'''. By use of the preceding transformation to Ψ'', we can thus obtain a direct $(1, 1)$ birational transformation between Ψ''' and the original curve $f = 0$; and in particular, the passage which we have made from the curve Ψ to Ψ''' in two steps, a projection from a $[r'-4]$, and a projection from a point O, may be made in one step, by projection from a suitably chosen space $[r'-3]$.

Remark. It will appear, from the greater detail of a subsequent chapter, that the linear system by which, after reduction, we pass to the curve Ψ, can be so taken that in the final system with which we make the transformation from $f = 0$ to Ψ, say $\mu_0\psi_0 + \ldots + \mu_{r'}\psi_{r'} = 0$, $(r' > 2)$ three of the polynomials ψ, say ψ_0, ψ_1, ψ_2, are of the forms $x\psi, y\psi, z\psi$, where x, y, z are the coordinates in $f = 0$. The corresponding coordinates ξ_0, ξ_1, ξ_2 of the space $[r']$ have then the ratios of x, y, z. In this case the transformation from $f = 0$ to Ψ is a projection on to the original plane from the space $[r'-3]$ for which

$\xi_0 = 0 = \xi_1 = \xi_2$. Thus any plane curve $f = 0$, however intricate its multiple points may be, may be regarded as a *projection*, from a curve having no multiple points, lying in higher space (or is in (1, 1) correspondence with a plane curve having no multiple points, if $r' = 2$).

For the preceding theory cf. Albanese, *Rend. Lincei*, XXXIII, 1924, pp. 13, 14; also Severi, *Trattato di geom. alg.* I, 1, 1927, p. 75. A valuable history is given on pp. 332–5 of the same volume.

Ex. 1. As a very simple application of the theory of this chapter we may take a plane quintic curve with 3 double points, and consider the system of all conics in the plane. By the process of reduction explained these lead to the contained system of conics through the double points, for which $n' = 4$, $r' = 2$. This is not further reducible, and leads to the transformation of the quintic curve to a plane quartic curve having no multiple points.

If however we take a plane quintic curve with 4 double points, the system of conics through these has $n' = 2$, $r' = 1$. The curve is therefore hyperelliptic, and cannot be transformed to a curve without multiple points with this system of conics. In this case we may take the cubic curves through the 4 double points, for which $n' = 7$, $r' = 5$; this system appears not to be further reducible in the sense of the text, and leads to a birational representation of the curve by a curve of order 7 in space of 5 dimensions, not having multiple points. But this curve can be projected from a line into a curve in ordinary space, and if the line be taken to be a general chord of this curve, this curve in [3] will be of order 5. It can be obtained directly from the original plane curve by considering cubic curves through the 4 double points and two arbitrary general points of the curve; the system of these cubic curves is of freedom 3 and grade 5; we can indeed consider cubic curves through the double points of the original curve and 3 arbitrary general points of the curve, forming a system of freedom 2 and grade 4, and so transform the original curve to a plane quartic curve; but this will have one double point, being obtained by projection of the quintic curve in [3] from a point of itself, and it can be shewn that there is one trisecant of the quintic curve through any point of itself. In fact it will appear that the quintic curve in [3] is the residual intersection of a quadric surface with a cubic surface containing one generator of the quadric, and the generator of the quadric surface of the same system through any point of the quintic curve is a trisecant. It will also appear that the septimic curve in [5] is the residual intersection of a rational ruled quartic surface with a quadric containing one generator of this.

Ex. 2. A more particular example, having however an intrinsic interest, is the curve $z(x^4 - y^3 z)^2 - 9x^5 y^4 = 0$. The system

$$\lambda_0 xy^3 + \lambda_1 zy^3 + \lambda_2 (x^4 - y^3 z) = 0$$

effects a birational transformation to the curve

$$3^6 X^7 (Y + Z)^2 - Y^3 Z^6 = 0,$$

the equations $\qquad X/xy^3 = Y/zy^3 = Z/(x^4 - y^3 z)$

leading to $\qquad x/X^2(Y + Z) = 9y/YZ^2 = z/XY(Y + Z)$,

but this system is reducible in the sense of the text, containing the system $\mu_0 xy^3 + \mu_2 (x^4 - y^3 z) = 0$ of grade $n_1 = 3$, for which $2r_1 - n_1 = -1$ is greater than $2.2 - 9$ or -5, of the original system. The new curve, in X, Y, Z, has accordingly multiple points.

If however we take the system
$$\lambda_0 x^2 y^3 + \lambda_1 x(x^4 - y^3 z) + 9\lambda_2 y(x^4 - y^3 z) = 0,$$
putting $\xi/x^2 y^3 = \eta/x(x^4 - y^3 z) = \zeta/9y(x^4 - y^3 z),$
which lead to $x/\eta(\eta^3 - \xi^2 \zeta) = 9y/\zeta(\eta^3 - \xi^2 \zeta) = z/\xi\eta^2 \zeta,$

we obtain a birational correspondence with the curve, without multiple points, $3^{-6}\eta\zeta^3 + \zeta\xi^3 - \xi\eta^3 = 0$. If we put $\xi = \xi_1$, $\eta = \omega\eta_1$, $\zeta = -\omega^3 \zeta_1$, with $\omega^7 = 3^6$, the last assumes a symmetric form $\eta_1 \zeta_1^3 + \zeta_1 \xi_1^3 + \xi_1 \eta_1^3 = 0$. For properties of this curve, the reader may consult Klein-Fricke, *Modulfunctionen*, I, 1890, p. 702, and also the writer's *Multiply-periodic Functions*, Cambridge, 1907, p. 269.

Ex. 3. We may employ the case of Ex. 2 to illustrate the expansion of the branches of a curve near a multiple point which has been referred to in the text. For the curve $3^{-6}\eta\zeta^3 + \zeta - \eta^3 = 0$, when t, $= \eta/3$, is sufficiently small, it can be proved that $\zeta = 3^3 t^3 - 3^4 t^{10} + 3^6 t^{17} - \ldots$, and hence, for the curve $(x^4 - y^3)^2 - 9x^5 y^4 = 0$, when x, y are sufficiently small, $x = t^6 - 6t^{13}\ldots$, $y = t^8 - 3^2 t^{15}\ldots$. Denoting the left side of the equation of this curve by $f(x, y)$, we may shew that, if ϕ denote a general sextic polynomial in x, y, 1, the number of linearly independent conditions for the coefficients of ϕ in order that the quotient $\phi x/\partial f/\partial y$ may vanish as we approach $x = 0 = y$ along the curve $f = 0$, is 21. Cf. Cayley, *Papers*, V, p. 524 and C. A. Scott, *Amer. J. of Maths.* xv, 1892, p. 318.

Ex. 4. It will appear subsequently that, for a curve of order μ which is not hyperelliptic or rational, and has only double points, the final reduced system, when we start with polynomials of order $\mu - 3$, is that determined by the curves of this order, passing through the double points; and that, for this system, the efficiency, $n' - 2r'$ vanishes. Any system contained in this final system has efficiency greater than zero (Clifford's theorem).

Ex. 5. If the fundamental curve be $y^2 = (x, y)_4$, a quartic with a self contact at $x = 0$, $y = 0$, it can be shewn that the system of conics is reducible, in the sense of the text, to conics touching the quartic curve at $x = 0$, $y = 0$, given by $y + (x, y)_2 = 0$. We therefore make a transformation to a space [3], by means of
$$\frac{x_0}{y} = \frac{y_0}{x^2} = \frac{y_1}{xy} = \frac{y_2}{y^2},$$
whereby the plane is transformed to the quadric cone in this space (x_0, y_0, y_1, y_2) which is expressed by
$$\frac{x_0}{\phi} = \frac{y_0}{\theta^2} = \frac{y_1}{\theta} = \frac{y_2}{1}$$
(where $\theta = x/y$, $\phi = 1/y$). The original curve, given by $\phi^2 = (\theta, 1)_4$, is represented by the intersection of this cone with the quadric surface $x_0^2 = (y_0, y_1, y_2)_2$, the multiple point being replaced by the two simple points where this quadric meets the line (x_0, y_0), or $y_1 = y_2 = 0$. The original curve is obtainable by projection, of the curve on the cone, from the point $x_0 = 0$, $y_1 = 0$, $y_2 = 0$. From an arbitrary point of the space (x_0, y_0, y_1, y_2) the curve on the cone is projected into a plane quartic curve with two double points.

Ex. 6. More generally, it will appear subsequently that the most general plane hyperelliptic curve can be given by an equation $y^{2p} = (x, y)_{2p+2}$; and that, if we begin with a general polynomial of order $n - 2$, that is $2p$, and effect the possible reduction explained in this chapter (due in fact to the character of the multiple point at $x=0$, $y=0$), then the final system obtained is expressed by an equation of the form $y^p(x, y)_{p-1} + (x, y)_{2p} = 0$, where $(x, y)_{p-1}$, $(x, y)_{2p}$ are homogeneous polynomials of the orders indicated, with arbitrary coefficients. In accordance with the text we therefore make a transformation to a space $[3p]$, with coordinates $x_0, \ldots, x_{p-1}, y_0, \ldots, y_{2p}$, by means of the equations

$$\frac{x_0}{x^{p-1}y^p} = \frac{x_1}{x^{p-2}y^{p+1}} = \ldots = \frac{x_{p-1}}{y^{2p-1}} = \frac{y_0}{x^{2p}} = \frac{y_1}{x^{2p-1}y} = \ldots = \frac{y_{2p}}{y^{2p}}.$$

Putting $x/y = \theta$ and $1/y = \phi$, these are the same as

$$\frac{x_0}{\phi\theta^{p-1}} = \frac{x_1}{\phi\theta^{p-2}} = \ldots = \frac{x_{p-1}}{\phi} = \frac{y_0}{\theta^{2p}} = \frac{y_1}{\theta^{2p-1}} = \ldots = \frac{y_{2p}}{1},$$

and represent in fact a rational ruled surface of order $3p-1$ obtained by joining every point in turn of the normal rational curve of order $p-1$ in space $(x_0, \ldots, x_{p-1})$ to the corresponding point (given by the same value of θ) of the normal rational curve of order $2p$ in space $(y_0, \ldots, y_{2p})$. This surface is the representative of the original plane. The curve of this plane corresponds to an equation $\phi^2 = (\theta, 1)_{2p+2}$, and its transformation on the ruled surface is the intersection of this with the quadric cone expressed by an equation $x^2_{p-1} = (y_{p-1}, y_p, \ldots, y_{2p})_2$. This quadric has, as base or vertex, the space $[2p-3]$ on which all the coordinates except $x_0, \ldots, x_{p-2}, y_0, \ldots, y_{p-2}$ are zero. The multiple point of the original plane curve is represented by the two simple points in which the curve on the ruled surface meets the generator $\theta = \infty$. This is the generator joining the vertices (x_0), (y_0) of the coordinate system, which lie respectively on the two rational curves spoken of; this generator lies on the base—or vertex—of the quadric cone specified. It can be shewn that this generator counts $2p-2$ times in the intersection of the quadric cone with the ruled surface, the remaining intersection, of order $4p$, being the transformation of the original curve. The original curve can be recovered by projection of the curve on the ruled surface, from any one of p spaces $[3p-3]$ such as

$$x_i = 0, \quad y_{p+i} = 0, \quad y_{p+1+i} = 0, \qquad (i = 0, \ldots, p-1).$$

In particular, we may take $x_0 = 0$, $y_p = 0$, $y_{p+1} = 0$, which contains, as its intersection with the ruled surface, the generator $\theta = 0$, taken $p-1$ times over, accounting therefore for $2p-2$ intersections with the curve of order $4p$, which thus projects into the original curve, of order $2p+2$. And the generator $\theta = \infty$ meets the space $x_0 = 0$, $y_p = 0$, $y_{p+1} = 0$, in the point (y_0), so that the two intersections of the curve with the generator $\theta = \infty$ project into the one multiple point in the plane, at $x=0$, $y=0$.

More generally, in space $[n-p]$, with $n \geqslant 2p+2$, a quadric containing $n-2p-2$ generators of a rational ruled surface of order $(n-p-1)$ meets this ruled surface further in a hyperelliptic curve of order n (Segre, *Math. Ann.* xxx, 1887, p. 203). This is easily proved by combining equations $x_0/\phi\theta^\lambda = \ldots = x_\lambda/\phi = y_0/\theta^\mu = \ldots = y_\mu/1$ with the equation of a general quadric in the coordinates $(x_0, \ldots, x_\lambda, y_0, \ldots, y_\mu)$.

It may be proved however that a hyperelliptic curve is equivalent with a curve of order $p+3$ on a quadric surface in ordinary space, obtained as the residual intersection with a surface of order $p+1$ passing through $p-1$ generators of the quadric, of the same system; this is without singularities. Or may be represented by a plane curve of order $p+2$ having a single p-fold multiple point of distinct tangents.

Ex. 7. For the curve expressed by $y^3 + y^2(x, y)_2 + y(x, y)_4 + (x, y)_6 = 0$, if we begin with polynomials of order 3, we can shew that the final reduced system is given by $y[y + (x, y)_2] = 0$. Putting then

$$x_0/y = y_0/x^2 = y_1/xy = y_2/y^2,$$

it is easily proved that the curve is transformed into the intersection of a quadric cone, in the space (x_0, y_0, y_1, y_2), expressed by $y_0 y_2 - y_1^2 = 0$ with a general cubic surface of that space. The multiple point at the origin is transformed into the three intersections of the surface with the generator $y_1 = 0$, $y_2 = 0$ of the quadric cone. At these points, as at the three intersections of the curve with any generator of the cone, the tangent lines of the curve are in one plane. The original curve can be obtained again by projection of the space curve from a point of this generator. By projection from a point of the space curve, there is obtained a plane quintic curve having one point of self contact. By projection from a general point of the threefold space there is obtained a plane sextic curve with 6 double points.

CHAPTER III

THE BRANCHES OF AN ALGEBRAIC CURVE; THE ORDER OF A RATIONAL FUNCTION; ABEL'S THEOREM

Introductory explanation. The present chapter deals with important fundamental conceptions in the theory of algebraic curves and functions, and, for greater precision, some familiarity with the elements of the theory of functions is assumed.

The preceding chapter incidentally brings out the want of definiteness in regarding a curve as defined by its points, each given by one set of values of the coordinates; for we have seen that a multiple point may be replaced, on another curve which is in $(1, 1)$ birational correspondence with it, by several distinct points. In the present chapter we are led to consider a point of a curve as belonging to a definite range of points, lying on, and forming all the points of a so-called *branch* of the curve. For clearness, a point so considered, in association with a branch of the curve to which it belongs, will be called a *place*. A point may belong to several branches, but when this is so it is accidental, the distinction of the branches being the essential fact. The point of view which is reached will ultimately be found to be of great importance.

The fundamental curve considered, with its equation expressed by non-homogeneous coordinates, $f(x, y) = 0$, may be spoken of as containing; (i), points for which both x and y are finite; (ii), points for which x is finite, but y is infinite; (iii), points for which x is infinite, but y is finite; (iv), points for which both x and y are infinite. If $f(x, y) = 0$ be regarded as arising from an equation $F(x_0, x_1, x_2) = 0$, homogeneous in x_0, x_1, x_2, by putting $x = x_0/x_2$, $y = x_1/x_2$, then points (i) are those for which x_2 is not zero; points (ii) are those for which $x_2 = 0$ and $x_0 = 0$; points (iii) are those for which $x_2 = 0$ and $x_1 = 0$; while points (iv) are those for which $x_2 = 0$ but neither x_0 nor x_1 vanishes. We may therefore deal with points (ii) by putting $\eta = 1/y$, and considering the equation $f(x, \eta^{-1}) = 0$, in x and η; with points (iii) by putting $\xi = 1/x$, and considering the equation $f(\xi^{-1}, y) = 0$, in ξ and y; and with points (iv) by putting $\xi = 1/x$, $\eta = 1/y$, and considering the equation $f(\xi^{-1}, \eta^{-1}) = 0$, in ξ and η.

If $x = a$, $y = b$ be a point of the curve, so that $f(a, b) = 0$, where a, b are finite, it will be proved that there are points of the curve given by $x = a + t^r$, $y = b + b_1 t + b_2 t^2 + \ldots$, for all sufficiently small, real, and complex, values of the parameter t, where the second expression is a power series, converging for such values of t, and r is a positive integer, and this in such a way that the value of t thus leading to any point (x, y) of the curve, sufficiently near to (a, b), is unique. Save for a finite number of values of a and b, this number r is unity, and the expression for y is a power series in $x - a$; otherwise, as we see by replacing t by ωt, where $\omega = \exp(2\pi i/r)$, the expression gives r values of y, corresponding to a single value of x; these values however can be continuously changed into one another by continuous variation of t, and form a *cycle*. The points (x, y) so obtained, for all sufficiently small values of t, are the points of a *branch* of the curve which are in the neighbourhood of (a, b). The branch does not cease to exist when t is so large that the power series for y fails to converge, but may be *continued*, by starting afresh from a new origin lying on the branch, instead of (a, b). It will further be proved that *all* values of x, y satisfying the equation $f(x, y) = 0$, in which x is sufficiently near to a, are given by a finite number of pairs of expressions, for x and y, of the same form as the pair above taken, the sum of the values of r which arise in all these pairs being n, the order in which y enters into the equation $f(x, y) = 0$. For points of the categories (ii), (iii), (iv), named above, there are similar pairs of expressions, respectively for (x, η), (ξ, y), (ξ, η).

Two simple examples may be given, to explain these statements:

(a) For the curve represented by $xy^2 = (x - 1)^3$, when x is near to 1, we find at once, with $x = 1 + t^2$, that $y = t^3 - \frac{1}{2}t^5 + \frac{3}{8}t^7 + \dots$, the two existing values of y being both included in this by change of t into $-t$; we say then that at $x = 1$, $y = 0$ there is only one place. When x is near to 0, by $x = t_1^2$ and $\eta = 1/y$, we find $\eta = it_1 + \frac{3}{2}it_1^3 + \frac{15}{8}it_1^5 + \dots$, there being again only one place. When x is infinite, and therefore also y is infinite, putting $x = 1/\xi$, $y = 1/\eta$, and $\xi = t_2$, there are two expansions of η when ξ is small,

$$\eta = t_2 + \tfrac{3}{2}t_2^2 + \tfrac{15}{8}t_2^3 + \dots, \quad \eta = -(t_2 + \tfrac{3}{2}t_2^2 + \tfrac{15}{8}t_2^3 + \dots),$$

which do not pass continuously into one another by continuous variation of t_2; they do not belong to a single cycle, and there are *two* places, at this single point of the curve.

(b) For the curve represented by $x^3y^2 = (x - 1)^3$, when x is infinite y is finite, and we put only $\xi = 1/x$, and thence $y^2 = (1 - \xi)^3$; near $\xi = 0$ there are two distinct expansions for y, and there are two places as this point of the curve.

Weierstrass's preparatory theorem. To prove the general result we enter into the proof of a theorem stated by Weierstrass (*Werke*, II, p. 135). Let $A_0 + A_1 y + A_2 y^2 + \dots$ be a power series in y, and $A_0, A_1, A_2, \dots$ be power series in x, so that the whole is a power series in x and y, which is supposed to converge for all sufficiently small values of x and y. It is further supposed that the function represented by this series vanishes for $x=0$, $y=0$. It is necessary then that A_0 vanish for $x=0$; for the sake of generality it is supposed that also A_1, A_2, ..., A_{n-1} vanish for $x=0$, but that A_n does not. It is then proved that the given power series may be regarded as a product ϖU, where U is a power series in x and y, converging for sufficiently small values of these, but *not vanishing for* $x=0$, $y=0$, while ϖ is of the form $\varpi = y^n + p_1 y^{n-1} + ... + p_n$, wherein p_1, p_2, ..., p_n are power series in x, converging for sufficiently small values of x, and all vanishing for $x=0$. In the application of the theorem which we make the power series is only a polynomial in x and y, namely $f(x, y)$, where $f(x, y) = 0$ is the curve considered; and the consequence of the theorem is that, in considering the small values of x and y which satisfy the equation $f(x, y) = 0$, where $f(0, 0) = 0$, it is sufficient to consider only the factor ϖ; the values of y satisfying the equation $f(x, y) = 0$, which reduce to zero when $x=0$, are then n in number, and given by $\varpi = 0$. The particular case when $n=1$, the equation $f(x, y) = 0$ being then of the form

$$ax + y + (x, y)_2 + (x, y)_3 + ... = 0,$$

where $(x, y)_r$ is a homogeneous polynomial of order r in x and y, is familiar; the only value of y satisfying this equation which reduces to zero when x vanishes is given by a power series $y = -ax + bx^2 + ...$.

In general, under the hypothesis made, the power series considered, which we may denote by $f(x, y)$, is of the form

$$f(x, y) = x\,(B_0 + B_1 y + ... + B_{n-1} y^{n-1}) + (C + xB_n)y^n + A_{n+1}y^{n+1} + ...,$$

where B_0, B_1, ..., B_n, A_{n+1}, ... are converging power series in x, *but C is a non-vanishing constant*. We denote $f(0, y)$ by ϕ, and $\phi - f(x, y)$ by ψ, so that ψ vanishes when $x=0$, for all values of y. The series ϕ will be of the form $\phi = Cy^n + Dy^{n+1} + ...$, in which C, D, ... are constants, and C is not zero. It may be assumed, as familiar, that the power series $f(x, y)$, converging for all sufficiently small values of x and y, by hypothesis, converges absolutely for sufficiently small values of these, and, if necessary, may be rearranged; and, further, that, for a power series in y, converging for all values near $y=0$, which vanishes at $y=0$, there is a finite neighbourhood of $y=0$ within which no other zero of the series exists. Thus, denoting the absolute value of y by $|\,y\,|$, as usual, we may suppose a (real positive) σ chosen to satisfy the two conditions, (a) that ϕ does not vanish for $0 < |\,y\,| \leqslant \sigma$; (b) that $f(x, y)$ converges, if x be sufficiently small, for $|\,y\,| < \sigma$. Here ϕ is a function of y only, and ψ vanishes when $x=0$ for every (small) y; if, therefore, we take x small enough, and take care that y is not too small, we can suppose ψ less in absolute value than ϕ. Precisely, a real positive number σ_1 exists, with $\sigma_1 < \sigma$, and a real positive ρ, such that, for $|x| < \rho$, and $\sigma_1 < |y| < \sigma$, we have $|\psi| < |\phi|$. Whence, for a definite value of x, since $f(x, y) = \phi(1 - \psi/\phi)$, we have (putting f for $f(x, y)$)

$$\frac{1}{f}\frac{\partial f}{\partial y} = \frac{1}{\phi}\frac{d\phi}{dy} - \frac{\partial}{\partial y}\sum_{\lambda=1}^{\infty}\frac{1}{\lambda}\left(\frac{\psi}{\phi}\right)^{\lambda};$$

here, since $1/\phi = 1/(Cy^n + Dy^{n+1} + \ldots)$, and $(\psi/\phi)^\lambda$ is expressible as a series of positive and negative powers of y, converging, with the definite x, for $\sigma_1 < |y| < \sigma$, while the series $\Sigma \lambda^{-1}(\psi/\phi)^\lambda$ is uniformly convergent, it follows, by a well-known theorem (Weierstrass, *Werke*, II, p. 205), that we can write the preceding equation in the form

$$\frac{1}{f}\frac{\partial f}{\partial y} = \frac{1}{\phi}\frac{d\phi}{dy} - \frac{\partial}{\partial y}\sum_{\mu=-\infty}^{\infty} G_\mu(x)y^\mu,$$

and, hence, in the form

$$\frac{1}{f}\frac{\partial f}{\partial y} = \frac{n}{y} - \frac{\partial}{\partial y}\sum_{\mu=-\infty}^{-1} G_\mu(x)y^\mu + G(y) - \frac{\partial}{\partial y}\sum_{\mu=0}^{\infty} G_\mu(x)y^\mu,$$

where $G_\mu(x)$ arises as a power series in x which vanishes for $x = 0$, and $G(y)$ is a power series in y.

Now let the number of values of y, less than σ in absolute value, which, for the definite value of x chosen, are zeros of $f(x, y)$, be m; as $f(x, y)$ is a power series in y, this number is finite. Denote these values of y, not necessarily all different, by $y_1, y_2, \ldots, y_m$; then, also for $|y| < \sigma$, the difference

$$\frac{1}{f}\frac{\partial f}{\partial y} - \left(\frac{1}{y - y_1} + \ldots + \frac{1}{y - y_m}\right)$$

is expressible as a power series in y; denote this by $\Phi(y)$. Thus, taking a value of y greater than all of $y_1, \ldots, y_m$ in absolute value, subject also to $\sigma_1 < |y| < \sigma$, we have

$$\frac{1}{f}\frac{\partial f}{\partial y} = \frac{m}{y} + \sum_{\nu=1}^{\infty} y^{-(\nu+1)}[y_1^\nu + \ldots + y_m^\nu] + \Phi(y).$$

This must agree with the previous expression of $f^{-1}\partial f/\partial y$ in powers of y; thus we have $m = n$, and $y_1^\nu + \ldots + y_m^\nu = \nu G_{-\nu}(x)$; defining then ϖ by $\varpi = (y - y_1)\ldots(y - y_n)$, $= y^n + p_1 y^{n-1} + \ldots + p_n$, and recalling an ordinary theorem of the theory of equations, we infer that $p_1 + G_{-1}(x) = 0$, $2p_2 + p_1 G_{-1}(x) + 2G_{-2}(x) = 0$, etc., whereby $p_1, p_2, \ldots$ are expressed as power series in x, vanishing for $x = 0$. Also we have

$$\frac{1}{f}\frac{\partial f}{\partial y} = \frac{1}{\varpi}\frac{\partial \varpi}{\partial y} + \Phi(y) = \frac{1}{\varpi}\frac{\partial \varpi}{\partial y} + G(y) - \frac{\partial}{\partial y}\sum_{\mu=0}^{\infty} G_\mu(x)y^\mu;$$

taking then U so that the right side is $\varpi^{-1}\partial\varpi/\partial y + U^{-1}\partial U/\partial y$, namely

$$U = A \exp\left[\int_0^y G(y)\,dy - \sum_{\mu=0}^{\infty} G_\mu(x)y^\mu\right],$$

where A is independent of y (for the definite value of x), we can infer the equation $f(x, y) = \varpi U$. The form of U is

$$U = A(\lambda_0 + \lambda_1 y + \lambda_2 y^2 + \ldots),$$

in which $\lambda_0, \lambda_1, \lambda_2, \ldots$ arise as power series in x, of which λ_0 reduces to unity for $x = 0$; comparing coefficients of y^n in the identity

$$A_0 + A_1 y + A_2 y^2 + \ldots = A(p_n + p_{n-1}y + \ldots + p_1 y^{n-1} + y^n)(\lambda_0 + \lambda_1 y + \ldots),$$

we see that A is given by $A = A_n(\lambda_0 + \lambda_1 p_1 + \ldots + \lambda_n p_n)^{-1}$; thus A is a power series in x, which reduces, when $x = 0$, to the constant term C in the power series A_n. For the values of x and y for which the two series $G(y)$ and $\sum\limits_{\mu=0}^{\infty} G_\mu(x) y^\mu$ converge, the factor U does not vanish, as its exponential form shews.

Ex. For the curve $y^2 = x^4 + 2Ax^3 y + Bx^2 y^2 + 2Cxy^3 + y^4$, prove that
$$\varpi = [y - Ax^3 - (AB + C)x^5 - \ldots]^2 - [x^2 + \tfrac{1}{2}(A^2 + B)x^4 + \ldots]^2.$$

The parametric expression of a branch of a curve We consider now, for small values of x and y, the form of the solution of the equation $\varpi(x, y) = 0$, where $\varpi(x, y) = y^n + p_1 y^{n-1} + \ldots + p_n$, wherein $p_1, \ldots, p_n$ are power series in x, converging when x is sufficiently small, all vanishing for $x = 0$. We suppose that $\varpi(x, y)$ is incapable of being written as a product

$$(y^m + q_1 y^{m-1} + \ldots + q_m)(y^h + r_1 y^{h-1} + \ldots + r_h),$$

in which $q_1, \ldots, q_m, r_1, \ldots, r_h$ are converging power series in x (which, as all the roots of $\varpi(x, y) = 0$ vanish when $x = 0$, would all vanish for $x = 0$); for instance $y^2 + x$ is so incapable. Otherwise we should deal with the factors separately.

Under this hypothesis, we shew that the n roots of $\varpi(x, y) = 0$ form a single cycle, being all given by one converging series, $y = a_1 t + a_2 t^2 + \ldots$, in the parameter t, defined by $x^n = t$, by allowing t to vary continuously in the neighbourhood of $t = 0$. Naturally, complex values are allowed for t; so that we can pass continuously from t to ϵt, where $\epsilon = \exp(2\pi i/n)$, and thence to $\epsilon^2 t$, and so on up to $\epsilon^{n-1} t$, and thence back to t; thereby we obtain n values of y corresponding to any value of x which is near enough to $x = 0$ to secure the convergence of the series.

For definiteness, let the Resultant of two integral polynomials in y, of respective degrees m and n, mean the determinant of $m + n$ rows and columns, whose rows consist first, for the first n rows, of n repetitions of the coefficients in the first polynomial, beginning in turn in the first, second, ..., n-th columns, the remaining elements being zeros, and then, for the remaining m rows, of m repetitions of the coefficients in the second polynomial, likewise beginning in turn in the first, second, ..., m-th columns, the other elements being zeros; for instance, for the polynomials $a_0 y^2 + a_1 y + a_2$, $b_0 y + b_1$, the rows of the determinant would be in turn a_0, a_1, a_2; b_0, b_1, 0; 0, b_0, b_1. Further, let the Discriminant of such a polynomial, of degree n in y, made homogeneous in y and z by inserting proper powers of z, so becoming $u(y,z)$, mean the Resultant of $\partial u/\partial y$ and $\partial u/\partial z$ (z being put equal to 1 after differentiation). The discriminant is a polynomial homogeneously of degree $2(n-1)$ in the coefficients of the original polynomial. Thus the discriminant of the polynomial in y which we have denoted by $\varpi(x, y)$ will be a homogeneous polynomial of degree $2(n-1)$ in the coefficients 1, p_1, ..., p_n; it will thus be capable of being written as a convergent power series in x, when x is sufficiently small. This discriminant of $\varpi(x, y)$ we may denote by Δ. In general, as is known, the vanishing of the resultant of two polynomials in y is the necessary and sufficient condition that they should have a common factor, linear in y, whose coefficients would then be rational in the coefficients of the two polynomials; and the vanishing of the discriminant of a polynomial in y is the necessary and sufficient condition that the polynomial contain as factor the square of a linear polynomial in y, likewise having therefore coefficients rational in the coefficients of the polynomial. In our case, as $\varpi(x, y)$ is incapable of being written as a product of factors of its own form, it cannot have such a square factor. Thus the discriminant Δ does not vanish for all small values of x. But Δ vanishes for $x=0$, for which $\varpi(x, y)$ reduces to y^n. Thus Δ is a power series in x vanishing for $x=0$; and, by a property of power series already remarked, there is a neighbourhood of $x=0$ within which no other zeros of Δ are found. We suppose the (real and complex) values of x represented on a Euclidean plane, in the familiar way; the neighbourhood in question will then be defined by a circle whose centre is the point $x=0$. For clearness we may speak of the interior of this circle as the *domain* of the origin, $x=0$; it is supposed that this circle is so taken that the series Δ, and all the series p_1, p_2, ... in ϖ, converge therein.

Now, let x_0 be a value of x within this domain, not at $x=0$; and let y_0 be one of the n roots of $\varpi(x_0, y)=0$. The polynomial of order $(n-1)$ in y, $\partial\varpi(x, y)/\partial y$, or, say, $\varpi_y'(x, y)$, will not vanish for $x=x_0$, $y=y_0$, since the two equations $\varpi(x_0, y_0)=0$, $\varpi_y'(x_0, y_0)=0$ would involve that Δ was zero for $x=x_0$, which is contrary to the definition of the domain. Next consider values $x=x_0+\xi$, $y=y_0+\eta$, which are near, respectively, to x_0 and y_0, $x_0+\xi$ being within the domain, and denote the values of $\partial\varpi(x, y)/\partial x$ and $\partial\varpi(x, y)/\partial y$, for $x=x_0$, $y=y_0$, respectively by A and B. The substitution of these values of x and y in $\varpi(x, y)$ then gives a result which, if ξ, η be sufficiently small, can be arranged in the form, a power series in ξ and η, $A\xi+B\eta+(\xi, \eta)_2+(\xi, \eta)_3+\ldots$. It is then a simple application of the general result proved earlier that, since B is not zero, there is, for small ξ, one and only one value of η, satisfying the equation $\varpi\,(x_0+\xi,\,y_0+\eta)=0$, which reduces to zero when $\xi=0$, this being expressible as a convergent power series in ξ. We may denote this solution of $\varpi\,(x,\,y)=0$, reducing to y_0 when x approaches x_0, by $y=y_0+C_1(x-x_0)+C_2(x-x_0)^2+\dots$. If y_1 be another root of $\varpi\,(x_0,\,y)=0$, there will similarly be a unique solution of $\varpi\,(x,\,y)=0$ which reduces to y_1 when $x=x_0$, likewise expressible by a convergent power series in $(x-x_0)$. Of the n series in $x-x_0$, so obtainable, representing the n roots of $\varpi\,(x,\,y)=0$ when x is near to x_0, let r be the least radius of convergence. This number r is not zero; it varies (presumably) as x_0 varies in the domain, and may be regarded as a real function of the real and imaginary parts of x_0. We desire to prove that as x_0 varies, this function r does not approach indefinitely near to zero, but has a lower bound greater than zero. The definition of r has assumed that x_0 is not at $x=0$; we suppose then a small circle put about $x=0$, and suppose x_0 to be without this. We likewise suppose another circle taken near to but within the original circle bounding the domain, also with centre at $x=0$. Then we restrict x_0 to the annulus, lying within the domain, which is bounded by these concentric circles.

We can then see that r varies continuously as x_0 varies within the domain. For, if x_0' be a point, within the annulus, in the immediate neighbourhood of x_0, while we could obtain the power series in $x-x_0'$, which give the roots of $\varpi\,(x,\,y)=0$ when x is near to x_0', directly from $\varpi\,(x,\,y)$, as we obtained the power series in $x-x_0$, we can also obtain them by the known process of analytical continuation from the power series in $x-x_0$, previously obtained, provided x_0' be sufficiently near to x_0. Hence, from the method of this continuation, if r' be the least radius of convergence for the n series in $x-x_0'$, we have $r'\geqslant r-|x_0'-x_0|$. If, however, $|x_0'-x_0|<r'$, we could equally begin with x_0' instead of x_0, and obtain the power series in $x-x_0$ by continuation from those in $x-x_0'$; thus we equally have $r\geqslant r'-|x_0-x_0'|$, and the inequalities

$$r-|x_0'-x_0|\leqslant r'\leqslant r+|x_0-x_0'|$$

shew that r varies continuously with x_0. As a continuous function of two real variables, in a limited domain, actually reaches its lower bound, and r is not zero at any point within the annulus, it follows that the lower bound of r is greater than zero. There exists therefore a real positive number ρ, greater than zero, such that for every point x_0, within the annulus, the radius of convergence of the series in $x-x_0$, which represents any root of $\varpi(x_0,\,y)=0$, is greater than ρ.

We start then with a particular x_0, within the annulus, and form the series in $x-x_0$ which represents the root of $\varpi\,(x_0,\,y)=0$ which reduces to y_0 when $x = x_0$, where y_0 is any of the n values satisfying $\varpi\,(x_0, y_0) = 0$. Then, within the circle of convergence of this series, and within the annulus, we take a point x_1, and form the analytic continuation of this particular series in powers of $x - x_1$. Since, for this derived series, the radius of convergence is $> \rho$, we can, by suitable choice of x_1, near to the circumference of the circle of convergence of the original series, secure that the circle of convergence of the derived series in $x - x_1$ contains a region lying outside the original circle of convergence. In this region, and within the annulus, we can take a point x_2, and transform the series in $x - x_1$ into a series in $x - x_2$, in such a way, by proper choice of x_2, that the derived series in $x - x_2$ contains a region outside the two former circles of convergence. It is thus clear that the original series in $x - x_0$ can be continued completely round the origin $x = 0$, the various centres $x_0, x_1, x_2, \ldots$, of the successive series, lying on a curve enclosing $x = 0$, which, after a finite number of steps, again reaches x_0. We thus obtain another series in $x - x_0$, giving a root of $\varpi\,(x_0, y) = 0$, obtained, as explained, by continuation of the original root, which was constructed to reduce to y_0 for $x = x_0$. This second root of $\varpi\,(x_0, y) = 0$ may not reduce to y_0, but to another root of $\varpi\,(x_0, y) = 0$, when $x = x_0$. In such case we again make the circuit of $x = 0$, and find a third root of $\varpi\,(x_0, y) = 0$, as a power series in $x - x_0$. As the number of roots of $\varpi\,(x_0, y) = 0$ is finite, we must, after, say μ circuits of $x = 0$, finally reach again the original root which reduces to y_0 for $x = x_0$.

We then take a variable t, similarly representable on a new plane of complex variables, such that $x = t^\mu$. As t makes a circuit about $t = 0$ in its own plane, the point x will make a circuit about $x = 0$ in the plane of x; but an unclosed path for t, by which it changes to $te^{2\pi i/\mu}$, will correspond to a complete circuit for x; and the complete circuit of t, about $t = 0$, will correspond to μ complete circuits by x, described in succession. Thus, if, instead of $\varpi\,(x, y) = 0$, we consider the equation $\varpi\,(t^\mu, y) = 0$, the root of the equation $\varpi(x_0, y) = 0$ which we have discussed, which reproduces its value after μ circuits about $x = 0$, will correspond to a root of $\varpi(t_0{}^\mu, y) = 0$, where $x_0 = t_0{}^\mu$, which reproduces itself by continuation, after one circuit of $t = 0$ in the plane of t. This root, being developable about every point of the annulus in the t-plane which corresponds to the annulus taken in the x-plane, as is clear from what has been said, is representable*, in a form valid for the whole of the annulus in the t-plane, by a series $\sum\limits_{h=-\infty}^{\infty} c_h t^h$. As, however, all the roots of $\varpi\,(x, y) = 0$ tend to zero when x tends to zero, the series can only contain terms for which $h > 0$,

---

* Weierstrass, *Werke*, I, p. 51. For the history of Laurent's theorem, cf. Mittag-Leffler, *Acta Math.* xxxix, 1923, p. 34.

and the root is representable over the whole of the t-annulus by a power series $y = c_1 t + c_2 t^2 + \dots$. Considering this root again in the annulus in the x-plane, we obtain $\mu - 1$ roots, into which it can successively be continued by circuit of $x = 0$, by replacing t in this series successively by ϵt, $\epsilon^2 t$, $\dots$, $\epsilon^{\mu-1} t$, where $\epsilon = \exp(2\pi i/\mu)$.

If then these roots, regarded as depending on x, be denoted by $\eta_1, \dots, \eta_\mu$, the product $(y - \eta_1) \dots (y - \eta_\mu)$ will be of the form $y^\mu + q_1 y^{\mu-1} + \dots + q_\mu$, where every coefficient q, symmetrical in $x^{1/\mu}$, $\epsilon x^{1/\mu}$, $\dots$, $\epsilon^{\mu-1} x^{1/\mu}$, is expressible by integral powers only of x, and arises as a power series; thus every coefficient q is a power series in x; and, as all the roots of $\varpi(x, y) = 0$ vanish when $x = 0$, this power series vanishes for $x = 0$. The function $y^\mu + q_1 y^{\mu-1} + \dots + q_\mu$ is thus a factor of $\varpi(x, y)$, which, however, by hypothesis, has no such factor, unless $\mu = n$. Thus, this is the case, and all the roots of $\varpi(x, y) = 0$ form a single cycle, and are all derivable from a single series, $y = c_1 t + c_2 t^2 + \dots$, where $x = t^n$, by continuous variation of t. The proof of the statements made at starting is thus completed.

Supplementary remarks. The expression of the points of a branch of a curve in terms of a parameter is ascribed to Puiseux (*Journ. de Math.* xv, 1850). Some further remarks, without complete proof, and suggestions for alternative methods of reaching the result, may be made:

(a) In the expressions $x = a + t^r$, $y = b + c_1 t + c_2 t^2 + \dots$, it is clear from the proof we have given that, to any values of x, y, in sufficient nearness to a, b, respectively, there corresponds only a single value of t. And it is not necessary for this that the exponent of the lowest power of t which is actually present in the series for $y - b$, should be unity, or even be prime to the exponent r. For instance, for the curve given by the equation $x^2 - 2xy^2 + y^4 - 4m^2 x^3 y - m^4 x^5 = 0$, there is only one place at $x = 0$, $y = 0$, with a cycle of four values, expressed by $x = t^4$, $y = t^2 + mt^5$; and t is actually given by

$$t = (-x + y^2 + m^2 x^2 y)/mx(2y + m^2 x^2).$$

It is clear too that, instead of t, we may use any other parameter τ which is expressible by a convergent power series in t, of the form $\tau = At + Bt^2 + \dots$, provided the coefficient A be not zero; then both x and y appear as expressed by a power series in τ.

(b) The number of pairs of values of a, b satisfying the equation $f(x, y) = 0$ for which, in the expression

$$x = a + t^r, \quad y = b + c_1 t + c_2 t^2 + \dots,$$

the value of r is greater than unity, is necessarily finite. For all such pairs equally satisfy the equation $\partial f/\partial y = 0$, and, by elimination of y between this equation and the equation of the curve, only a finite number of values of x arises. If we take account also, as is proper, of the associated equations $f(x, \eta^{-1}) = 0$, $f(\xi^{-1}, y) = 0$, $f(\xi^{-1}, \eta^{-1}) = 0$, as explained above, when this is necessary, the number of places for which $r > 1$ is still finite. It is convenient to have a name for the number $r - 1$, at any place, when it is not zero; it may perhaps, for reasons which appear below, be called the *winding index* at the place. The sum of the values of this index, for $f(x, y) = 0$ and the associated equations such as $f(x, \eta^{-1}) = 0$, etc., at the finite number of places where this index is not zero, will be found to be of great importance.

(*c*) The number of pairs of expressions

$$x = a + t^r, \quad y = b + c_1 t + c_2 t^2 + \dots,$$

with $r = 1$ and $r > 1$, taken with the analogous expressions arising for the associated equations such as $f(x, \eta^{-1}) = 0$, which are *necessary*, in order to give every pair of values of x and y (or of (x, η), (ξ, y), (ξ, η)) which satisfy the equation of the curve (or its associated equations), *is only finite*. This may be seen by regarding the complex variable x as represented upon a Riemann sphere (replacing the plane on which x is represented); every one of the expressions in question has then a region of existence on this sphere of which the radius has a lower bound not infinitesimally small.

(*d*) Moreover, all the pairs of expressions such as those in (*c*), which are necessary to represent all pairs of associated values of x and y satisfying the equation of the curve, are derivable by analytic continuation from any one such pair of expressions (provided we may assume all the coefficients in this one expression to be known). Thus any one such pair of expressions theoretically contains the equation of the curve. For this however the condition that the equation $f(x, y) = 0$ is incapable of being obtained by multiplication of other such equations, equally rational in x and y, or, that the curve $f(x, y) = 0$ is irreducible, is fundamental.

(*e*) It may be proved directly that, for the neighbourhood of a simple point $(a, b, c, \dots)$ on an algebraic curve in space in which the (non-homogeneous) coordinates are $x, y, z, \dots$, the values of $y - b, z - c, \dots$ are expressible as converging power series in $x - a$ (if x be chosen appropriately among the coordinates). If therefore we assume, from Chapter II above, that any given curve can be put in (1, 1) birational correspondence with a non-singular curve, it follows that any place of the given curve can be represented by power series in a parameter; and further that, for this parameter we may choose one which is rationally expressible by the coordinates of the point of the given curve. This is evidently the simplest way in which the general theorem may be reached. But also, conversely, it appears at once on consideration that, in a birational transformation between two curves, a branch of one necessarily changes into a single branch of the other, and consequently a place of one into a single place of the other. Thus, in the birational transformation of a curve into one without multiple points, the number of points of the latter curve which correspond to a multiple point of the original, is the number of places which exist at this multiple point.

(*f*) But the object of representing the neighbourhood of a particular multiple point of a plane curve by an aggregate of power series would also be achieved if any transformation, birationally reversible, were found which replaced this point by a number of simple points on the transformed curve, even if thereby new multiple points were introduced not corresponding to the neighbourhood of the original multiple point. There exists an extensive theory of this possibility, in which the transformations employed are birational in the whole plane (so-called Cremona transformations); and this suffices to establish the general theorem referred to. The theory was first elaborated in detail by Noether, using a succession of quadratic transformations to build up the necessary Cremona transformation. In regard to the theory in question, which will not be fully treated in this volume, the reader may consult: The Brill-Noether *Bericht*; *Deut. Math. Ver.* 1894, p. 377, etc. (cf. Noether, *Math. Ann.* ix, 1876; and Segre, *Ann. d. Mat.* xxv, 1897, p. 1); also Segre and Castelnuovo, *Atti...Torino*, xxxvi, 1901, pp. 645 and 861; and, thereto Alexander, *Trans. Amer. Math. Soc.* xvii, 1916, p. 295; and further, the very full account in Enriques-Chisini, *Teoria geometrica*, ii, with the bibliography, pp. 535 ff. As a simple example of the application of the method we may quote the case of the curve represented by $y^{2p} = (x, y)_{2p+2}$, for which the neighbourhood of $x = 0$, $y = 0$ may be discussed by making the transformation $x = \xi\eta^{p-1}$, $y = \xi\eta^p$, reversible by $\xi = x^p/y^{p-1}$ and $\eta = y/x$, whereby the curve becomes $\eta^2 = \xi^2(1, \eta)_{2p+2}$, with the two branches of a double point at $\xi = 0$, $\eta = 0$.

(*g*) It follows from the theory given here that the number of *places* existing at a multiple point of a plane curve is the same as the number of irreducible factors of the function denoted above, in the text, by $\varpi(x, y)$. Indications of real value, giving also the effective terms in the relative factor, to the first approximation, are very easily found by use of Newton's polygon. As a simple example, consider the curve represented by the equation

$$y^3 + y^2 u_2 + y u_4 + u_6 = 0,$$

where u_r is homogeneous of order r in x and y. The Newton polygon indicates at once that there are three (branches) places for this curve at $x = 0$, $y = 0$, each given to the first approximation by an equation of the form $y = mx^2$; namely, there are three ordinary branches with a common tangent. The multiple point is described in the Noether theory, referred to under (f), as consisting of a triple point, and a further triple point in the immediate neighbourhood of this. In regard to the use of Newton's polygon reference may be made to a paper, *Trans. Camb. Phil. Soc.* xv, Part iv, Oct. 1893, p. 403.

Ex. The transformation $\xi/xy^{n-1} = \eta/y^n = \zeta/[zy^{n-1} + (x, y)_n]$ is uniquely reversible, leading to $x/\xi\eta^{n-1} = y/\eta^n = z/[\zeta\eta^{n-1} - (\xi, \eta)_n]$. The curve $y^{n-1} + (x, y)_n = 0$ has one place at $x = 0$, $y = 0$, representable by a pair $x = t^{n-1}$, $y = mt^n + \dots$; the curve is in $(1, 1)$ correspondence with a curve in space $[n+1]$ obtainable by a prime section of the cone which joins an arbitrary point, O, to the rational normal curve given by

$$y_0/\theta^n = y_1/\theta^{n-1} = \dots = y_n,$$

the single place at $x = 0$, $y = 0$ corresponding to a simple point on the line joining O to the point $\theta = \infty$ of the rational curve. Any two curves $axy^{n-1} + by^n + c[zy^{n-1} + (x, y)_n] = 0$, for different values of a, b, c, have, beside a single variable intersection, common points at $x = 0$, $y = 0$ which are described in Noether's phraseology as consisting of a $(n-1)$-ple point, followed by $(2n-2)$ simple points, on a branch of order $(n-1)$, in successive neighbourhoods (Segre, *Atti...Torino*, xxxvi, 1901). In fact $n^2 = (n-1)^2 + 2n - 2 + 1$.

General theorem for infinities of a rational function. We pass now to make application of the parametric expression we have established, to the theory of the rational functions of x and y, these variables being supposed to be connected by the equation of the fundamental curve, $f(x, y) = 0$. These functions will be briefly referred to as rational functions on the curve. For clearness, we repeat that, in speaking of the places belonging to the equation $f(x, y) = 0$, we mean not only those for whose neighbourhood there is an expression of the form $x - a = t^r$, $y - b = c_1 t + c_2 t^2 + \dots$, but also those, derived from the associated equations, for which, in this expression, $x - a$ is to be replaced by $1/x$, or $y - b$ replaced by $1/y$, or both.

A function, explicitly rational in x and y (cf. the Note at the end of the chapter), is expressible in the neighbourhood of any place belonging to the curve $f(x, y) = 0$, after substitution for x and y of the appropriate series in the parameter t, in the form $t^\lambda P(t)$, where $P(t)$ denotes a power series in t, not vanishing when t is zero, and λ is an integer, which may be positive, zero, or negative. If λ is positive, the rational function is said to vanish to order λ at this place; if λ is negative, equal to $-\mu$, the function is said to have a *pole* of order μ at this place. In general λ will be zero; it will indeed appear that there is only a finite number of places for which λ is > 0, and only a finite number at which $\mu > 0$; and that the sum of the values of λ at the former places (often called the number of zeros of the function) is equal to the sum of the values of μ at the latter places (often called the number of poles of the function). Either of these sums is then called the *order of the rational function*. Further, if the function be denoted by Φ, and C be an arbitrary constant, the difference $\Phi - C$, at a place which is not a pole of Φ, may vanish to a certain order; for brevity, this is often spoken of as the number of times for which Φ is equal to C at this place. It will appear that the sum of the numbers of times for which Φ is equal to C, at all the places where this happens (including, if necessary, those for which x, or y, is infinite), is equal to the order of the function Φ.

In order to prove these results, and others of still greater importance, we proceed to prove a theorem which we express in the form

$$\left[\Phi \frac{dx}{dt} \right]_{t^{-1}} = 0.$$

For the neighbourhood of any place satisfying $f(x, y) = 0$ we suppose Φ expressed, in the manner explained, in terms of the parameter t; and the expression multiplied by dx/dt; where $x = a + t^r$, $y = b + c_1 t + c_2 t^2 + \ldots$, this dx/dt will be rt^{r-1}; but, where $1/x = t^{r_1}$, and $y - b$, or $1/y$, is $c_1 t + c_2 t^2 + \ldots$, this dx/dt will be $- r_1 t^{-r_1-1}$. In the product $\Phi \, dx/dt$, so formed, arranged as an ascending series in t, there will be only a finite number of negative powers of t; if the power t^{-1} occur, we take the coefficient of this. It will appear that there is only a finite number of places where negative powers of t arise, in this expansion of $\Phi \, dx/dt$. The theorem expresses that the sum of the coefficients of t^{-1}, at all such places, is zero.

We first remark a corresponding, but much simpler, theorem, for a rational function, ϕ, of a *single independent* variable x. For convenience we express this in the same form $[\phi \, dx/dt]_{t^{-1}} = 0$. Any such function, ϕ, is necessarily of the form

$$Ax^m + Bx^{m-1} + \ldots + Mx + N + \sum_{i=1}^{k} [P_i (x - a_i)^{-n_i}$$
$$+ Q_i (x - a_i)^{-n_i+1} + \ldots + R_i (x - a_i)^{-1}],$$

where k is the number of places at which the function becomes infinite. In the neighbourhood of a finite value of x, say $x = c$, we can put $x = c + t$, and expand the function in ascending powers of t. For general values of c, no negative powers of t will arise; for $c = a_i$, there will be the negative powers $P_i t^{-n_i} + Q_i t^{-n_i+1} + \ldots + R_i t^{-1}$, in which the coefficient of t^{-1} is R_i. In the neighbourhood of $x = \infty$, we can put $x = 1/t$, and similarly develop the function in ascending powers of t; the result will again be a power series with the addition of terms involving negative powers of t, the number of such negative powers being only finite; in this case, if we multiply the function by dx/dt, or $- 1/t^2$, the coefficient of t^{-1} in the product will be the negative of the coefficient of t in the function ϕ itself, namely will be $- \Sigma R_i$. Thus, it appears that the sum of the coefficients of t^{-1} in the expansions of the product $\phi \, dx/dt$, for all the places where this contains negative powers, is zero. This is what is expressed by the equation put down, for this case of a rational function of a single variable.

Now take the theorem for the rational function $\Phi(x, y)$, where $f(x, y) = 0$. Consider a place at which $x = a + t^r$, and $y - b$, or $1/y$, is a power series in t. Denoting the values of y, belonging to the cycle, when x is near to a, by $y_1, \ldots, y_r$, form the sum

$$\Phi(x, y_1) + \ldots + \Phi(x, y_r),$$

regarded as a function of t. Each constituent of this sum is a series of integral powers of t, with only a finite number of negative powers; if ϵ denote $\exp(2\pi i/r)$, the series are the same in the respective quantities $t, \epsilon t, \epsilon^2 t, \ldots, \epsilon^{r-1}t$. Thus the sum is a series of integral powers of t^r or $x - a$. In this series, the coefficient of $(x-a)^{-1}$, or t^{-r}, if not zero, arises by equal contributions from each constituent of the sum, and is thus equal to the coefficient of t^{-1} in the series of powers of t given by $\Phi(x, y) r t^{r-1}$, or $\Phi(x, y) dx/dt$, where y denotes any one of $y_1, \ldots, y_r$. Next consider a place arising for an infinite value of x, for the neighbourhood of which $x = t^{-k}$, the appropriate corresponding values of y, as series in t, being $y_1, \ldots, y_k$ (which may become infinite for $t = 0$). The sum $\Phi(x, y_1) + \ldots + \Phi(x, y_k)$ is then, similarly, a series of integral ascending powers of x^{-1}, in which the term in t^k, or x^{-1}, if present, arises by equal contributions from each constituent of the sum. This is then equal to the coefficient of t^{-1} in $\Phi(x, y) k t^{-k-1}$, where y denotes any one of $y_1, \ldots, y_k$. The *negative* coefficient of x^{-1} in the sum is thus equal to the coefficient of t^{-1} in $\Phi(x, y) dx/dt$.

Lastly consider, for any value of x, finite or infinite, the sum $\phi(x)$ defined by $\phi(x) = \Phi(x, y_1) + \ldots + \Phi(x, y_n)$, where $y_1, \ldots, y_n$ are all the values of y satisfying the equation $f(x, y) = 0$, for the value of x taken. As a symmetrical function of these, $\phi(x)$ is expressible as a rational function of x only. For a finite value of x, say $x = a$, if the sum be regarded as an aggregate of partial sums, each extended to all the values of y belonging to one place, and forming a cycle, the coefficient of $(x-a)^{-1}$ in the complete sum is, by what we have shewn, equal to the sum of the coefficients of t^{-1} in $\Phi(x, y) dx/dt$, evaluated in turn at the various places arising for this value of x, the symbol t denoting in turn the parameters at these places. Likewise for infinite x; the negative coefficient of x^{-1} in $\phi(x)$ is equal to the sum of the coefficients of t^{-1} in $\Phi(x, y) dx/dt$, at the various places arising for $x = \infty$. As the number of values of x, giving contributions, when we are considering the function $\phi(x)$, is finite, it is clear that the number of places providing a term to be taken into account is also finite for $\Phi(x, y) dx/dt$. And, from the theorem proved for $\phi(x)$, the sum of the coefficients of t^{-1}, at this finite number of places, in the appropriate expressions for

$\Phi(x, y) \, dx/dt$, evidently vanishes. Which is the theorem to be proved.

Order of a rational function on the curve. An immediate corollary is the theorem justifying the definition of the order of an algebraic function which we have given. Let $\Psi(x, y)$ be any rational function, and K any constant. Then the function

$$\Phi(x, y) = [\Psi(x, y) - K]^{-1} \frac{d}{dx}[\Psi(x, y) - K],$$

where d/dx means total differentiation, equal to

$$\partial/\partial x + (-\partial f/\partial x \div \partial f/\partial y)\partial/\partial y,$$

is a rational function. Also, at a place where $\Psi(x, y)$ has a pole of order μ, being expressible in the neighbourhood of this place in the form $\Psi(x, y) = t^{-\mu}P(t)$, where $P(t)$ is a power series not vanishing for $t = 0$, the expression $\Phi(x, y) \, dx/dt$, say $(\Psi - K)^{-1} d(\Psi - K)/dt$, has the form

$$\frac{d}{dt} \log(\Psi - K), \quad \text{or} \quad \frac{d}{dt} \log[(P(t) - Kt^{\mu})/t^{\mu}];$$

and in this, as the power series $P(t)$ does not vanish for $t = 0$, the coefficient of t^{-1} is $-\mu$. Again, at a place where $\Psi(x, y) - K$ vanishes to order ρ, having an expression for the neighbourhood of this place, if the parameter here be also denoted by t, which we may write $t^{\rho}P(t)$, the coefficient of t^{-1} in $d \log(\Psi - K)/dt$ is ρ. Hence, by the theorem $[\Phi(x, y) \, dx/dt]_{t^{-1}} = 0$, we see that the sum of the values of μ which arise, for all finite and infinite values of x, is equal to the sum of the values of ρ; and this, therefore, is the same for all values of the constant K. This is the theorem in question, justifying the definition of the order of the rational function $\Psi(x, y)$. That the sum of these values of ρ is finite has been remarked; it is obvious also because the places where $\Psi(x, y) = 0$, subject to $f(x, y) = 0$, are obtainable by algebraic elimination, say of y, from these two equations, leading to an equation for x of finite order.

The theorem that a rational function assumes, on $f(x, y) = 0$, any assigned value at a number of places which is independent of the value, is, in its simpler cases at least, a well-known theorem for the intersection of curves. If the explicit expression of the function be $u(x, y)/v(x, y)$, where u, v are polynomials, the theorem is that the curve $u(x, y) - Kv(x, y) = 0$ intersects $f(x, y) = 0$ in a number of points which is independent of K. But it is clear that this familiar result may be difficult of interpretation when some of the intersections are at multiple points of $f(x, y) = 0$, or if, at an intersection which is ordinary on $f(x, y) = 0$, there is multiple contact. The intersections at infinity present no special difficulty if homogeneous coordinates be employed. The computation of the exact number of intersections, in any case of difficulty, can only be made by some method equivalent to that we have employed, with the precedent expression of the places of the curve in terms of a representative parameter.

Ex. 1. As a very simple illustration, suppose $f(x, y)$ to be given by $f(x, y) = y^2 - 4(x-a)(x-b)(x-c)$. On the curve $f(x, y) = 0$ there is then a single place at each of $x = a$, $x = b$, $x = c$, $x = \infty$, there being a cycle of two values; for all other values of x there are two places. Thus the rational function $x - m$, for general values of the constant m, is infinite to the second order at $x = \infty$, and vanishes to the first order at each of the two places for which $x = m$. But these zeros coincide in one zero of the second order if $m = a$, or b, or c. Or again, the rational function y is infinite to the third order at the place given by $x = \infty$, and vanishes to the first order at each of the places given by $x = a$, or $x = b$, or $x = c$. Or, further, if z be not a, nor b, nor c, and s be either of the two values for which $f(z, s) = 0$, the rational function, of x and y, $(y+s)/(x-z)$, is infinite to the first order at one (and only one) of the two places where $x = z$, and is infinite to the first order at $x = \infty$; its zeros are for the two values of x satisfying the quadratic equation

$$[(x-a)(x-b)(x-c) - (z-a)(z-b)(z-c)]/(x-z),$$

each of these values of x being associated with *one* of the two corresponding values of y which satisfy the equation $f(x, y) = 0$. The function $(y+s)/(x-z)$ is thus of the second order.

Ex. 2. In further illustration of the order of a rational function, we may prove that a curve for which there exists a rational function of order 2 is of special character, being in fact hyperelliptic, see Chap. II; or is rational if there be a rational function of order 1. It is clear that if there exist a rational function of order 1, the explicit expression of this function being $u(x, y)$, then the elimination of y from the two equations $u(x, y) = \xi$, $f(x, y) = 0$, must lead to an equation for x, giving only one value, and hence, as the process is rational, x may be expressed rationally in terms of ξ; and, similarly, so may y; and ξ by definition is rational in x and y. This is the definition of a rational curve.

If there exists a rational function $u(x, y)$, of order 2, the two equations $u(x, y) = \xi$, $f(x, y) = 0$, in which ξ is a general value, must, on elimination of y, lead to a quadratic equation $x^2 - 2xr + s = 0$, where r and s are rational in ξ, unless the two places, where $u(x, y) = \xi$, have the same value of x for all values of ξ. If this exception arises, these two places cannot also have the same value of y for all values of ξ, because then, taking x_0, y_0 arbitrarily, subject to $f(x_0, y_0) = 0$, the function $u(x, y) - u(x_0, y_0)$ vanishes to the second order at the arbitrary place (x_0, y_0), and hence the differential coefficient of $u(x, y)$ vanishes at every place (x_0, y_0), or $u(x, y)$ is a constant. Thus we can use, instead of x, a coordinate $x + \lambda y$, where λ is constant; and, that done, we can make the statement that the elimination of y between $u(x, y) = \xi$, and $f(x, y) = 0$, leads to a quadratic equation $x^2 - 2xr + s = 0$, in which $r^2 - s$ vanishes only for particular values of ξ. Take now the rational function of x and y given by $\eta = x - r$, the value of ξ in r being expressed by x and y. Then, when ξ and η are given, we have an expression for x in terms of these, namely $x = \eta + r$, with also $\eta^2 = r^2 - s$, in which $r^2 - s$ is rational in ξ. Further, if (x_1, y_1) be one of the two places where $u(x, y) = \xi$, the equations $u(x_1, y) = \xi$, $f(x_1, y) = 0$, must on elimination of x_1, lead to $y = y_1$. If $x_1 = \eta + r$, the value of x at the other place at which $u(x, y) = \xi$ must be $x_2 = -\eta + r$;

thus the equations $u(\eta+r, y)=\xi$, $f(\eta+r, y)=0$ lead to a rational expression of y in terms of ξ and η. There exists therefore an equation, $\eta^2=$ rational function of ξ, to which the original equation $f(x, y)=0$ is birationally equivalent. If this equation be $\eta^2=p(\xi)/q(\xi)$, in which $p(\xi)$, $q(\xi)$ are polynomials, by taking $\eta_1=\eta q(x)$, we obtain $\eta_1^2=p(\xi)q(\xi)$. If the polynomial $p(\xi)q(\xi)$ be of the first or second order, it is familiar that ξ and η_1 can be rationally expressed by a proper rational function of themselves, and hence the original curve $f(x, y)=0$ is rational; if $p(\xi)q(\xi)$ be a cubic or quartic polynomial, the original curve is elliptic (see Chap. I).

Ex. 3. More generally the following may also be stated, but without complete proof at this stage. Let ξ and η be two rational functions of x and y for the curve $f(x, y)=0$; let σ be the order of ξ. To any general value a there are σ places of $f(x, y)=0$ at which $\xi=a$, and, save for a finite number of values of a, these places will be different. Let $b_1, \ldots, b_\sigma$ be the values of η at these places. Then there are two possibilities; either these values $b_1, \ldots, b_\sigma$ are different from one another except for a finite number of values of a; or, these values consist, *for every value of a*, of q different values, each repeated p times, where $\sigma=pq$. In the former case there is a relation connecting ξ and η, rational in both, and irreducible, of order σ in η; in the latter case there is also such an irreducible rational relation, but of order q in η. In the former case, this relation may be obtained by eliminating x, y between $f(x, y)=0$, $u(x, y)=\xi$, $v(x, y)=\eta$, where u, v represent the explicit forms of ξ and η; in the latter case, the corresponding result of the elimination appears as the p-th power of the single irreducible relation connecting ξ and η. In the former case, the rational relation connecting ξ and η may be regarded as that of a curve in $(1, 1)$ birational correspondence with $f(x, y)=0$; in the latter case, the new curve is such that to every point of it there correspond p points of $f(x, y)=0$; or, the new curve regarded as repeated p times may be considered as in $(1, 1)$ correspondence with the original. The former case arises certainly if there be a single value of a for which the values $b_1, b_2, \ldots, b_\sigma$ are all different.

Abel's theorem for a sum of algebraic integrals.
We now deduce, from our general theorem, another consequence of great importance.

Let $\Phi(x, y)$, $\Psi(x, y)$ be two rational functions, on $f(x, y)=0$. If C be a constant, the function $\Psi(x, y)-C$ has a definite number of zeros, say m zeros, for every value of C. As C is taken differently, the positions of these zeros on $f(x, y)=0$ change, and these zeros may thus be regarded as depending on C. We shall suppose that these zeros are all of the first order, for all the values of C which we consider, and that none of them coincides with any of the places where the expansion of $\Phi(x, y)\,dx/dt$ contains negative powers of t. We shall also put

$$I=\int_{(a)}^{(x)}\Phi(x, y)\,dx,$$

the integral being extended along a continuous range, on $f(x, y)=0$, given by a variation of x, and the consequent variation of an associated value of y satisfying $f(x, y)=0$, from a place (a), which is fixed, to the place (x). Further, denoting the m places where $\Psi(x, y) = C$ by $(x_1), \ldots, (x_m)$, and the m places where $\Psi(x, y)$ has a particular value C_0 by $(a_1), \ldots, (a_m)$, it will be convenient to denote the sum

$$\int_{(a_1)}^{(x_1)} \Phi(x, y)\, dx + \ldots + \int_{(a_m)}^{(x_m)} \Phi(x, y)\, dx$$

by a symbol, say Σ. Denoting $\Phi(x, y)$, $\Psi(x, y)$ by Φ, Ψ, the function $[\Psi(x, y) - C]^{-1} \Phi(x, y)$ is a rational function, and, by what we have proved, the equation expressed by

$$\left[\frac{1}{\Psi - C} \Phi \frac{dx}{dt} \right]_{t^{-1}} = 0$$

holds. This we can write in the form

$$\left[\left(\frac{1}{\Psi - C} \right) \frac{dI}{dt} \right]_{t^{-1}} = -\left[\left(\frac{dI}{dt} \right) \frac{1}{\Psi - C} \right]_{t^{-1}},$$

in which the sum on the left refers to the m places where $\Psi(x, y) = C$, and only to these, and the sum on the right is for places where dI/dt contains negative powers of t. We have assumed that, for all the values of C considered, these two sets of places are different. For a point $(x_i + dx_i, y_i + dy_i)$, near to a zero (x_i, y_i) of $\Psi(x, y) - C$, and on a particular branch of $f(x, y) = 0$ issuing from (x_i, y_i), for which the points are given in terms of the parameter t, the value of $\Psi(x, y)$, expressed in powers of t, will be $C + tA + \ldots$, where A is not zero because we have assumed that the zeros of $\Psi(x, y) - C$ are simple for the values of C which we consider. We may denote this value by $C + dC$, regarding x_i and y_i as functions of C. The contribution to the sum on the left side, corresponding to this zero, will then be the value at (x_i, y_i) of

$$\frac{1}{A} \frac{dI}{dx_i} \cdot \frac{dx_i}{dt}, \text{ or } \frac{dI}{dx_i} \cdot \frac{dx_i}{dC},$$

and the whole value of the left side of the equation will be $d\Sigma/dC$, where Σ is defined above. On the right side, the places to be considered are those for which dI/dt contains negative powers of t; at all of these, by hypothesis, $\Psi(x, y) - C$ differs from zero, while dI/dt does not depend on C. For the neighbourhood of such a place, at which $\Psi(x, y)$ is Ψ_0, suppose, we have an expression of the form

$$-\frac{1}{\Psi - C} = \frac{d}{dC} \log(\Psi - C) = \frac{d}{dC} [\log(\Psi_0 - C) + Pt + Qt^2 + \ldots],$$

where $P, Q, \ldots$ depend on C, and the series converges, uniformly for C, when t is small enough. Here

$$-\left[\left(\frac{dI}{dt} \right) \frac{1}{\Psi - C} \right]_{t^{-1}} = \frac{d}{dC} \left[\left(\frac{dI}{dt} \right) \log(\Psi - C) \right]_{t^{-1}}.$$

Thus, varying C from C_0 to C, we have the equation

$$\int_{(a_1)}^{(x_1)} \Phi(x, y)\,dx + \ldots + \int_{(a_m)}^{(x_m)} \Phi(x, y)\,dx = \left[\left(\frac{dI}{dt}\right) \log \frac{\Psi - C}{\Psi - C_0}\right]_{t^{-1}}.$$

Here $(\Psi - C)/(\Psi - C_0)$ is a rational function, say $\Theta(x, y)$, whose zeros are the upper limits $(x_1), \ldots, (x_m)$ of the integrals on the left, and whose poles are the lower limits, $(a_1), \ldots, (a_m)$ of these integrals; the right side can then be written

$$\left[\left(\frac{dI}{dt}\right) \log \Theta(x, y)\right]_{t^{-1}}.$$

This is the result, equivalent to the famous theorem due to Abel, at which we desired to arrive. A simple example will serve to make the matter clear: Suppose the curve $f(x, y) = 0$ is given by $y^2 - 4x^3 + 8ax + b = 0$; that the function $\Phi(x, y)$ is x/y; and the function $\Theta(x, y)$ is $(y - mx - c)/(y - m_0 x - c_0)$, where m, c, m_0, c_0 are constants of general values. This function $\Theta(x, y)$ evidently vanishes at three places, say $(x_1, y_1), (x_2, y_2), (x_3, y_3)$, and has poles at three places, say $(x_1{}^0, y_1{}^0), (x_2{}^0, y_2{}^0), (x_3{}^0, y_3{}^0)$, so that the sum on the left may be denoted by

$$\int_{(x_1{}^0)}^{(x_1)} \frac{x\,dx}{y} + \int_{(x_2{}^0)}^{(x_2)} \frac{x\,dx}{y} + \int_{(x_3{}^0)}^{(x_3)} \frac{x\,dx}{y}.$$

To evaluate the right side, notice that dI/dt, or $y^{-1}x\,dx/dt$, can only have negative powers of t in its expansion when either y vanishes or x becomes infinite. At a place, $x = e$, where y vanishes, the equation $f(x, y) = 0$, of the form $y^2 = 4(x - e)(x^2 + a'x + b')$, shews that the two values of y form a cycle, and we must put $x = e + t^2$, giving for y a form $y = 2t(A + Bt + \ldots)$, so that no negative power of t arises in $y^{-1}x\,dx/dt$. At $x = \infty$, the two values of y equally form a cycle, and the appropriate expression is to be found by putting $x = t^{-2}$, giving $y = -2t^{-3}(1 - at^4 - \ldots)$, and hence $dI/dt = t^{-2}(1 - at^4 - \ldots)$. Also, if, in Θ, $= (y - mx - c)/(y - m_0 x - c_0)$, we substitute $x = t^{-2}$, and

$$y = -2t^{-3}(1 - at^4 - \ldots),$$

we obtain

$$\log \Theta = \log(1 + \tfrac{1}{2}mt + \tfrac{1}{2}ct^3 + \ldots)/(1 + \tfrac{1}{2}m_0 t + \tfrac{1}{2}c_0 t^3 + \ldots),$$
$$= \tfrac{1}{2}(m - m_0)t + \ldots.$$

Wherefore $[(dI/dt) \log \Theta]_{t^{-1}} = \tfrac{1}{2}(m - m_0)$. The three equations $y_i = mx_i + c$ give $m = (y_1 - y_2)/(x_1 - x_2)$; thus finally the sum of the three integrals is equal to $\tfrac{1}{2}(y_1 - y_2)/(x_1 - x_2) - \tfrac{1}{2}(y_1{}^0 - y_2{}^0)/(x_1{}^0 - x_2{}^0)$.

It can be shewn, as will appear, that, except when the curve $f(x, y) = 0$ is rational, there are rational functions $\Phi(x, y)$, such that, for the integral $I = \int \Phi(x, y)\,dx$, there are no places at which dI/dt contains negative powers of t. Then the right side, in our general theorem, vanishes. The integral I is then spoken of as an everywhere finite integral; and for every one such integral we have an equation

$$\int_{(a_1)}^{(x_1)} \Phi(x, y)\, dx + \dots + \int_{(a_m)}^{(x_m)} \Phi(x, y)\, dx = 0,$$

where $(a_1), \dots, (a_m)$ are the places at which an arbitrary rational function $\Psi(x, y)$ takes any assigned value, for instance ∞, and $(x_1), \dots, (x_m)$ are the places at which the same function takes any other assigned value, for instance 0; the paths of integration are defined by continuous variation of $\Psi(x, y)$ from one of these values to the other. It has been assumed for simplicity in the proof given here that, for the range considered, no two of the places, at which $\Psi(x, y)$ takes any of the values passed over, coincide with one another; but this appears evidently unnecessary for the final result obtained. It should be remarked however that, if the path of integration be not assigned, the integral $\int_{(a)}^{(x)} \Phi(x, y)\, dx$ is not definitely determined by the extreme points (a) and (x); it is ambiguous by additive multiples of certain constants, the so-called periods of the integral. The theory of such periods is considered in subsequent chapters (V, VI).

It can also be shewn that there exists a rational function $\Phi(x, y)$ such that, for the integral $I = \int \Phi(x, y)\, dx$, there are two places only at which dI/dt contains negative powers of t, each of the first order, so that the expansions at these places are of the forms $dI/dt = \alpha t^{-1} + P(t)$, $dI/dt = \beta t^{-1} + P_1(t)$, where $P(t)$, $P_1(t)$ denote power series in the (unconnected) parameters t belonging to these places. The theorem $[\Phi(x, y)\, dx/dt]_{t^{-1}} = 0$, or $[dI/dt]_{t^{-1}} = 0$ shews then that $\beta = -\alpha$; so that, multiplying $\Phi(x, y)$ by a constant, we may suppose $\alpha = 1$, $\beta = -1$. It will be shewn later how to construct such an integral, with its infinities at any two assigned places of $f(x, y) = 0$. If $\Phi(x, y)$ be such that these infinities are at (z) and (c), whereat we have respectively $dI/dt = t^{-1} + P(t)$, $dI/dt = -t^{-1} + P_1(t)$, and Θ denote an arbitrary rational function with simple zeros at $(x_1), \dots, (x_m)$, and simple poles at $(a_1), \dots, (a_m)$, the general formula leads to

$$\int_{(a_1)}^{(x_1)} \Phi(x, y)\, dx + \dots + \int_{(a_m)}^{(x_m)} \Phi(x, y)\, dx = \log[\Theta(z)/\Theta(c)].$$

In this formula it is assumed that the paths of the m places under the integrals on the left do not pass through either of the places (z), (c); but, further, an appropriate integral multiple of $2\pi i$ must be added on the right. For the integral $\int_{(a)}^{(x)} \Phi(x, y)\, dx$, regarded as a function of (x), when (x) is near to (z), is of the form $\log t + P(t)$.

It can also be shewn that there exists a rational function $\Phi(x, y)$, so that for the resulting integral $I = \int \Phi(x, y)\, dx$, there is only one place, which may be arbitrarily taken, at which dI/dt has negative powers of t in its expansion, if this power be the second or a higher power; in particular, for which $dI/dt = -t^{-2} + P(t)$.

The possibilities and relations of the algebraic integrals for an elliptic curve (represented by a quartic plane curve with a point of self-contact) were investigated, after Abel had given the general theorem obtained above, by Legendre, whose work, extending over forty years, was entirely algebraical. The classification of the algebraic integrals, for any curve, by reference to the nature of their possible infinities, is essentially due to Riemann; the theory of the conduction of heat in a metal plate depends on a potential function which may either have two logarithmic infinities (for the case of a single source and a single sink) or one algebraic infinity (for the case of a single doublet); the two possible kinds of algebraic integrals with infinities which we have referred to above, have analogous behaviour; there is ground for believing that the analogy was in Riemann's mind. The exact, and simple, treatment of the theory which we have given (entirely different from Riemann's) is essentially due to Weierstrass.

A very important character of a curve, indeed the most important, is the number, p, of algebraic integrals $\int \Phi_i(x, y)\,dx$ existing for it which are everywhere finite in the sense we have explained, under the condition that there exists no linear equation, with constant coefficients, c_i, of the form $c_1\Phi_1(x, y) + \ldots + c_p\Phi_p(x, y) = 0$, satisfied in virtue of $f(x, y) = 0$. To emphasise this character we remark as follows: If an arbitrary rational function $\Psi(x, y)$, of order m, take the same value at the places $(x_1), \ldots, (x_m)$ of $f(x, y) = 0$, and the increments $dx_1, \ldots, dx_m$ refer to the passage to a consecutive set of such places, for the same function $\Psi(x, y)$, there exist, by what we have proved, p equations $\Phi_i(x_1, y_1)\,dx_1 + \ldots + \Phi_i(x_m, y_m)\,dx_m = 0$; it will be proved later, however, that, conversely, the existence of these p equations, with the p definite functions Φ_i, is sufficient to shew that there exists a rational function assuming the same value at all of them. Thus the problem of integrating this set of p total differential equations is that of finding the most general rational function of order m belonging to the given curve $f(x, y) = 0$. It will appear that when $m > p$ the number of arbitrary constants entering in the general integral of these equations is p.

Ex. 1. Suppose it be required to integrate the single differential equation
$$[u(x_1)]^{-\frac{2}{3}}\,dx_1 + [u(x_2)]^{-\frac{2}{3}}\,dx_2 + [u(x_3)]^{-\frac{2}{3}}\,dx_3 = 0,$$
where $u(x)$ denotes a given cubic polynomial in x. It can be proved that, for the cubic curve represented by the equation $y^3 - u(x) = 0$, there is one everywhere finite integral, namely $\int y^{-2}\,dx$, which is $\int [u(x)]^{-\frac{2}{3}}\,dx$. The problem of integrating this differential equation is therefore that of finding the general rational function of order 3 for this curve.

A particular function of order 3 is given by $(y - mx - n)/(y - m_0x - n_0)$, where m, n, m_0, n_0 are constants; and the intersections $(x_1, y_1), (x_2, y_2), (x_3, y_3)$ of the cubic curve with any line are such that
$$y_1(x_2 - x_3) + y_2(x_3 - x_1) + y_3(x_1 - x_2) = 0.$$

This equation then is a particular integral of the given differential equation. If $u(x) = a + 3bx + 3cx^2 + dx^3$, this equation, expressed in terms of x_1, x_2, x_3 only, which are the roots of an equation of the form $u(x) - (mx + n)^3 = 0$, is $(a + bp_1 + cp_2 + dp_3)^3 = u(x_1)\, u(x_2)\, u(x_3)$, where $p_1 = x_1 + x_2 + x_3$, $p_2 = x_2 x_3 + x_3 x_1 + x_1 x_2$, $p_3 = x_1 x_2 x_3$. It is not necessary, however, that the points (x_1, y_1), (x_2, y_2), (x_3, y_3), of the curve $y^3 = u(x)$, which satisfy the differential equation, should lie in line. For if $U = 0$, $V = 0$ represent two conics having three common intersections with the curve, the latter meeting the curve also in (a_1), (a_2), (a_3), while $U = 0$ meets the curve also in (x_1), (x_2), (x_3), then U/V is a rational function of third order with (x_1), (x_2), (x_3) as its zeros. Conversely, it will be proved in the next chapter that any rational function with given poles, (a_1), (a_2), (a_3), is capable of expression, with help of the equation of the given cubic curve, in this form U/V. Thus, the most general set (x_1, y_1), (x_2, y_2), (x_3, y_3) satisfying the differential equation are the intersections with the given cubic curve of a conic having three fixed intersections therewith; of these three fixed points, however, two may be taken to be the same for all sets (x_1, y_1), (x_2, y_2), (x_3, y_3), and the coordinates of these may be regarded as absolutely fixed, the single arbitrary constant of the general integral of the differential equation being the position of the third base point of the conics. This will be clearer after the discussion of coresidual sets of points in the next chapter. The particular integral already found arises when the third base point of the conics is in line with the two absolutely fixed points. To express the general integral, suppose that the cubic curve is met by $x = 0$ in $(0, p)$, $(0, \omega p)$, $(0, \omega^2 p)$; then the general conic through the two latter of these points has an equation

$$Ax^2 + 2Hxy + 2Gx + y^2 + yp + p^2 = 0.$$

The equation which expresses that this conic contains the four points (x_1, y_1), (x_2, y_2), (x_3, y_3), (x_0, y_0), of the cubic curve, is the general integral in question, with (x_0, y_0) as the arbitrary constant of integration. In particular when $(x_0, y_0) = (0, p)$, this integral, after removal of the factor $x_1 x_2 x_3$, reduces to the particular integral found above. For this example, the reader may compare Cayley, *Papers*, XII, p. 30.

Ex. 2. The curve given by $y^3 = az^3 + 3bz^2 x + 3czx^2 + dx^3$ has inflexions at its three intersections with $y = 0$. If $ez + x$ be any one of the linear factors of the cubic polynomial on the right, prove that $y(ez + x)^{-1}$ and $z(ez + x)^{-1}$ are rational functions, on the cubic curve, of orders respectively 2 and 3. Hence reduce the equation of the cubic curve, birationally, to the form $\eta^2 = \xi^3 + m$.

Ex. 3. If $y^2 = 4x^3 + ax + b$, shew that the general integral of the differential equation $y_1^{-1} dx_1 + y_2^{-1} dx_2 + y_3^{-1} dx_3 = 0$, is obtained by expressing that the conic $Ax^2 + Bx + Cy + D = 0$ contains the four points (x_1, y_1), (x_2, y_2), (x_3, y_3), (x_0, y_0), the last point furnishing the arbitrary constant of integration. Obtain the particular integral when $(x_0, y_0) = (\infty, \infty)$.

Ex. 4. For the curve represented by $y^2 = u(x)$, where $u(x)$ is a polynomial in x of order $2p + 1$, having only simple roots, shew that the complete integral of the p simultaneous differential equations

$$y_1^{-1} x_1^{i-1} dx_1 + \ldots + y_{p+1}^{-1} x_{p+1}^{i-1} dx_{p+1} = 0, \qquad (i = 1, 2, \ldots, p),$$

is expressed by the p equations

$$F(\alpha_i) \left[\sum_{r=1}^{p+1} y_r / (\alpha_i - x_r)\, F'(x_r) \right]^2 = C_i, \qquad (i = 1, 2, \ldots, p),$$

where $F(\theta)$ denotes $(\theta - x_1)\dots(\theta - x_{p+1})$, and $\alpha_1, \dots, \alpha_p$ are any p roots of $u(x)$, the $C_1, \dots, C_p$ being the arbitrary constants of the integration. The general everywhere finite integral for the given curve is in fact $\int \phi(x)\,dx/y$, where ϕ is an integral polynomial in x of order $p-1$.

Ex. 5. The expression of the branches of a plane curve in terms of a parameter, and the formula for Abel's theorem, may be extended to a curve in space.

As a very simple example, prove that, for the quartic curve given by the two equations $y^2 = x(1-x)$, $z^2 = x$, wherein x is regarded as the independent variable, there are: (i), four places on the curve for any value of x other than $x = 0$, $x = 1$, $x = \infty$; (ii), two places for $x = 0$ (where the curve has a double point), expressed by the expansions

$$x = t^2, \quad y = t(1 - \tfrac{1}{2}t^2 - \dots), \quad z = t, \quad \text{and} \quad x = t^2, \quad y = -t(1 - \tfrac{1}{2}t^2 - \dots), \quad z = t;$$

(iii), two places for $x = 1$, expressed by the expansions

$$x = 1 + t^2, \quad y = it + \dots, \quad z = 1 + \tfrac{1}{2}t^2 + \dots,$$

and

$$x = 1 + t^2, \quad y = it + \dots, \quad z = -(1 + \tfrac{1}{2}t^2 + \dots);$$

(iv), two places for $x = \infty$, expressed by

$$x = t^{-2}, \quad y = it^{-2} + \dots, \quad z = t^{-1} \quad \text{and} \quad x = t^{-2}, \quad y = -it^{-2} + \dots, \quad z = t^{-1}.$$

Prove also that the integral $I = \tfrac{1}{2}\int y^{-1}z^{-1}dx$ is such that dI/dt contains a negative power of t on each of the two branches at $x = 0$, but not elsewhere; further that the sum of the two integrals I, taken from the two places, other than $x = 0$, $y = 0$, $z = 0$, where the plane $z + ax + by = 0$ meets the curve, to the two places where $z + Ax + By = 0$ meets it, is

$$\log[(1 + B)(1 - b)/(1 - B)(1 + b)].$$

Verify the identity

$$\int^{x_1} \frac{dx}{2x(1-x)^{\frac{1}{2}}} + \int^{x_2} \frac{dx}{2x(1-x)^{\frac{1}{2}}} = \log[(\phi + x_2^{\frac{1}{2}} - x_1^{\frac{1}{2}})/(\phi - x_2^{\frac{1}{2}} + x_1^{\frac{1}{2}})] + \text{const.},$$

where $\phi = x_1^{\frac{1}{2}}(1 - x_2)^{\frac{1}{2}} - x_2^{\frac{1}{2}}(1 - x_1)^{\frac{1}{2}}$.

Note as to definition of a rational function on the curve.

We have defined as a rational function on the curve $f(x, y) = 0$ one which is given explicitly rationally in x and y. But, with the help of a fundamental theorem for functions of one independent variable, we can prove that if there be given a function R which has a perfectly definite value for every point of the curve (including those for which x, or y, is infinite), and on every branch at such point, being expressible in terms of the parameter appropriate to the branch, for the immediate neighbourhood of the origin of the branch, in the form $t^\lambda P(t)$, where $P(t)$ is a power series in t not vanishing for $t = 0$, and λ is a finite integer, positive, zero, or negative, then this function is expressible explicitly as a rational function of x and y. In brief, the conditions are that the function be single valued on the curve, and analytic with no discontinuities other than poles. There is then a finite neighbourhood about every pole within which the function is finite.

For first, for the values of x in the neighbourhood of a finite value $x = a$, within which all the values of R are finite, save possibly at $x=a$ itself, let $y_1, y_2, ..., y_n$ be the values of y satisfying $f(x, y) = 0$, not necessarily all different, and $R_1, ..., R_n$ be the values of the function R at the places (x, y_1), (x, y_2), ..., (x, y_n). Consider the sum $y_1{}^k R_1 + ... + y_n{}^k R_n$, where k is a positive integer. This is the sum of portions, such as $y_1{}^k R_1 + ... + y_r{}^k R_r$, where $y_1, ..., y_r$ belong to a cycle, there being a single place, on the curve $f(x, y) = 0$, for which $x = a$, in the neighbourhood of which the values of x, y are given by $x = a + t^r$, $y_i = b + c_1 t + c_2 t^2 + ...$, for $i = 1, 2, ..., r$. The portion $y_1{}^k R_1 + ... + y_r{}^k R_r$ is thence expressible by a series in t, involving only integer powers, and only a finite number of negative powers. As, however, $y_1, ..., y_r$ and the values $t, \epsilon t, ..., \epsilon^{r-1} t$, (where $\epsilon = \exp(2\pi i / r)$) enter symmetrically in this partial sum, the series will only involve the powers of t^r, namely will be a series in integer powers of $x - a$, with a finite number of negative powers. The whole sum $y_1{}^k R_1 + ... + y_n{}^k R_n$ is thence a function of x, expressible by series of integer powers about every finite value of x, single valued, and having no singularities other than poles. For the neighbourhood of $x = \infty$, by a similar argument, this same sum is single valued, and has no singularity other than perhaps a pole. Hence by the theory of functions of a single variable, this sum has only a finite number of poles, and is a rational function of x. Thus we may put $y_1{}^k R_1 + ... + y_n{}^k R_n = U_k(x)$. From the equations of this form for $k = 0, 1, ..., n-1$, we can express R_1 as M_1 / Δ, where Δ is the determinant of n rows whose general row is $y_1{}^k, y_2{}^k, ..., y_n{}^k$, and M_1 is the determinant obtainable from Δ by replacing the first column by $U_0(x), U_1(x), ..., U_{n-1}(x)$. Put in the form $M_1 \Delta / \Delta^2$, the denominator may be expressed as a rational function of x only, and $M_1 \Delta$ as a rational function of y_1 and x. This rational expression of R_1 by x and y_1, involves the like expression of R by x and y, which we desired to establish.

CHAPTER IV

THE GENUS OF A CURVE. FUNDAMENTALS
OF THE THEORY OF LINEAR SERIES

Main objects of the chapter. The present chapter has the main purposes of establishing the definition of that most important number called the *genus* of a curve, which is invariant in all birational transformations of the curve; and of obtaining the fundamental results of the theory of linear series of sets of points upon the curve. Both these are intimately related to the theory of the rational functions existing on the curve, and to the theory of the algebraic integrals belonging thereto, especially the so-called everywhere finite integrals. But it is desired that the account given shall be logically sound, shall be brief, and shall be simple; these conditions seem best satisfied by employing, together, ideas from several different modes of approach, due mainly to Abel, Riemann, Weierstrass, and Brill and Noether. A further method, developed by Kronecker and by Dedekind, in extension of the arithmetical theory of integer numbers, is explained at length in a subsequent chapter (VII). Unless the contrary is stated, the curve considered is supposed to be a plane curve.

We distinguish provisionally between what we may call the *integral genus*, and the *arithmetical genus*; the former is easy to explain in general, the latter can be computed for a curve of sufficient simplicity; it is part of our task to shew that these are the same.

As regards the former, it has already been remarked that, if the fundamental curve $f(x, y) = 0$ be not rational, there exists at least one algebraic integral attached thereto, $\int R(x, y) dx$, where $R(x, y)$ is rational in x and y, which is everywhere finite on the curve. It is clear that if the given curve be transformed birationally into another curve, such an integral gives rise to an algebraic integral for the other curve, likewise finite everywhere on this. The number of such integrals, $\int R_i(x, y) dx$, for which there exists no linear equation $\Sigma \lambda_i R_i(x, y) = 0$, with constant coefficients λ_i, is thus the same for two curves which are in $(1, 1)$ birational correspondence. It is this number which is the integral genus. For a rational curve this number is zero, every integral $\int r(t) dt$, in which $r(t)$ is a rational function of t, becoming infinite for a finite or infinite value of t; for an elliptic curve this number is unity.

This definition suggests at once the enquiry, in fact begun by Abel in his great paper *Mémoire sur une propriété générale d'une classe très étendue de fonctions transcendantes* (1826, cf. *Oeuvres complètes*), as to the behaviour, at infinity and at the multiple points of the curve $f(x, y) = 0$, necessary for the function $R(x, y)$ in order that the integral should be everywhere finite; and this enquiry is intimately related with the further one of the behaviour of the polynomials which form the numerator and denominator in the expression of a rational function, at the multiple points of a curve, in order that the function should be quite general. It has been seen that it may be possible to transform the given curve birationally so that a multiple point is replaced by several points of the new curve; it appears probable then that a rational function belonging to the curve should assume different values at these; this can only happen when, in the expression of the function by the coordinates on the original curve, both the denominator and numerator vanish at the multiple point of the original curve. The different values of the function thereat may then be obtainable by a limiting process, approaching the multiple point on the different branches of which it is the origin. For the sake of brevity, we answer this enquiry by defining what is meant by saying that a polynomial, in the coordinates, is *adjoint* at a multiple point of the given curve $f(x, y) = 0$; then we shew that any rational function is expressible as a quotient of two such adjoint polynomials. The definition we give covers all cases; later we shall see that for simple multiple points it can be replaced by an easy geometrical definition.

In order that it may be clear that the definition is not limited to points of the curve of which the coordinates are finite, we suppose the curve first given in homogeneous coordinates x_0, x_1, x_2, say $F(x_0, x_1, x_2) = 0$, of order n. If, beside (x_0, x_1, x_2), the consecutive point $(x_0 + dx_0, x_1 + dx_1, x_2 + dx_2)$ also belong to the curve, we have the two equations $F_0 dx_0 + F_1 dx_1 + F_2 dx_2 = 0$, $x_0 F_0 + x_1 F_1 + x_2 F_2 = 0$, where

$$F_i = \frac{1}{n} \partial F / \partial x_i;$$

and hence

$$(x_1 dx_2 - x_2 dx_1)/F_0 = (x_2 dx_0 - x_0 dx_2)/F_1 = (x_0 dx_1 - x_1 dx_0)/F_2,$$

so that, if c_0, c_1, c_2 be quite arbitrary, each of these fractions is equal to $(c, x, dx)/F_x^{n-1}F_c$, where (c, x, dx) denotes the determinant whose rows are (c_0, c_1, c_2), (x_0, x_1, x_2), (dx_0, dx_1, dx_2), and the ordinary symbolical notation is used, $F(x_0, x_1, x_2)$ being written F_x^n. We say now that any integral form $\Psi(x_0, x_1, x_2)$, homogeneously of order m in x_0, x_1, x_2, is *adjoint* to F when the curve $\Psi = 0$ intersects $F = 0$ in such a way that the differential

$$\Psi . (cxdx)/F_x^{n-1}F_c$$

is everywhere finite. In other words, Ψ must vanish appropriately at the zeros of $F_x{}^{n-1}F_c$ upon $F = 0$, in so far as these are not compensated by the zeros of $(cx\,dx)$.

To see the meaning of this, consider first a point of the curve $F = 0$ at which the coordinate x_2 does not vanish. The quotient $x_2{}^{-(m-n+3)}\Psi \cdot (cx\,dx)/F_x{}^{n-1}F_c$ is a functional differential, unaltered by replacing x_0, x_1, x_2 by any constant multiples of themselves. Herein put $x = x_0/x_2$, $y = x_1/x_2$, and take $c_0 = 0$, $c_2 = 0$; denote $x_2{}^{-m}\Psi$ by $\psi(x, y)$, or ψ, and $x_2{}^{-n}F$ by $f(x, y)$, so that $nF_x{}^{n-1}F_1$ is $x_2{}^{n-1}\partial f/\partial y$, or, say, $x_2{}^{n-1}f'(y)$. Save for the factor n, the differential is then $\psi\,dx/f'(y)$, and the meaning is that, in the neighbourhood of any point for which x is finite, given by such a pair of formulae as $x = a + t^r$, $y - b = c_1 t + c_2 t^2 + \dots$ (as in the preceding chapter), the expression $[\psi/f'(y)]\,dx/dt$, written as an ascending series of powers of t, must contain no negative powers of t, this being so for every branch of the curve having origin at the point, (a, b). As $dx/dt = rt^{r-1}$, this is exactly the same as saying that $(x - a)\psi/f'(y)$ must vanish at every finite place for which $x = a$.

Consider next a point at which $x_2 = 0$. If also $x_0 = 0$, we can choose a constant λ so that $x_0 - \lambda x_1$ does not vanish, and then use $x_0' = x_0 - \lambda x_1$, $c_0' = c_0 - \lambda c_1$ instead of x_0 and c_0; we suppose then x_0 is not zero. Then the quotient $x_0{}^{-(m-n+3)}\Psi \cdot (cx\,dx)/F_x{}^{n-1}F_c$ is a functional differential. Herein put $\xi = x_2/x_0$, $\eta = x_1/x_0$, and take $c_0 = 0$, $c_2 = 0$; denote $x_0{}^{-n}\Psi$ by $\psi_1(\xi, \eta)$, or ψ_1, and $x_0{}^{-n}F$ by $f_1(\xi, \eta)$, so that $nF_x{}^{n-1}F_1$ is $x_0{}^{n-1}\partial f_1/\partial \eta$, or, say, $x_0{}^{n-1}f_1'(\eta)$. Save for the factor n, the differential is then $-\psi_1\,d\xi/f_1'(\eta)$, and the meaning is that, for every place in the neighbourhood of $\xi = 0$, given by such a pair of expansions as $\xi = t^r$, $\eta - b = c_1 t + c_2 t^2 + \dots$ (as in the preceding chapter), the expression of $[\psi_1/f_1'(\eta)]\,d\xi/dt$ as an ascending series in t must contain no negative powers; or, what is the same, that $\xi\psi_1/f_1'(\eta)$ must vanish. Still supposing x_0 not zero, the equation $f_1(\xi, \eta) = 0$ is derived directly from $f(x, y) = 0$ by $\xi = x^{-1}$, $\eta = yx^{-1}$, which lead to $x^{-(m-n+3)}\psi\,dx/f'(y) = -\psi_1\,d\xi/f_1'(\eta)$; thus the condition of adjointness for $x = \infty$ is that the differential on the left here must be finite. The case of a point for which $x_2 = 0$ and $x_0 = 0$, of which we have evaded the detailed consideration, arises, for the non-homogeneous equation $f(x, y) = 0$, when $x = 0$, $y = \infty$.

The quotient $(x - a)\psi(x, y)/f'(y)$ certainly vanishes, and the condition of adjointness is nugatory, at every finite value of x, unless $f'(y) = 0$. The condition is thus effective only, (i), at a point of the curve $f(x, y) = 0$ at which x is finite but y is infinite, namely for a value of x which reduces to zero the coefficient of the highest power of y, when this power is not y^n. We can easily evade this by writing for x, in the equation, $x + hy$, where h is a constant; and may thus suppose the term y^n to be present; or (ii), when $f'(y) = 0$ for a

finite value of x. This happens, as we see by eliminating y between
$f'(y) = 0$ and $f(x, y) = 0$, only for a finite number of values of x. The
equation $f'(y) = 0$ expresses that two, or more, of the n values of y
corresponding to the same value of x coincide.

It is easy to give a geometrical interpretation in the simplest
cases. For instance, if at $x = 0$, $y = 0$ there be a multiple point of
order k, with distinct tangents; so that the terms of lowest order
in $f(x, y) = 0$ are of order k, having k distinct linear factors, which,
first, we suppose to be all of the form $y - mx$: then in $f'(y)$, the terms
of lowest order are of order $k - 1$, and it is sufficient that the terms
of lowest order in ψ should equally be of order $k - 1$, namely, that
$\psi = 0$ should have a multiple point of order $k - 1$ at $x = 0$, $y = 0$. In
this case the quotient $\psi/f'(y)$ is finite on all branches of $f(x, y) = 0$
as we approach the origin. If, however, next, one of the tangents of
$f(x, y) = 0$ be $x = 0$, and along the corresponding branch, to the
first approximation, $x \propto y^2$, it is still true that the condition of
$x\psi/f'(y)$ vanishing is satisfied if $\psi = 0$ have a multiple point of order
$k - 1$ at the origin. Another simple case is that of an ordinary cusp;
when the equation of the curve near the origin, $(x, y)_r$ denoting a
homogeneous polynomial in x, y of order r, is $y^2 + (x, y)_3 + \ldots = 0$, the
term in x^3 being present; in this case $f'(y)$, to the first approxima-
tion, $\propto x^{\frac{3}{2}}$, and it is sufficient that $\psi/x^{\frac{1}{2}}$ should vanish, or that the
curve $\psi = 0$ should pass through the cusp.

Ex. 1. When in $f(x, y) = 0$, the terms of lowest order are of the form
$y^2 + y(x, y)_2 + y(x, y)_3 + \ldots + y(x, y)_{r-2} + (x, y)_r + \ldots$ with $r > 3$, the term
in yx^2 being present, prove that there are two branches of the curve at
the origin, but it is sufficient for adjointness that the curve $\psi = 0$ should
pass through the origin with $y = 0$ for tangent.

Ex. 2. When in $f(x, y) = 0$, the terms of lowest order are of the form
$x + x(x, y)_1 + x(x, y)_2 + \ldots + x(x, y)_{r-2} + (x, y)_r + \ldots$, with $r > 2$, the term
in yx being present, it is not necessary for adjointness that $\psi = 0$ should pass
through the origin, though $f'(y)$ there vanishes.

Ex. 3. For the sextic curve $y^3 + y^2(x, y)_2 + y(x, y)_4 + (x, y)_6 = 0$, the
condition of adjointness of $\psi = 0$, at the origin, is that $\psi = 0$ should have
a self-contact, with $y = 0$ as the tangent.

Further, it is capable of proof, though we do not enter into this in
detail at this stage, that the conditions of adjointness of a polynomial ψ, at a multiple point of the curve $f(x, y) = 0$, are those
which we should employ in carrying out the process of reduction
described in Chap. II, so far as this reduction is made possible by
the existence of the multiple point. Consider the case of a multiple
point of $f(x, y) = 0$, of multiplicity k, with k distinct tangents. In
order that $\psi = 0$ should have there a multiple point of order λ, it
is necessary to impose $\frac{1}{2}\lambda(\lambda + 1)$ linear conditions for ψ; by this
we secure λk intersections of $\psi = 0$ with $f = 0$; the condition that
this should be at least twice the number of conditions, is that

$\lambda k \geqslant \lambda(\lambda + 1)$, or $\lambda \leqslant k - 1$; and we have shewn that for $\psi = 0$ to be adjoint at this point, we have $\lambda = k - 1$. It is an incidental consequence of the theory that a polynomial ψ satisfying the conditions of adjointness wherever, on $f(x, y) = 0$, they are effective, cannot be further reduced in the sense of Chap. II.

Expression of rational function by adjoint polynomials. Now let $R(x, y)$ be any rational function for the curve, of order n, $f(x, y) = 0$; and suppose, (a), that the term in y^n is present in $f(x, y)$, a substitution of the form $x + hy$ for x being made, if necessary, to secure this; (b), that the terms of order n in $f(x, y)$ are the product of n distinct linear factors, as may be secured by a suitable (fractional) linear substitution for x and y. Further, suppose that the poles of the function R are all at places for which x is finite, as can also be secured in a similar way. Next choose a polynomial $\psi(x, y)$ quite arbitrarily subject to the condition that the product $R\psi$ is finite at all the poles of R; or that ψ vanishes on $f(x, y) = 0$, at any pole of R, to an order at least equal to the order of this pole, and subject further to the condition that ψ is adjoint at all the places of $f(x, y) = 0$, arising for finite values of x, at which $f'(y)$ vanishes, where $f'(y) = \partial f / \partial y$. For a place at which $x = a$, the second condition is that $(x - a)\psi / f'(y)$ shall vanish at this place. If it happen that $R(x, y)$ becomes infinite at a place where $f'(y) = 0$, the two conditions for ψ are understood to require that $(x - a)R\psi / f'(y)$ shall vanish at this place. By taking ψ of sufficiently high order, the conditions can be satisfied. Consider then the value, on $f(x, y) = 0$, of the sum

$$\sum_{i=1}^{n} \frac{(\theta - y_1) \dots (\theta - y_n)}{\theta - y_i} M \frac{R(x, y_i) \psi(x, y_i)}{f'(y_i)},$$

where M is the product of all the different factors $x - a$ at all the finite values of x for which $f'(y) = 0$, θ is an undetermined quantity, and, for a general value of x, the $y_1, \dots, y_n$ are the n values of y which satisfy $f(x, y) = 0$, all of which, by the hypothesis, are finite. This sum, from its symmetry in regard to $y_1, \dots, y_n$, is expressible as a rational function of x only; and it can only become infinite, for a finite value of x, when $f'(y_i)$ vanishes, or $R(x, y_i)$ becomes infinite. By the conditions for ψ, $MR\psi / f'(y)$ vanishes for every such value of x. The sum is therefore a rational function of x with no poles for finite values of x, and is therefore a polynomial in x; moreover it vanishes for all the values of x which make $M = 0$, so that it divides by M. Thus, on $f(x, y) = 0$, it is capable of the form

$$M(\theta^{n-1} u_1 + \theta^{n-2} u_2 + \dots + u_n),$$

where $u_1, u_2, \dots, u_n$ are integral polynomials in x. If we now put y_1 for θ, and divide by M, since $f'(y_1) = (y_1 - y_2) \dots (y_1 - y_n)$, we obtain $R(x, y_1) = (y_1^{n-1} u_1 + y_1^{n-2} u_2 + \dots + u_n) / \psi(x, y_1)$, and in this we may put y for y_1. Hence it is proved that any rational function R, for the curve $f(x, y) = 0$, is expressible as the quotient of two polynomials of which the denominator is an arbitrary adjoint polynomial ψ chosen to vanish at the poles of R, so that $R\psi$ is finite at such a pole, or, if such a pole be also a point, with $x = a$, for which $f'(y)$ vanishes, chosen to compensate this zero, so that

$$(x - a) R\psi/f'(y)$$

vanishes. The numerator polynomial must needs vanish at all the zeros of the denominator which are not poles of the function, and thus in particular must also be adjoint; and since the function is taken here not to have a pole for infinite values of x, the order of the numerator polynomial will not be higher than that of the denominator.

We have defined adjointness with use of homogeneous variables; thus the references to infinite values of x in this enunciation can easily be eliminated.

Expression of most general everywhere finite integral. Before applying the preceding result to discuss further the properties of rational functions, we apply an almost identical reasoning to prove that all the everywhere finite integrals belonging to the curve $f(x, y) = 0$ can be expressed in the form $\int \phi \, dx/f'(y)$, where ϕ is an integral polynomial of suitably limited order. Suppose, first, as before, that the equation $f(x, y) = 0$ is taken so that y becomes infinite only when x is infinite, and that all the places arising for $x = \infty$ are distinct; let u denote any everywhere finite integral; then du/dx is a rational function only becoming infinite, for finite x, at a place where the parametric expression of the form $x = a + t^r$, $y = b + c_1 t + c_2 t^2 + \ldots$, has $r > 0$; at such a place

$$du/dx = \frac{1}{r} t^{-(r-1)} \, du/dt,$$

and du/dt is finite; thus $(x - a) du/dx$ vanishes. Hence if M denote the product of all the different factors $x - a$ so arising, and $(du/dx)_i$ denote the value of du/dx for $y = y_i$, we have, as above, the equation

$$\sum_{i=1}^{n} \frac{(\theta - y_1) \ldots (\theta - y_n)}{\theta - y_i} M \left(\frac{du}{dx} \right)_i = M (\theta^{n-1} u_1 + \ldots + u_n),$$

and can infer an equation of the form $du/dx = (y^{n-1} u_1 + \ldots + u_n)/f'(y)$. At any one of the places arising for $x = \infty$, however, we have $x = t^{-1}$, $y = mt^{-1} + \ldots$; thus du/dx vanishes to the second order. From this it appears that the aggregate order of the polynomial $y^{n-1} u_1 + \ldots + u_n$ in x and y does not exceed $n - 3$. The deduction shews that this polynomial is adjoint at all places where $f'(y) = 0$. Conversely, any such adjoint polynomial, of aggregate order $n - 3$, is easily seen to give rise to an everywhere finite integral.

And as before, the references to infinite values of x are easily eliminated, the everywhere finite integrals belonging to the curve $F(x_0, x_1, x_2) = 0$ being given by $\int \phi(x_0, x_1, x_2)(cxdx)/F_x{}^{n-1}F_c$, where $\phi(x_0, x_1, x_2)$ is an everywhere adjoint polynomial homogeneously of order $n-3$. The number we have defined as the integral genus is thus the number of undetermined coefficients in such an everywhere adjoint polynomial.

Equivalent or coresidual sets of points on the curve. We now introduce another technical term, and say that two sets, each of m places, on the curve $f(x, y) = 0$, are *equivalent*, or *coresidual*, when there exists a rational function for this curve, say $K(x, y)$, which has one value, say A, at the places of one of these sets (and not elsewhere), and has another value, say B, at the places of the other set (and not elsewhere). As we may consider the rational function $[K(x, y) - A]/[K(x, y) - B]$, we see that this is the same as saying that two sets of places are coresidual when they constitute respectively the zeros and poles of the same rational function. Hence, the most general set, N, of places coresidual with a given set D, is obtained by finding the most general rational function of which the places of the latter set, D, are the poles, and taking the zeros, N, of this function. By what we have proved above, this is done by first taking a perfectly arbitrary adjoint polynomial (D) which vanishes in the places of the set D; this polynomial (D) will in general vanish, on $f(x, y) = 0$, at places other than those of the set D, say in places of a set O, which is then said to constitute a set residual to D. The formation of the rational function is then to be completed by finding the most general adjoint polynomial, (N), vanishing in the set O, in each case to the same order as does (D), which is of the same order as (D). The rational function is then given by $(N)/(D)$. In general the polynomial (N) is not completely determined by the conditions imposed upon it, but the most general rational function with given poles contains a number of linearly entering arbitrary coefficients.

The statements made seem easiest of application if we think of the curve $f(x, y) = 0$ as given by the homogeneous equation

$$F(x_0, x_1, x_2) = 0;$$

then the conditions of adjointness are effective only at the zeros of the polar $F_x{}^{n-1}F_c$ which are independent of the arbitrary point (c_0, c_1, c_2), that is, at the multiple points of the curve; they require the adjoint curve to have a certain multiplicity at such a point; in general the set O spoken of will consist of the residual intersections of the curve $(D) = 0$, other than those which this has at the multiple points of $F = 0$.

Complete linear series of sets of points on the curve. Now suppose there are $(r+1)$ homogeneously entering arbitrary coefficients in the numerator polynomial (N). Then the sets of places determined on the curve $F=0$ by $(N)=0$, other than those required by the conditions of adjointness (at the multiple points of the curve), and other than the set O which are also zeros of $(D)=0$, are said to form, on $F=0$, a linear series of sets of places, of freedom r; one such set exists for every definite set of ratios of the undetermined coefficients in (N); thus one such set, of the linear series of sets coresidual with the given zeros of (D), exists with arbitrary positions for r places of the set. This number r is necessarily less than the whole number, m, of zeros of (D). Further, we may form a partial series, of sets each of m places, of freedom less than r, defined by the zeros (other than at the multiple points of $F=0$) of a polynomial found from (N) by assigning definite values (say zero) to some of the undetermined coefficients therein. But the series of sets of m places which we have found is said to be *complete*, because, by the manner of its construction, it is not contained in any other series, of sets of the same number, m, of places, having freedom greater than r. It may be remarked that among the sets of this complete series, the set of zeros of (D) occurs, one of the curves $(N)=0$ through the points O being evidently $(D)=0$. We may say then that a complete series of sets of coresidual places is uniquely determined by any one of its sets, which may be taken arbitrarily.

Ex. 1. Consider a plane quartic curve with two double points, and the set D determined by the intersections of the curve with a line u; coresidual sets of four points, forming a series *of freedom* 2, are evidently obtained by the intersections of the curve with a variable line v. But the complete series is to be obtained by taking an adjoint curve through the points of the set u, meeting the quartic again in a set O, and then taking the most general adjoint curve of the same order through the points of the set O. If, for instance, we take the adjoint conic through the set u, which breaks up into the line u and the line joining the double points, in which case there exists no residual set O, then the most general adjoint conic is conditioned only by passing through the double points, and determines on the quartic curve a coresidual series of sets of four points *of freedom* 3. Likewise if we take an adjoint cubic curve through the original set u, the residual set O consists of four points lying on a particular conic through the double points; the series of coresidual sets is then determined by cubic curves through the double points and the four points O, and is *of freedom* 3 (since the equation of a cubic curve contains ten terms).

If, however, upon the original quartic curve with two double points we start with a set of 4 points which do not lie on a conic containing the double points of the curve, we determine a residual set O of four points by a cubic curve through the double points and the original set of 4 points; and the coresidual series, of sets of 4 points, is obtained by cubic curves through the double points and the set O. This series is likewise of freedom 3, but it is distinct from the series found before; it cannot contain a set of 4 points which lie on a line unless there is a conic through the set O and the double points, which is not generally the case.

Ex. 2. Prove that on a plane quartic curve with one double point, the complete series of sets coresidual with a set of four collinear points of the curve is given by the intersections of the curve with a variable line. Examine whether this series is distinct from that of points coresidual with four general points.

Theorem of coresiduation. The theorem we have proved, that in the expression of a rational function the denominator polynomial is arbitrary, so long as it is adjoint and vanishes at the poles of the function, is a remarkable result, and leads to an important geometrical property, namely: Suppose there be, upon the curve $f(x, y) = 0$, two sets N, D, of the same number of places, which have the same residual set of places O, so that there is an adjoint curve $\psi = 0$ whose aggregate set of intersections with $f = 0$, other than at the multiple points, consists of the sets O, D; and, also, another adjoint curve $\phi = 0$, of the same order as $\psi = 0$, whose aggregate set of intersections, other than at the multiple points, consists of the sets O, N. Let then any other adjoint curve $\psi' = 0$ be taken, which has the same intersections D with $f = 0$ as has $\psi = 0$; let its residual set of intersections with $f = 0$, other than at the multiple points, be O'; *then* there exists an adjoint curve $\phi' = 0$, of the same order as $\psi' = 0$, whose aggregate intersections, other than the multiple points, consist of the sets O', N together. This is called the *theorem of coresiduation*. It is sometimes important to bear in mind that in this theorem there may be points which are common to the sets N and D.

The equality of the ratios ϕ/ψ and ϕ'/ψ', which holds in virtue of $f(x, y) = 0$, must involve a polynomial identity of the form $\phi\psi' = \phi'\psi + \theta f$, where θ is equally a polynomial in the coordinates. In a later chapter a theorem of Noether's, in regard to the expression of the equation of a curve which passes through the common points of two given curves, will be considered (Chap. VIII), which can be used to furnish an alternative proof of the theorem of coresiduation. The method we have followed has a wide bearing; it is used in the exposition of Galois' theory of the invariants of a group of permutations of the roots of an algebraic equation (cf. Kronecker, *Werke*, II, p. 202).

Sum and difference of complete linear series. The sets of a linear series are given by the intersections with $f = 0$ of the curves of a system with equation $\lambda_0\phi_0 + \lambda_1\phi_1 + \ldots + \lambda_r\phi_r = 0$, wherein $\lambda_0, \ldots, \lambda_r$ are variable parameters, and $\phi_0, \ldots, \phi_r$ are definite polynomials not connected, in virtue of $f = 0$, by any linear equation with constant coefficients. The places of a set of this series are the intersections, with $f = 0$, other than those which are independent of $\lambda_0, \ldots, \lambda_r$, namely, other than those common to all of $\phi_0 = 0, \ldots, \phi_r = 0$ (among which are the intersections at the multiple points). It is sometimes convenient however to adjoin, to the variable sets thus defined, certain of the intersections which are not variable. If a particular set of the series is denoted by A, it is usual to denote by

$|A|$ the complete series determined thereby. If $|B|$ be another complete series, given by a system $\lambda_0\psi_0 + \dots + \lambda_s\psi_s = 0$, there exists a further complete series $|A+B|$, defined by the aggregate of a set A taken with a set B. This contains all sets given by the system $\Sigma\lambda_{ij}\phi_i\psi_j = 0$, but may contain also other sets not so given. Conversely, given two complete series $|C|$, $|A|$, it may be that there are sets of $|C|$ which contain all the places of a particular set A_0 of $|A|$; if $|C|$, of freedom k, be given by $\mu_0\vartheta_0 + \dots + \mu_k\vartheta_k = 0$, the curves of $|C|$ which contain the places A_0 will be obtained by imposing proper linear conditions upon $\mu_0, \dots, \mu_k$. These curves define a complete system $|C-A_0|$, which can be shewn, utilising the theorem of coresiduation proved above, to be equally definable from any other set of $|A|$ instead of A_0. The resulting series is therefore denoted by $|C-A|$. If this be denoted by $|B|$, we have $|C| = |A+B|$. It is clear that these notions, of the sum or difference of two linear series, can be explained in another way, speaking of the product or quotient of two rational functions belonging to the curve $f = 0$.

Riemann–Roch formula for linear series, and for rational functions. There exists an important relation connecting the freedom of a linear series, upon a curve of given genus, with the number of places in a set of the series, which we develop in connexion with a numerical definition of the genus; such a definition is possible when the fundamental curve $F = 0$ has no multiple points other than those with distinct tangents. The genus defined by the number of everywhere finite integrals belonging to the curve, has been proved (p. 65) to be the same as the number of linearly independent curves of order $n-3$ (if n be the order of $F = 0$) which are everywhere adjoint to $F = 0$; but, even when $F = 0$ has the simple form referred to, this is not final without an examination of whether the conditions of adjointness at all the multiple points of $F = 0$ are linearly independent. We proceed then with care.

Let $F = 0$ be a curve of order n, *which at every multiple point, say of multiplicity k, has k distinct tangents.* We define the number P by the equation $P = \frac{1}{2}(n-1)(n-2) - \frac{1}{2}\Sigma k(k-1)$, where the summation extends to all the multiple points. It has already been shewn (Chap. I, p. 10) that this number is not negative.

Now consider a curve, $\psi = 0$, of order m, subject to no conditions other than of having a $(k-1)$-ple point at every k-ple point of $F = 0$; it is clearly possible to construct such a curve if m be great enough. Let N be the whole number of intersections of $\psi = 0$ with $F = 0$ other than these prescribed ones, and $R+1$ the number of homogeneously entering coefficients which remain arbitrary in ψ after satisfying the conditions. We cannot obtain R by adding the number of conditions imposed upon ψ at the separate multiple points, because we do not know that these conditions are linearly independent for a given limited value of the order m of ψ. When the R ratios, of the variable coefficients in ψ, vary, some, possibly all, of the N intersections of $\psi = 0$ with $F = 0$ will also vary; a very particular instance when this is not so is when ψ is a line through the two double points of a quintic curve—the remaining intersection is then fixed; but such a thing is possible with $R > 0$. As before we call R the freedom of the series determined on $F = 0$ by the curves $\psi = 0$ when the $R + 1$ coefficients vary; and we call N the *complete grade* of the series, even when some of the N intersections are fixed.

Suppose first $m \geqslant n$, and that $F(x_0, x_1, x_2)$ contains the term in $x_2{}^n$; regarding F and ψ as polynomials in x_2, we can write, if ψ be a general polynomial of order m,

$$\psi = F(cx_2{}^{m-n} + u_1 x_2{}^{m-n-1} + \ldots + u_{m-n}) + v_{m-n+1} x_2{}^{n-1} + \ldots + v_m,$$

where $u_1, \ldots, v_m$ are polynomials in x_0, x_1, and c is a constant. The intersections of $\psi = 0$ with $F = 0$ are therefore the same as the intersections with $F = 0$ of the curve $\psi_1 = 0$, where

$$\psi_1 = v_{m-n+1} x_2{}^{n-1} + \ldots + v_m.$$

In general ψ contains $\frac{1}{2}(m+1)(m+2)$ coefficients, but ψ_1 contains only $\frac{1}{2}(m+1)(m+2) - \frac{1}{2}(m-n+1)(m-n+2)$, or, say, $\rho + 1$ coefficients. Conversely, ψ_1 is a polynomial of order m. Whence, a general polynomial of order m, with $m \geqslant n$, contains, so far as its intersections with $F = 0$ are concerned, precisely $\rho + 1$ homogeneously entering arbitrary coefficients. Thus the number, $R + 1$, of homogeneously entering arbitrary coefficients in such a polynomial, when it is further conditioned by having a $(k-1)$-ple point at every k-ple point of $F = 0$ (equivalent to $\frac{1}{2}k(k-1)$ conditions, as we may see by supposing the point to be at $x_0 = 0$, $x_1 = 0$), is given by

$$R + 1 \geqslant \tfrac{1}{2}(m+1)(m+2) - \tfrac{1}{2}(m-n+1)(m-n+2) - \tfrac{1}{2}\Sigma k(k-1);$$

here the right side is $mn + 1 - \frac{1}{2}(n-1)(n-2) - \frac{1}{2}\Sigma k(k-1)$; also we have, *by definition*, $N = mn - \Sigma k(k-1)$; thus, recalling the definition of P, we have, for $m \geqslant n$, $R \geqslant N - P$.

Next suppose $m < n$, and write m as $n - 3 + \alpha$. Then, as before, we have $R + 1 \geqslant \frac{1}{2}(m+1)(m+2) - \frac{1}{2}\Sigma k(k-1)$, which is the same as $R \geqslant \frac{1}{2}\alpha(\alpha + 2n - 3) + P - 1$; thus, if α be not negative (so that it is 0, or 1, or 2), $R + 1$ is certainly positive for $P \geqslant 1$, $n \geqslant 3$, and a curve $\psi = 0$ exists passing through the multiple points in the way prescribed. In particular, if $\alpha = 0$, $R \geqslant P - 1$. Also we have

$$N = mn - \Sigma k(k-1),$$

which is equal to $\alpha n + 2P - 2$, and is not negative if $P \geqslant 1$, $\alpha \geqslant 0$. Thus $R - N \geqslant \frac{1}{2}\alpha(\alpha - 3) - P + 1$. Wherefore, when $\alpha = 0$, namely for curves $\psi = 0$ of order $n - 3$, we have $R > N - P$. But, as

$$\tfrac{1}{2}\alpha(\alpha - 3) + 1 = 0,$$

when $\alpha = 1$ or 2, for curves $\psi = 0$ of order m, greater than $n-3$, including curves for which $m \geqslant n$, as we have seen above, our result is $R \geqslant N - P$.

The sign of inequality arises from the possibility that the linear conditions to be satisfied by the coefficients in ψ, at the multiple points of $F = 0$, may not be linearly independent. The number of such conditions depends however only on $F = 0$, and not on $\psi = 0$; and if the number of coefficients in ψ be great enough, that is, if ψ be of sufficiently high order, these linear conditions must certainly be linearly independent. Thus we may say:

For curves $\psi = 0$ of sufficiently high order, we have $R = N - P$; for curves $\psi = 0$ of less order, but greater than $n-3$, we have $R \geqslant N - P$; finally, for curves $\psi = 0$ of order $n - 3$, we have $R > N - P$.

Consider in particular the last case, $m = n - 3$. Then $N = 2P - 2$, and we have $R > P - 2$. We can prove in an instructive way that $R = P - 1$, as follows: Denote $R + 1$ by p, so that $p > P - 1$; say $p = P + \epsilon$, with $\epsilon \geqslant 0$. Denote the general adjoint polynomial of order $n - 3$ by

$$\lambda_1 \phi_1 + \ldots + \lambda_p \phi_p,$$

where $\lambda_1, \ldots, \lambda_p$ are arbitrary coefficients; the zeros on $F = 0$, other than those prescribed at the multiple points, of any one of the polynomials $\phi_1, \ldots, \phi_p$ are $2P - 2$ in number. Further,

$$N - R = 2P - 2 - (p-1) = p - 1 - 2\epsilon,$$

so that $N - R < R$ if $\epsilon > 0$. Now, first, suppose that, save for the prescribed points, there is on $F = 0$ no zero common to all the polynomials $\phi_1, \ldots, \phi_p$. Take R, or $p - 1$, arbitrary general points, $M_1, \ldots, M_{p-1}$, on $F = 0$, and determine the particular curve $\lambda_1 \phi_1 + \ldots + \lambda_p \phi_p = 0$ which passes through these, say Φ; let its remaining intersections with $F = 0$, other than at the multiple points, be $A_1, A_2, \ldots, A_{N-R}$. When $\epsilon > 0$ we have $N - R < p - 1$, so that, beside Φ, at least one other curve $\lambda_1 \phi_1 + \ldots + \lambda_p \phi_p = 0$, linearly independent of Φ, can be drawn through the derived points $A_1, \ldots, A_{N-R}$. We prove that any such curve, say $\Psi = 0$, necessarily contains also the points $M_1, \ldots, M_{p-1}$; this we prove by shewing that $\Psi = 0$ contains M_{p-1}, any one of $M_1, \ldots, M_{p-1}$. To this end remark, that there exists, for $F = 0$, a rational function having M_{p-1} as a pole of the first order, and having also $A_1, \ldots, A_{N-R}$, or some of these, as poles of the first order. For, since $M_1, \ldots, M_{p-1}$ are arbitrary and general, we can suppose that the most general curve $\lambda_1 \phi_1 + \ldots + \lambda_p \phi_p = 0$ through $M_1, \ldots, M_{p-2}$, only, has an equation of the form $\mu \Phi + \mu_1 \Phi_1 = 0$, where Φ is as above, passing through $M_1, \ldots, M_{p-1}$, and μ, μ_1 are arbitrary parameters, and can suppose that this curve does not pass through M_{p-1} for all values of μ, μ_1, that is, that $\Phi_1 = 0$ does not pass through M_{p-1}. Then the quotient $(\mu\Phi + \mu_1\Phi_1)/\Phi$ is a rational function, for $F=0$, with M_{p-1} as pole, which has also for poles such of $A_1, ..., A_{N-R}$ as do not happen to be zeros of Φ_1. On the other hand, to form such a rational function with pole at M_{p-1}, and other poles occurring only among $A_1, ..., A_{N-R}$, all of the first order, we may, by the theorem of coresiduation, proceed thus: draw through M_{p-1} an arbitrary line $L=0$, meeting $F=0$ in $n-1$ other points (B); draw through $A_1, ..., A_{N-R}$ a general adjoint curve of order $n-3$, say $\Psi=0$, meeting $F=0$ in remaining points (C); thus $L\Psi=0$ is an adjoint curve of order $n-2$ containing the points $M_{p-1}, A_1, ..., A_{N-R}$, with residual intersection consisting of (B), (C) together; then draw an adjoint curve of order $n-2$ through these points (B), (C); we have proved that the function exists, so that such a curve is possible, other than $L\Psi=0$. This curve, taken as numerator, with $L\Psi$ as denominator, will express the most general rational function of the description. But a curve of order $n-2$ containing the $n-1$ points (B) will contain the line L as factor; the function will therefore have for numerator a curve of order $n-3$, and have Ψ for denominator. The function has thus for poles only the zeros of Ψ. But by hypothesis it has M_{p-1} for pole. Whence $\Psi=0$, which is any curve $\lambda_1\phi_1 + ... + \lambda_p\phi_p = 0$ drawn through $A_1, ..., A_{N-R}$, contains M_{p-1}; and, therefore, contains all of $M_1, M_2, ..., M_{p-1}$. This, however, is inconsistent with $R > P-1$, or $\epsilon > 0$, and shews that $p = P$. For the $2P-2$ points $M_1, ..., M_{p-1}, A_1, ..., A_{N-R}$ are determined by the $p-1$ perfectly arbitrary points $M_1, ..., M_{p-1}$; while the points $A_1, ..., A_{N-R}$, in number $p-1-2\epsilon$, are not only fewer than $p-1$ if $\epsilon > 0$, but are, we may say, less general in character than $M_1, ..., M_{p-1}$, as being derived from these, and not assigned arbitrarily. It is therefore impossible that the whole $2P-2$ points, which depend upon $p-1$ quite arbitrary general points, should be determined by these fewer and less general points $A_1, ..., A_{N-R}$, as they would be if every curve $\lambda_1\phi_1 + ... + \lambda_p\phi_p = 0$, through these, also contained $M_1, ..., M_{p-1}$. We can only conclude therefore that $\epsilon = 0$, and $p = P$; in which case $N-R=R$, and the points $A_1, ..., A_{N-R}$ are as many as $M_1, ..., M_{p-1}$. This conclusion is reached on the hypothesis that the curves $\phi_1=0, ..., \phi_p=0$ have no common intersection with $F=0$, save at the multiple points. If there are such points, they will be among the points $A_1, ..., A_{N-R}$, the derived zeros of the curve $\lambda_1\phi_1 + ... + \lambda_p\phi_p = 0$ drawn through $M_1, ..., M_{p-1}$; say they are $A_1, ..., A_k$. These points can then be excluded from consideration, the same argument proceeding with $A_{k+1}, ..., A_{N-R}$ instead of $A_1, ..., A_{N-R}$. The conclusion now will be that $N-R-k=p-1$, leading to $2P=2p+k$. This can only be consistent with $p \geqslant P$, if $p=P$, $\epsilon=0$, and $k=0$.

The sign of inequality, for the freedom of adjoint curves of order

$n-3$, in the formula $R \geqslant P-1$, is thus unnecessary; and the conditions of adjointness, at the multiple points of $F=0$, for curves of this order, are linearly independent. These conditions are therefore independent for all adjoint curves of order $m \geqslant n-3$. *This is a capital result.* The formulae we found are thus replaced by $R = N - p$ for all adjoint curves of order $> n-3$, while for adjoint curves of order $n-3$ we have $R = p - 1$. This is $R = N - P + 1$ and replaces $R > N - P$, because $N = 2p - 2$, $P = p^*$.

The preceding argument is for curves of assigned order, intersecting the given curve $F=0$, which are subject to no conditions

---

* Some remarks of importance may be placed in a footnote, so as not to interrupt the general argument. (*a*) The general adjoint curve of order $n-3$, $\lambda_1 \phi_1 + \ldots + \lambda_p \phi_p = 0$, though not having zeros on $F=0$, save at the multiple points, which are common to $\phi_1 = 0, \ldots, \phi_p = 0$, may have such zeros elsewhere in the plane. For example, for a plane sextic curve with 8 double points, the adjoint cubic curves $\lambda_1 \phi_1 + \lambda_2 \phi_2 = 0$ all have a point in common; unless the sextic curve is further specialised this ninth point does not lie thereon. (*b*) The curves $\lambda_1 \phi_1 + \ldots + \lambda_p \phi_p = 0$ may all have a common part; by the theorem of the text, this part is then a curve intersecting $F=0$ only at its multiple points. For instance, for the sextic curve given by an equation of the form

$$y^3 z^3 + y^2 z^2 u_2 + y z u_4 + u_6 = 0,$$

where u_i is homogeneous of order i in y, x, the adjoint curves of order $n-3$ have an equation of the form $y(yz + v_2) = 0$, where v_2 is homogeneous of order 2 in y and x; the partial curve $y=0$ meets the sextic curve only at the multiple point $x=0$, $y=0$. The linear series, of sets of $2p-2$ points, with freedom $p-1$, is then given by variable curves of order less than $n-3$, constituting the *pure*, or *reduced*, adjoint system. It may be shewn that, in any birational transformation *of the plane*, the pure adjoint system is transformed into the pure adjoint system of the curve into which the given curve is transformed. This is clear here from the association with the everywhere finite algebraic integrals; a simple independent proof is given by Bertini, *Ann. d. Mat.* XXII, 1894, p. 14. Conversely, it may be shewn that, when the multiple points of the given curve $F=0$ are such that it is possible to put through them a curve γ, having no other intersections with the given curve, then this curve γ is part of the general adjoint curve of order $n-3$, provided the multiplicities of γ at the multiple points of $F=0$ are such as would constitute independent conditions in defining γ (Castelnuovo, *Memorie Torino*, XLII, 1892). For instance, for a sextic curve with two triple points, the variable part of the adjoint system of cubics is given by conics passing simply through each triple point; but, for a sextic curve having six double points which lie on a conic, the system of adjoint cubics has no common part. (*c*) Reference may be made here to a general theorem for a linear system of curves in a plane, $\lambda_0 \phi_0 + \ldots + \lambda_r \phi_r = 0$, not necessarily adjoint to any given curve, that if the variable part of every curve of the system is composite, the general curve of the system is given by an equation of the form

$$\psi [\lambda_0 u^r + \lambda_1 u^{r-1} v + \ldots + \lambda_r v^r] = 0,$$

where $u=0$, $v=0$ are curves of the same order (Bertini, *Rend. Lombardo*, XV, 1882; Lüroth, *Math. Ann.* XLII, XLIV; 1893, 1894). For the system consisting of the adjoint curves of order $n-3$ of a curve $F=0$ of order n, this happens if $F=0$ be hyperelliptic (see Chaps. II, III, above), but only then. For instance, for the curve expressed by $y^{2p} z^2 = (y, x)_{2p+2}$, the adjoint system in question is given by $y^p (y, x)_{p-1} = 0$.

save the conditions of adjointness at the multiple points of $F = 0$. Now consider adjoint curves which, beside, are required to pass through certain specified simple points of $F = 0$. Let the number of these points be k, and let v, $= N - k$, be the number of remaining intersections which are not assigned (other than at the multiple points of $F = 0$). If the k prescribed points give linearly independent conditions for the coefficients of the curve to be obtained, this curve will define a linear series of freedom $\rho = R - k$; of this, however, we cannot be sure unless the curve be of sufficiently high order; when the conditions are not independent, we shall have $\rho > R - k$. Thus we conclude that, for the linear series defined by adjoint curves with prescribed fixed base points on $F = 0$, we have, for the freedom, $\rho = v - p$, for all curves of sufficiently high order; but for curves of less order ($> n - 3$ however), $\rho \geqslant v - p$; and, for curves of order $n - 3$, $\rho \geqslant v - p + 1$, that is $\rho > v - p$.

It can however be shewn, for adjoint curves of order $> n - 3$, with k fixed base points, that, if these k points do not furnish independent conditions, so that, for the linear series of sets of v points, of freedom ρ, formed by the sets of residual intersections with $F = 0$, we have $\rho > v - p$, then the points of any one of the sets of the linear series lie upon an adjoint curve of order $n - 3$; and that the series can be given by variable adjoint curves of order $n - 3$, with possibly fixed prescribed base points; of these curves there will then be v intersections not prescribed, and the general curve of the system will contain $\rho + 1$ homogeneously entering arbitrary coefficients. This we proceed to prove. In saying that the k prescribed base points for the adjoint curves in question are not independent, or that some are determined by the others, we have not said that there may not be other points of $F = 0$ also necessarily lying on all the adjoint curves through the k prescribed points. The linear system of adjoint curves of the assigned order through the k prescribed points, say (Ψ), given by an equation $\lambda\Psi + \lambda_1\Psi_1 + \ldots + \lambda_\rho\Psi_\rho = 0$, may be such that there are, say, l points of $F = 0$, common to all the curves $\Psi = 0, \ldots, \Psi_\rho = 0$, in addition to the k prescribed base points. Choose now ρ arbitrary general points of $F = 0$, say $M_1, \ldots, M_\rho$, and let the definite curve (Ψ) passing through these, and the k prescribed base points, meet $F = 0$ further in points $A_1, \ldots, A_{v-\rho}$ (beside the multiple points of $F = 0$); the l common zeros of $\Psi = 0, \ldots, \Psi_\rho = 0$, if existent, will be among $A_1, \ldots, A_{v-\rho}$. Since $v - \rho < p$, an adjoint curve of order $n - 3$, or several linearly independent such curves, can be put through $A_1, \ldots, A_{v-\rho}$. We proceed to prove that every such adjoint curve, of order $n - 3$, contains M_ρ, any one of $M_1, \ldots, M_\rho$, and hence contains all of these. The argument is the same as one used above. As $M_1, \ldots, M_\rho$ are independent points for curves (Ψ), there are such curves, given, say, by

$\mu\Psi' + \mu_1\Psi'_1 = 0$, containing only $M_1, \ldots, M_{\rho-1}$, where μ, μ_1 are arbitrary, and $\Psi' = 0$, suppose, is the (Ψ') curve containing $M_1, \ldots, M_\rho$, $A_1, \ldots, A_{\nu-\rho}$. Hence there is a rational function, $(\mu\Psi' + \mu_1\Psi'_1)/\Psi'$, having M_ρ as pole, having beside, for poles, such of the points from $A_1, \ldots, A_{\nu-\rho}$ as are not common zeros of Ψ' and Ψ'_1. By the theorem of coresiduation, however, such function must be capable of expression by taking an arbitrary adjoint curve through $M_\rho, A_1, \ldots, A_{\nu-\rho}$, and then an adjoint curve of the same order through the residual intersections (other than the multiple points) of this curve with $F = 0$. Thus, draw a line $L = 0$ through M_ρ, meeting $F = 0$ in $(n-1)$ points (B); and take an adjoint curve of order $n-3$ through $A_1, \ldots, A_{\nu-\rho}$, say $\phi = 0$, whose residual intersections with $F = 0$ are denoted by (C). The denominator in the expression of the function may then be taken to be $L\phi$, the numerator being found from an adjoint curve of order $n-2$ containing the sets (B), (C). But such a curve, containing the $(n-1)$ points (B) upon the line $L = 0$, is of the form $L\psi = 0$, where $\psi = 0$ is an adjoint curve of order $n-3$. The function is then ψ/ϕ; and as it has M_ρ as pole, the curve $\phi = 0$ must pass through this point (and the curve $\psi = 0$ will not). Thus we see that an adjoint curve of order $n-3$ contains all the points $M_1, \ldots, M_\rho$, $A_1, \ldots, A_{\nu-\rho}$; let $\phi = 0$ be such a curve, and the general adjoint curve $\psi = 0$ be taken passing through the residual intersections (other than the multiple points) of $\phi = 0$ with $F = 0$; the function ψ/ϕ must be the same, by the theorem of coresiduation, save for a constant multiplier, as the function $(\Psi')/\Psi'$, where $\Psi' = 0$ is a particular curve of the original system, containing $M_1, \ldots, M_\rho$, $A_1, \ldots, A_{\nu-\rho}$, and $(\Psi') = 0$ is the general adjoint curve, of the same order as $\Psi' = 0$, formed, similarly to $\psi = 0$, from the residual set on $\Psi' = 0$. And as the common points, other than the k prescribed points, of all the curves $\Psi' = 0, \Psi'_1 = 0, \ldots,$ $\Psi'_\rho = 0$ are eliminated in considering the rational function (being common to numerator and denominator), and are zeros of $\phi = 0$, they will also be zeros of $\psi = 0$. Thus, the system $\psi = 0$ will have the same fixed points as the system $(\Psi') = 0$. In the argument given, we have allowed, as possible, that there may be l points among $A_1, \ldots, A_{\nu-\rho}$ which are zeros of all the adjoint curves $(\Psi') = 0$ put through the k points originally prescribed; if we denote $\nu - l$ by m, the function $(\Psi')/\Psi'$ will be of the m-th order, and ρ will, *a fortiori*, be $> m - p$. But, conversely, if a function with m given poles have a number $\rho + 1$ of homogeneously entering arbitrary coefficients in its expression, where $\rho > m - p$, the function can be expressed as a quotient of adjoint polynomials of order $n - 3$; for we can suppose the l points spoken of to be included among the k prescribed points, and then apply the preceding argument. From this investigation we conclude:

(*a*) *That a linear series of sets of ν points, obtained by general variable adjoint curves of order* $> n-3$*, passing through certain prescribed points of* $F = 0$*, has a freedom ρ given by* $\rho = \nu - p$*, unless every one of the sets of ν points lies entirely upon an adjoint curve of order* $n-3$*; and*

(*b*) *For a series of sets of ν points, determined by general adjoint curves of order* $n-3$*, which pass through k prescribed points, not necessarily all independent, the freedom ρ is such that* $\rho > \nu - p$*. There may be points common to all the sets of ν points of the series, and, hence, common to all the adjoint curves of order* $n-3$ *through the k prescribed points.*

A linear series of sets of ν points in which all the ν points of any set lie upon an adjoint curve of order $n-3$ is said to be *special*. Such a series can, by what we have proved, be determined by the free intersections, with $F = 0$, of a system of adjoint curves of order $n-3$, in general with prescribed fixed points, say k in number. We proceed to prove that, if the number of linearly independent adjoint curves of order $n-3$, conditioned only by their adjointness, which pass through the points of any set of the series be $\rho' + 1$, then the freedom of the series is given by $\rho = \nu - p + \rho' + 1$. This formula then gives definiteness to the result (*b*) above, and may be understood also to contain the result (*a*), with $\rho' + 1 = 0$.

Take the particular set of the series which contains ρ independent general points of $F = 0$, say $M_1, \ldots, M_\rho$; let $A_1, \ldots, A_{\nu-\rho}$ be the other points of this set. In the first place define $\rho' + 1$ as the number of linearly independent adjoint curves of order $n-3$, conditioned only by their adjointness, which contain $A_1, \ldots, A_{\nu-\rho}$; as $\nu - \rho < p$ such curves exist. We have proved that all these curves pass through $M_1, \ldots, M_\rho$. We clearly have $\rho' + 1 \geqslant p - (\nu - \rho)$, the sign of inequality being unnecessary if the points $A_1, \ldots, A_{\nu-\rho}$ furnish independent conditions for such curves. These $\rho' + 1$ adjoint curves of order $n-3$, which contain all of $M_1, \ldots, M_\rho, A_1, \ldots, A_{\nu-\rho}$, define a new linear series of sets of ν' points, where $\nu' = 2p - 2 - \nu$, which is of freedom ρ'. It follows from the theorem of coresiduation that this new series is complete, namely that there is no series of sets of ν' points, of freedom greater than ρ', which contains all the sets of this series. Likewise, the whole original complete series of freedom ρ, of sets of ν points, is determined by all the adjoint curves of order $n-3$ which can be drawn through any one of the sets of ν' residual points. These two complete series are thus reciprocal; and the number of linearly independent adjoint curves of order $n-3$, conditioned only by their adjointness, which pass through the particular set of ν' residual points which we select, is $\rho + 1$. Thus, beside $\rho' + 1 \geqslant p - (\nu - \rho)$, we also have $\rho + 1 \geqslant p - (\nu' - \rho')$. Since $\nu' + \nu = 2p - 2$, these lead to the two equations

$$\rho = \nu - p + \rho' + 1, \quad \rho' = \nu' - p + \rho + 1.$$

A particular consequence, also, is that the $\nu - \rho$ points $A_1, \dots, A_{\nu - \rho}$ used in the argument furnish independent conditions for adjoint curves of order $n - 3$, conditioned only by their adjointness, which pass through them. In the nature of the case, and it must be borne in mind, the equation $\rho = \nu - p + \rho' + 1$ is proved on the hypothesis that the linear series is complete.

We have introduced the theory of linear series by considering the numerator polynomial of a rational function with given poles, the freedom of the series depending on the number of arbitrary constants in the function; and we have obtained a relation connecting the freedom of the series with the number of points in a set of the series. But the order of a rational function may prove less than was intended owing to undesigned coincidences in the zeros of the numerator and denominator. It is necessary then to examine whether the relation found for a complete linear series is the same as that between the number of constants in the expression of a rational function of given poles, and the order of this function.

Consider a linear series, of sets of ν points, of freedom ρ, given by a system of adjoint curves of order $n - 3$, $\lambda\phi + \lambda_1\phi_1 + \dots + \lambda_\rho \phi_\rho = 0$, having k prescribed common zeros (other than at the multiple points of the given curve $F = 0$); and suppose that all of $\phi = 0$, $\phi_1 = 0, \dots, \phi_\rho = 0$, have, beside, l common zeros on $F = 0$. Then the quotient $(\lambda\phi + \dots + \lambda_\rho \phi_\rho)/\phi$ is a rational function of order $\nu_0 = \nu - l$, having poles only at the unprescribed zeros of $\phi = 0$ other than these l common zeros. By what we have proved, the number, $\rho + 1$, of the homogeneously entering arbitrary coefficients in the function is given by $\rho = \nu_0 - p + \rho' + 1 + l$, where $\rho' + 1$ is the number of linearly independent adjoint curves, of order $n - 3$, otherwise unconditioned, which pass through the points of a set of the linear series. We have shewn that if the particular curve $\lambda\phi + \dots + \lambda_\rho \phi_\rho = 0$ which passes through ρ independent general points of $F = 0$, $M_1, \dots, M_\rho$, have for its remaining zeros (other than the multiple points, and the k prescribed points) the set $A_1, \dots, A_{\nu - \rho}$, then the adjoint curves of order $n - 3$ which pass through $A_1, \dots, A_{\nu - \rho}$, and are otherwise unconditioned, all pass through $M_1, \dots, M_\rho$, and that $A_1, \dots, A_{\nu - \rho}$ furnish independent conditions for these curves. Among $A_1, \dots,$ $A_{\nu - \rho}$ will be found the l undesigned common zeros spoken of, say these are $A_1, \dots, A_l$; and, from the independence of $A_1, \dots, A_{\nu - \rho}$, there will be $\rho' + 1 + l$ linearly independent adjoint curves of order $n - 3$, otherwise unconditioned, which pass through $A_{l+1}, \dots, A_{\nu - \rho}$. These $\rho' + 1 + l$ curves will all contain $M_1, \dots, M_\rho$. The proof of this is the same as the proof of the former statement, and depends on the fact that there exists a rational function having M_ρ for pole and having its other poles among the points $A_{l+1}, \dots, A_{\nu - \rho}$. Thus there are $\rho' + 1 + l$ adjoint curves of order $n - 3$, otherwise unrestricted, passing through $M_1, \ldots, M_\rho, A_{l+1}, \ldots, A_{\nu-\rho}$. Put now

$$\rho_0' + 1 = \rho' + 1 + l;$$

then we have the conclusion that a rational function of order ν_0 contains $\rho + 1$ homogeneously arbitrary constants, where $\rho = \nu_0 - p + \rho_0' + 1$, and $\rho_0' + 1$ is the number of adjoint curves of order $n - 3$, otherwise unconditioned, which pass through the points which are the poles of the function. The same result can be proved to hold for a rational function expressible as the quotient of adjoint polynomials of order $> n - 3$, the number $\rho_0' + 1$ being replaced by zero.

Ex. 1. A linear series of sets of points being said to be *simple* when it is not the case that every set which contains an arbitrary general point of the given curve necessarily contains one or more points of the curve, prove that a simple series which is *special*, that is, determined by adjoint curves of order $n - 3$, which has freedom ρ, is such that *any* ρ points of a set of the series suffices to individualise the set (cf. Bertini, *Ann. Mat.* XXII, 1894, p. 25). Thus, if the number of points in a set be ν, *every* $\nu - \rho$ points of a set are independent conditions for adjoint curves of order $n - 3$.

Ex. 2. The series, of freedom $p - 1$, with $2p - 2$ points in a set, which is determined by unconditioned adjoint curves of order $n - 3$, is always simple, unless the given curve $F = 0$ be hyperelliptic. This series is called the *canonical series* on $F = 0$. The theorem is proved below (p. 80).

In the preceding theory the number p was effectively defined as the number of linearly independent everywhere finite algebraic integrals belonging to the given curve $F = 0$; for it was shewn previously that these integrals are those of the form $\int \psi \, dx / f'(y)$ (or the equivalent of this in homogeneous variables), in which $\psi = 0$ is an adjoint curve of order $n - 3$. It has now been shewn, when the multiple points of the curve are of the simplest kind, k-ple points with k distinct tangents, in which case the condition of adjointness for another curve is the possession of a $(k - 1)$-ple point at such a multiple point, that the conditions of adjointness at all the multiple points, for curves of order $n - 3$, or more, are independent; and accordingly that the value of p in this case is given by

$$\tfrac{1}{2}(n-1)(n-2) - \tfrac{1}{2}\Sigma k(k-1).$$

Further, arguing with this simple case in view, we have found a relation connecting the number of points in a set of a linear series and the freedom of the series, and, correspondingly, a relation connecting the order of a rational function belonging to the given curve and the number of constants left arbitrary in such a function when the poles of the function are given. It is clear, however, when two curves are in $(1, 1)$ birational correspondence, that not only do everywhere finite integrals associated therewith transform into such integrals, but also rational functions and linear series; in the latter cases the number of arbitrary constants, or the freedom of the

linear series, is unaltered by such transformation; so also is the order
of a rational function, and likewise, with due consideration of fixed
points of a linear series which may undesignedly become merged
with a multiple point of a transformed curve, is the number of
points in a set of a linear series. Thus we may say that the equation
$\rho = \nu - p + \rho' + 1$, as applied to a linear series or a rational function,
with inclusion of the case when $\rho' + 1 = 0$, is universally true, p being
defined by the number of everywhere finite integrals, and *computed*,
for instance, from a simple transformed equation. This relation is
known as the Riemann-Roch theorem, having been given, in its
application to rational functions, by Riemann, with neglect of the
possibility expressed by the term $\rho' + 1$, which was supplied by
Roch. The interpretation in terms of linear series was given by
Brill and Noether. While, in Riemann's theory, rational functions
are built up from algebraic integrals having algebraic infinities (as
is explained in Chap. VI below), in Brill and Noether the theorem of
coresiduation is based on a theorem for the general curve of
assigned order which passes through the common points of two
given curves. (Riemann, *Ges. Werke*, 1876, p. 100, etc.; Roch,
Crelle, LXIV, 1864, p. 372; Brill u. Noether, *Math. Ann.* VII, 1873,
p. 269. For an elementary exposition of Riemann's theory, see
C. Neumann, *Riemann's Theorie*, 1884; for the transition to the
geometrical theory, see Clebsch u. Gordan, *Abelsche Functionen*,
1866; for a comprehensive history, Brill u. Noether, *Entwicklung
der Theorie der algebraischen Functionen*, *Deutsch. Math. Ver.
Bericht*, III, 1894, pp. 109–566.)

Applications of the Riemann–Roch formula. We now develop
in turn various consequences of the theory which has been established. Most often we denote by r the freedom of a linear series of
sets of points upon the given fundamental curve, and by n the
number of points in a set of the series; when it is necessary to refer
to the order of the fundamental curve, a symbol different from n
will be employed, say N; the adjoint curves of order $N - 3$ will
often be called simply ϕ-curves. The series itself will often be
denoted* by $g_r{}^n$. There may be points common to every set of the
linear series (beside the multiple points of the fundamental curve);
to allow for this we may speak of n as the *complete grade* of the
series, denoting the number of points of a set which actually vary
from set to set as the *grade*. When every set of the series consists of
points which lie upon a ϕ-curve, so that $r > n - p$, the series is called
a *special* series; the series of sets of $2p - 2$ points, of freedom $p - 1$,
which is determined by the complete series of ϕ-curves, is called

---

* The established notation is $g_n{}^r$, which however conflicts with the equally
established notation whereby a manifold of order n and dimension r is denoted
by $M_r{}^n$.

the canonical series; the Riemann-Roch theorem, which we now write in the form $r = n - p + r' + 1$, enables us to prove easily that the canonical series is the only existing series g_{p-1}^{2p-2}. This series is very fundamental, the canonical series of any curve changing into the canonical series of any other curve obtained from the former by (1, 1) birational transformation. We have proved in the foregoing that, if $\phi_1 = 0, \dots, \phi_p = 0$ be linearly independent ϕ-curves, there is no point of the fundamental curve common to all these; thus, a rational function expressed by $(\lambda_1 \phi_1 + \dots + \lambda_p \phi_p)/\phi_i$ is of order $2p - 2$ for general values of the constants $\lambda_1, \dots, \lambda_p$; or, we may say, the canonical series has no 'fixed' points. The theory has been developed on the hypothesis that the fundamental curve is not rational; when this is so, however, we have $p = 0$, and there exists no canonical series, while for every linear series $r = n$ (as we have seen in earlier chapters). Conversely, it is easy to see that the existence of a series g_r^r involves that the curve is rational. Upon such a curve there exist series g_1^1; there also exist series of sets of 2 points of freedom 1 (quadratic involutions), but such a series is not complete, being contained in a series of sets of 2 points of which both are arbitrary. When $p = 1$, there exists one everywhere finite integral, but the canonical series has no freedom, the existing ϕ-curve having no intersections with the fundamental curve other than at the multiple points (for instance, for a plane quartic curve with two double points, the ϕ-curve is the line joining these). Upon a curve for which $p = 1$ there exist complete series g_1^2 (for instance, upon a plane cubic curve, the lines drawn through a fixed point of the curve determine such a series); likewise upon a curve for which $p = 2$, there exists one complete series g_1^2, the canonical series (for instance, upon a plane curve of order m with a single $(m-2)$-ple point, this series is given by the lines drawn through this point). In general, however, for $p > 2$, the existence of a linear series of sets of two points with freedom 1, involves that the fundamental curve is not the most general curve of its genus; it involves the existence of a rational function of order 2; and we have shewn (Chap. III, p. 50) that then the fundamental curve is hyperelliptic, and can be birationally transformed to a simple form. The Riemann-Roch theorem, applied to a complete series g_1^2 on a curve of genus p, giving $1 = 2 - p + r' + 1$, shews that $p - 1$ of the ϕ-curves pass through every set of the series g_1^2, namely that every ϕ-curve through one point of this necessarily contains the other. For a curve which is not hyperelliptic, the canonical series is simple, that is, it is not the case that every ϕ-curve, otherwise unconditioned, which is made to pass through an arbitrary general point of the fundamental curve, necessarily passes then through one or more other points of the curve, determined by the arbitrary point. For suppose it possible that every

ϕ-curve through an arbitrary point contained in consequence $(\mu-1)$ others; such a curve can be put through $(p-1)$ perfectly general independent points, and would therefore contain $(\mu-1)(p-1)$ others, or, in all, $\mu(p-1)$ points of the fundamental curve. But a ϕ-curve has only $2p-2$ zeros; thus μ, if >1, is 2. Suppose this is so. Then the ϕ-curves through $(p-2)$ arbitrary general points (which are expressible in the form $\mu\phi+\mu_1\phi_1=0$, where ϕ, ϕ_1 are definite and μ, μ_1 are arbitrary constants) will contain $p-2$ other definite points; the curves $\mu\phi+\mu_1\phi_1=0$ have therefore two intersections with the fundamental curve variable with μ and μ_1. There exists therefore a rational function, expressible in the form ϕ_1/ϕ, which is of the second order. The fundamental curve is therefore hyperelliptic. Suppose now that the fundamental curve is not hyperelliptic. Then the canonical series is not only simple, as we have just proved, but is further incapable of being *reduced*, in the sense of Chap. ii; namely, it is not possible, by considering the ϕ-curves with k prescribed zeros, to obtain a linear series of sets of ν points, with freedom ρ, for which $\nu-2\rho$ is less than the value $[2p-2-2(p-1),$ or] zero, arising for the original canonical series defined by the ϕ-curves without prescribed zeros. The ϕ-curves through k general prescribed points may, let us suppose, all pass through l other points determined by the k points (beside the multiple points of the curve $F=0$); these curves will then determine a linear series of sets of $2p-2-k-l$, say ν, variable points, with freedom $\rho=p-1-k$. We can prove that, when the curve is not hyperelliptic, and $k>0$, we always have, for such a special series, $\nu>2\rho$; that is $k>l$. This result, in the form $\nu\geqslant 2\rho$, is usually called Clifford's theorem (see Clifford, *Math. Papers*, 1882, p. 329, where general results are assumed which are obtained here much later). The argument is one we have employed before: when the series g_ρ^ν is established, a particular set is identified by assigning ρ points thereof, namely, of the ϕ-curves conditioned by the fixed points belonging to all of them, there is one, determined by ρ points, which contains all the points of a set. On the other hand, we have shewn above that ϕ-curves conditioned only by their adjointness which pass through $\nu-\rho$ arbitrary points of a set of the series likewise contain all the points of the set. These unconditioned ϕ-curves must evidently require at least as many conditions, in order that they may contain all the points of a set of the series, as are required by the conditioned ϕ-curves. Thus $\nu-\rho\geqslant\rho$. This argument is under the hypothesis that the series g_ρ^ν is complete; the result holds *a fortiori* for a series g_σ^ν contained therein with $\sigma<\rho$. That the series g_ρ^ν, obtained as described, is in fact complete is easily proved. From the fact that the canonical series g_{p-1}^{2p-2} is simple, it follows, as in Chap. II, above, that, with coordinates $(\xi_1, ..., \xi_p)$, of a space of $p-1$ dimensions $[p-1]$, the equations $\xi_1/\phi_1 = ... = \xi_p/\phi_p$ transform the original *non-hyperelliptic* curve $F = 0$, with a $(1, 1)$ birational correspondence, into a curve of order $2p-2$ in this space. And from the inequality $\nu \geqslant 2\rho$, for any series g_ρ^ν contained in the canonical series, it follows that the curve so obtained is without multiple points. Moreover this curve is not the projection of another curve of the same order, in a space of dimension $p+q-1, > p-1$; for such a curve, being in $(1, 1)$ correspondence with this, and therefore with the original curve, would have the coordinates of a point expressible by equations

$$\xi_1/u\phi_1 = ... = \xi_p/u\phi_p = \xi_{p+1}/\psi_1 = ... = \xi_{p+q}/\psi_q,$$

where $u, \psi_1, ..., \psi_q$ are polynomials in the coordinates of the original plane; the canonical series on the original curve would then be incomplete, being contained in a series of sets of $2p-2$ points, of freedom $p+q-1$, given by an equation

$$u(\lambda_1\phi_1 + ... + \lambda_p\phi_p) + \mu_1\psi_1 + ... + \mu_q\psi_q = 0,$$

the zeros of u, on the original curve $F = 0$, being also zeros of all of $\psi_1, ..., \psi_q$. This is contrary to what we have proved. The curve, of order $2p-2$, in space of dimension $p-1$, thus obtained as representative of any non-hyperelliptic curve, is called the *canonical curve*; and (like any other curve with the same property) is said to be *normal* in the space $[p-1]$ because it cannot be obtained by projection from another curve of the same order in space of higher dimension. This curve has the great simplicity that the canonical series thereon, that is the series of sets corresponding to the canonical sets of the original curve, are determined by its intersections with the *prime spaces* of the space $[p-1]$, given by linear equations $\lambda_1\xi_1 + ... + \lambda_p\xi_p = 0$; and thence, any special series contained in the canonical series, is given by its intersections with primes passing through fixed points of the canonical curve. In particular, consider a special series g_ρ^ν; we have shewn that all the points of a set of this series lie on unconditioned ϕ-curves passing through any $\nu - \rho$ points of the set. In the space $[p-1]$, this number of points of the canonical curve determine a linear space $[\nu-\rho-1]$. The sets of the series g_ρ^ν on the canonical curve thus lie each in such a space. For instance, for $p = 4$, there is a canonical curve of order 6 lying in ordinary space; and thereon there is a special series g_1^3 of which the sets lie upon trisecant lines of the curve; it will appear in fact that the curve is the complete intersection of a cubic surface and a quadric surface; the trisecants in question are one system of generators of the quadric surface (and there is another g_1^3). If we now assume that, for a curve c^m, of order m, in space of k dimensions $[k]$, a space of dimension $s-1$, put through s *arbitrary* points of the curve, has no further intersections with the curve, unless $s=k$, and the space is a prime (for instance, a curve, not lying in a plane, cannot be such that every chord is a trisecant), then we can infer that Clifford's theorem $\nu \geqslant 2\rho$, for a special series g_ρ^ν, can be replaced by $\nu > 2\rho$, except when $\rho = p-1$, $\nu = 2p-2$, the special series being then the whole canonical series. For when $\nu - \rho = \rho$, a set of the series g_ρ^ν being determined by ρ quite arbitrary points of the canonical curve, a space $[\nu - \rho - 1]$ put through $\nu - \rho$ quite arbitrary points will wholly contain a set of g_ρ^ν, and will thus meet the curve again. By the assumption made, this is so only when $\nu - \rho = \rho = p-1$.

Remark I. A prime through a set of a special series g_ρ^ν, on the canonical curve, meets the curve again in a set of ν' points $(\nu' = 2p - 2 - \nu)$, determining a complementary special series $g_{\rho'}^{\nu'}$ where $\rho = \nu - p + \rho' + 1$, so that $\nu' - 2\rho' = \nu - 2\rho$; and any set of this complementary series lies in a space $[\nu' - \rho' - 1]$. The two spaces $[\nu - \rho - 1]$, $[\nu' - \rho' - 1]$, containing complementary sets, as they both lie in a prime $[p-2]$, meet one another in a space of dimension $\nu - \rho - 1 + \nu' - \rho' - 1 - (p-2)$, which is a $[\nu - 2\rho - 1]$ or $[\nu' - 2\rho' - 1]$. When $\rho > 0$ and $\rho' > 0$, this space is not coextensive either with the $[\nu - \rho - 1]$ or the $[\nu' - \rho' - 1]$. For instance when $p = 4$, two complementary sets of the two series g_1^3 lie in generators of opposite systems of a quadric surface. These have a point in common (not lying on the canonical curve).

Remark II. If $p - 3$ arbitrary points be taken on the canonical curve, the space $[p-4]$ determined thereby does not meet the curve further. By means of the ∞^2 spaces $[p-3]$ through this $[p-4]$, the canonical curve can be projected into a plane curve of order $p+1$. Any curve of genus p, which is not hyperelliptic, can thus be transformed birationally into a plane curve of order $p+1$ (with appropriate multiple points). It has already been remarked that a hyperelliptic curve of genus p can be represented by a plane curve of order $p+2$ (with one p-ple multiple point).

Remark III. It was proved above that unless a curve be hyperelliptic there exists no rational function of order 2. There exists, however, always a rational function of order $p+1$, with its poles at quite arbitrary places, as is easy to see. And this is the lowest order for which a rational function can always be constructed with quite arbitrary poles. For particular positions of the poles, a rational function may be constructed of order $\frac{1}{2}p+1$, or $\frac{1}{2}(p+1)+1$ (according as p is even or odd); and this is the absolutely lowest order for a rational function unless the curve be in some way less general than the most general curve of genus p. This will be proved.

Remark IV. We have proved that the canonical series g_{p-1}^{2p-2} is simple, and not further reducible in the sense of Chap. II. We can prove that the same is true of any series g_r^n in which $n > 2p$ (so that this series is not special); by such a series then the curve can be put in $(1, 1)$ birational correspondence with a curve of order n, in space $[r]$, which is normal in this space if the series g_r^n is complete, as we suppose. As the series is not special we have $r = n - p$; thus $n > 2p$ is the same as $n < 2r$. That the series is simple, follows because, if every set containing an arbitrary point of the fundamental curve contained $\mu - 1$ other points of this curve, we could, by taking r arbitrary points, find a set of the series containing μr points, so that $\mu r \leqslant n$, and this, by $n < 2r$, involves $\mu = 1$. Again there are no points of the curve $F = 0$ common to all the sets of n points of the series; for if there were only $n - l$ variable points in these sets, then $n - l < n < 2r$, shewing (by Clifford's theorem) that the series g_r^{n-l}, of these sets of variable points, is not special; so that $r = n - l - p$, or $l = 0$. Lastly, if we fix a single point, and consider the residual series g_{r-1}^{n-k}, then $k = 1$; for, if $k \geqslant 2$, the condition $n < 2r$ involves $n - k < 2(r-1)$, and the series g_{r-1}^{n-k} (by Clifford's theorem) is not special, so that $r - 1 = n - k - p$. Hence, as in Chap. II, if the system of curves determining the series be

$$\lambda_0 \psi_0 + \ldots + \lambda_r \psi_r = 0,$$

the equations $\xi_0 / \psi_0 = \ldots = \xi_r / \psi_r$ define a $(1, 1)$ birational transformation to a non-singular curve. This argument holds if the original curve be hyperelliptic. By projection on to a plane, from $r - 2$ general points of this curve in $[r]$, we thus obtain a plane curve of order $p + 2$ in $(1, 1)$ correspondence with the original.

Proof of a formula for the genus in terms of the branch points. Jacobian series of a linear series. There is a very important formula, illustrating the relation between the theory of linear series, and the functional point of view, which seems to find its appropriate place here. The formula is moreover often of practical use for determining the genus of a given curve.

Let ξ be any rational function upon the fundamental curve, say of order m. At a place of the curve where ξ has a finite value, say α, the expression of ξ, in terms of the usual local parameter t, may be of the form $\xi = \alpha + \alpha_1 t^r + \alpha_2 t^{r+1} + \ldots$, where α_1 is not zero. At a place where ξ is infinite, with local parameter t_0, the expression may be of the form $\xi^{-1} = k_1 t_0^{r_0} + k_2 t_0^{r_0+1} + \ldots$, where k_1 is not zero. We have then the formula $\Sigma(r-1) + \Sigma(r_0 - 1) = 2m + 2p - 2$, where p is the genus of the fundamental curve, and the summation extends to all places of this at which $r > 1$, or $r_0 > 1$.

To illustrate, first, the geometrical meaning of this result, suppose that every r which is >1 is 2; and let the expression of ξ as a quotient of adjoint polynomials be $\xi = \vartheta/\psi$. Consider the linear series of sets of m variable points on the fundamental curve given by $\vartheta - \lambda\psi = 0$, as λ varies. For any set of this series, corresponding to a definite value of λ, there may be one or more coincidences of 2 of the m points of the set, the curve $\vartheta - \lambda\psi = 0$ having (in general) contact with the fundamental curve at such a coincidence. The aggregate of the points, on the fundamental curve, at which such coincidences occur, in all the sets of the linear series, is called the *Jacobian set* of the series, and the formula states that the number of points in this set is $2m + 2p - 2$. It will in fact be proved that the Jacobian set, on the fundamental curve, is equivalent with, or coresidual to, a set constituted by the sum of two sets of the linear series, and a set of the canonical series. Conversely, this fact enables us to define the canonical series when we have determined the Jacobian set of any given linear series of freedom 1, on the fundamental curve, and the complete series defined by this set. For if this complete series be given by the system of adjoint curves, $\lambda_0\Phi_0 + \lambda_1\Phi_1 + \ldots = 0$, the curves of this system which pass through the points of any two sets of the given series (or pass doubly through the points of one such set) have, for their residual intersections with the fundamental curve, the (complete) canonical series.

For instance, if the fundamental curve have for multiple points only nodal double points, the complete series on this curve determined by the points of contact of tangents to the curve from an arbitrary point, O, of the plane, is given by the intersections, other than at the nodes, of a linear system of curves all passing through the nodes. The curves of this system which contain the intersections of the curve with two arbitrary lines through O, have, as residual intersection, the sets of the canonical series. When the curve has cusps, as well as nodes, the Jacobian set of the series determined by lines through O, will include the cusps. In this case, the system of curves employed must not only be adjoint at each cusp (or pass simply through this), but must have 3 intersections, namely touch the cuspidal tangent.

To prove the theorem in general, let u be any everywhere finite integral of the fundamental curve. Then $du/d\xi$ is a rational function, and, by what was shewn as to the construction of u, this function has a canonical set of the fundamental curve as part of the set of its zeros. Its only other zeros arise when ξ is infinite; at a place where, in the notation above, $\xi = k_1^{-1}t_0^{-r_0} + \ldots$, we have, effectively, $du/d\xi = -k_1 r_0^{-1} t_0^{r_0+1} du/dt_0$, so that there is a zero of order $r_0 + 1$. The poles of $du/d\xi$ arise only from places whereat, in the notation above, $\xi = \alpha + \alpha_1 t^r + \ldots$, with $r > 1$; this form gives in fact, effectively,

$du/d\xi = (r\alpha_1 t^{r-1})^{-1} du/dt$, or a pole of $du/d\xi$ of order $r-1$. As the number of zeros of the rational function $du/d\xi$ is equal to the number of poles, and a canonical set consists of $2p-2$ simple points, we thus have $\Sigma(r-1) = \Sigma(r_0+1) + 2p-2$. By hypothesis ξ is of order m, so that $\Sigma r_0 = m$. Hence we obtain the formula originally stated $\Sigma(r-1) + \Sigma(r_0-1) = 2m + 2p - 2$. If K denote a canonical set on the fundamental curve, C denote a set where the function ξ has a given arbitrary value, and J denote a set of places whose orders are given, each with its appropriate multiplicity, by the terms of $\Sigma(r-1) + \Sigma(r_0-1)$, and the sign $\equiv$ denote that two sets of places are coresidual or equivalent, the proof we have given establishes that $K + 2C \equiv J$. The set J is the aggregate of all coincidences in the sets given by $\xi = $ constant (including $\xi = \infty$), a coincidence of r points of a set being regarded as $r-1$ coincidences; or J is the generalised form of the Jacobian set.

It may be remarked here that if, from any linear series g_r^n with $r > 1$, defined by a system of curves $\lambda\psi + \lambda_1\psi_1 + \ldots + \lambda_r\psi_r = 0$, we choose a series of freedom 1, which is given, with variable μ, and fixed $a, \ldots, a_r, b, \ldots, b_r$, by curves $a\psi + \ldots + a_r\psi_r + \mu(b\psi + \ldots + b_r\psi_r) = 0$, and take the Jacobian set of this series g_1^n, then all the Jacobian sets so obtained by different choice of $a, \ldots, a_r, b, \ldots, b_r$ are equivalent with one another, or belong to the same linear series. In fact the Jacobian set of a series given by curves $\vartheta - \lambda\psi = 0$ is at once seen to be found from the intersections of the fundamental curve $F = 0$ with the curve expressed by the vanishing of the Jacobian determinant of the three polynomials ϑ, ψ, F (other than at the multiple points of $F = 0$); and if in this determinant ϑ, ψ are replaced respectively by $a\psi + \ldots + a_r\psi_r$ and $b\psi + \ldots + b_r\psi_r$, the curve expressed by the vanishing of the determinant is one of a linear system of curves, in which the part of the variable parameters is played by binary determinants of the form $a_r b_s - a_s b_r$.

When the curve $F = 0$ has only multiple points with distinct tangents, it is easy to prove directly that the intersections with $F = 0$, other than at the base points common to all curves $\vartheta - \lambda\psi = 0$, of the curve represented by the vanishing of the Jacobian determinant $(\vartheta_1, \psi_2, F_3)$, are in number $2m + 2p - 2$, where m is the number of intersections, other than at the base points, of any curve of the system $\vartheta - \lambda\psi = 0$ with $F = 0$. For let $F = 0$ be of order N, and $\vartheta = 0$, $\psi = 0$ of order M, and have, beside a $(k-1)$-ple point at every k-ple point of $F = 0$, a number q of simple intersections with $F = 0$. It is easy to prove that, at a k-ple point of $F = 0$, the curve represented by the vanishing of the Jacobian determinant has a multiple point of order $3k - 4$, with k tangents coinciding with those of $F = 0$, and has therefore $3k(k-1)$ intersections with $F = 0$; and that this curve has two intersections with $F = 0$ at each of the q

simple base points of $\vartheta - \lambda\psi$. The determinant is of order $2M + N - 3$. Thus the number of intersections, not at the multiple points, or common points of $\vartheta = 0$, $\psi = 0$ or $F = 0$, is

$$N(2M + N - 3) - 3\Sigma k(k-1) - 2q.$$

This, however, is $2m + 2p - 2$, because $m = NM - \Sigma k(k-1) - q$, $p = \frac{1}{2}(N-1)(N-2) - \frac{1}{2}\Sigma k(k-1)$.

Ex. 1. Supposing $F = 0$ without multiple points, the degree in the coefficients of a curve $\Theta = 0$, of order M, of the function whose vanishing expresses that $\Theta = 0$ touches $F = 0$ (the *tact invariant*), is $N(2M + N - 3)$. And, if these curves be written symbolically as $F_x{}^N = 0$, $\Theta_x{}^M = 0$, and $u_x = 0$ be an arbitrary line, this tact invariant is the product of the values, at all the NM intersections of these curves, of the expression

$$(F\Theta u)F_x{}^{N-1}\Theta_x{}^{M-1}/u_x.$$

For a line to touch $F = 0$ is a condition of degree $N(N-1)$ in the coefficients of the line, and degree $2(N-1)$ in the coefficients of $F = 0$.

Ex. 2. The relation $\Sigma(r-1) + \Sigma(r_0-1) = 2m + 2p - 2$ is often convenient for computing the genus of a curve whose equation is given. For instance, when the equation of the curve is of the form

$$y^m = (x - a_1)^{m_1}...(x - a_k)^{m_k},$$

the rational function x is of order m, and the values of $r-1$, r_0-1 are obtained by considering only the places for which $x = a_1, ..., x = a_k, x = \infty$. For example, the curve represented by $y^{10} = x(x-a)^4(x-b)^5$ is of genus 2.

Note I. The existence of a rational function with assigned poles. Weierstrass's gap theorem.

It seems worth while to give greater definiteness to some of the results of this chapter by reconsidering them in connexion with the question whether, given a set of points on the fundamental curve $F = 0$, there exists a rational function having every one of these as a pole of the first order.

It was shewn that if $T_1, ..., T_k$ be points lying on a ϕ-curve (an adjoint curve of order $N-3$, if N be the order of $F = 0$), then it is impossible to construct a rational function having T_k as an actual pole of the first order, with its other poles lying among $T_1, ..., T_{k-1}$, unless it is the case that every ϕ-curve passing through $T_1, ..., T_{k-1}$ also passes* through T_k. Thus, where there are ϕ-curves through $T_1, ..., T_{k-1}$ which do not contain T_k, the linear series defined by $T_1, ..., T_{k-1}, T_k$ is such that T_k belongs to every set of the series. When $k \leqslant p - 1$ a ϕ-curve can always be put through $T_1, ..., T_k$; but the remark remains equally true for $k = p, ..., 2p - 2$, if

---

* The argument, depending on the theorem of coresiduation, proceeded by the attempt to construct a function of the form $\psi/L\phi$, where $L = 0$ is a line through T_k. The deduced theorem of Riemann-Roch shews that the freedom $k - p + r' + 1$, of the sets coresidual with $T_1, ..., T_k$, can only be as great as 1, when, with $k \leqslant p$, we have $r' + 1 > p - k$, shewing that the k points are not independent conditions for unconditioned ϕ-curves.

$T_1, \ldots, T_k$ lie on a ϕ-curve. If $\phi_1 = 0, \ldots, \phi_p = 0$ denote independent ϕ-curves, the condition that the ϕ-curves containing $T_1, \ldots, T_{k-1}$, which we now denote by $(c_1), \ldots, (c_{k-1})$, should contain (c_k), or T_k, is the existence of the p equations

$$\phi_i(c_k) = \mu_1 \phi_i(c_1) + \ldots + \mu_{k-1} \phi_i(c_{k-1}), \qquad i = 1, \ldots, p,$$

where $\mu_1, \ldots, \mu_{k-1}$ are certain constants, the same for all values of i. For $k < p+1$, these equations are necessary for the existence of a rational function having (c_k) as a simple pole with its other poles among $(c_1), \ldots, (c_{k-1})$. For $k = p+1$ a rational function can be constructed with (c_k) as pole and its other poles among $(c_1), \ldots, (c_{k-1})$ (a function with $p+1$ arbitrary poles, by the formula $\rho = \nu - p$, always exists, of the form $\lambda + \mu\zeta$, where λ, μ are arbitrary constants); and, for $k = p+1$, values of $\mu_1, \ldots, \mu_{k-1}$ can be found to satisfy the p equations whatever $(c_1), \ldots, (c_{k-1})$, (c_k) be, it being understood, in both these statements, that $(c_1), \ldots, (c_{k-1})$ do not themselves lie on a ϕ-curve. Similarly for $k > p+1$. Now suppose we construct an array of p rows and k columns, in which the elements of the sth column are $\phi_1(c_s)$, $\phi_2(c_s)$, $\ldots$, $\phi_p(c_s)$. Then, without further statement as to the value of k, we may express our conclusion by saying that a *necessary* condition for the existence of a rational function having (c_k) as pole, with its other poles among $(c_1), \ldots, (c_{k-1})$, is that the kth *column* of this array should be linearly expressible in terms of the other columns. The converse of this result is true in the form: If the kth column of this array is expressible linearly by certain of the other preceding columns, so that there exist the p equations

$$\phi_i(c_k) = A_m \phi_i(c_m) + A_t \phi_i(c_t) + \ldots + A_s \phi_i(c_s), \qquad i = 1, \ldots, p,$$

where $(c_m), (c_t), \ldots, (c_s)$ are certain points from $(c_1), \ldots, (c_{k-1})$, taken for convenience in their natural order, and $A_m, A_t, \ldots, A_s$ are constants, *all different from zero*, having the same values for all values of i, *then* there exists a rational function having (c_k) and (c_m), $(c_t), \ldots, (c_s)$ as actual poles of the first order, and no other poles. It is understood that the columns containing $(c_m), (c_t), \ldots, (c_s)$ are linearly independent; otherwise the equations put down could be expressed without one or more of these.

When the number, say q, of the points $(c_m), (c_t), \ldots, (c_s)$, which precede (c_k), is at most $p-1$, we can determine $p-q$ ϕ-curves passing through these q points—these points furnishing independent conditions for ϕ-curves because the columns $(c_m), (c_t), \ldots, (c_s)$, of the array, are independent by hypothesis. In virtue of the equations, all these $p-q$ ϕ-curves pass through (c_k), and $p-q$ is the total number of linearly independent ϕ-curves passing through the $q+1$ points. Thus the linear series of sets of $q+1$ points determined by the set $(c_m), (c_t), \ldots, (c_s), (c_k)$ has freedom $q+1-p+p-q$, or 1. There is, therefore, just one rational function having these $q+1$

points as poles of the first order; and the function is actually infinite at every one of these $q+1$ points—for, if the function were infinite only at some of these, the ϕ-curve drawn through all but one of this partial set would pass through this one, and the linear equations put down would involve less than the $q+1$ points.

When the number of points (c_m), (c_t), ..., (c_s) is p, there is no ϕ-curve passing through these points, since this would involve a linear equation connecting the p corresponding columns of the array. The linear series determined by (c_m), (c_t), ..., (c_s), (c_k) thus has freedom $q+1-p$, or 1, and there is one rational function with these as poles; the function is actually infinite at every one of these, for the same reason as before.

When we take more than p columns (c_m), ..., (c_s), we cannot maintain the hypothesis that the columns are independent, any $p+1$ columns being necessarily connected linearly. But we have the result: If $p+f$ points be given, there being no rational function having simple poles chosen from the first p points (and no other poles), then there are f rational functions, all having actual poles of the first order at the first p points, having respectively each a single pole at one of the other f points. The sum of these functions, each multiplied by an arbitrary constant, is a function effectively infinite at each of these f points, with also poles among the p first points. This function contains $1+f$ homogeneously entering arbitrary constants.

This result can be stated differently: As the p ϕ-curves $\phi_1(x)=0$, ..., $\phi_p(x)=0$, do not all vanish for any point (x), on the fundamental curve $F=0$, other than at the multiple points, as we have proved, and the ratios of the functions $\phi_1(x)$, ..., $\phi_p(x)$ have definite values at any point (x) of $F=0$, these functions can be regarded as the homogeneous coordinates of a point in space $[p-1]$, of $p-1$ dimensions. Any set of points (c_1), ..., (c_h), of the curve $F=0$, thus give rise to points, say C_1, ..., C_h, in this space $[p-1]$. The expression of the column (c_k) in terms of the columns (c_m), (c_t), ..., (c_s) involves that the linear space determined by C_m, C_t, ..., C_s contains the point C_k (cf. p. 79 preceding). When $F=0$ is hyperelliptic, the points C_m, C_t, ... lie on a rational curve; of the two points of $F=0$ corresponding to any one of these, only one is represented by a column in the array.

There are theorems, analogous to those preceding, which relate to rational functions having their poles all at one place of the curve $F=0$, this being a multiple pole. A function, when expressed by the local parameter for the neighbourhood of the place where it is infinite, contains negative powers, $A_r t^{-r}+A_{r-1}t^{-r+1}+...+A_1 t^{-1}$; when we say that the function is infinite to order r, we mean that the coefficient A_r is not zero, but we make no statement in regard to the coefficients $A_{r-1}, \ldots, A_1$. Consider now rational functions having (multiple) poles only at one place (x) of the curve. If $u_1, \ldots, u_p$ denote linearly independent everywhere finite integrals, and t the parameter for the neighbourhood of (x), we can consider the limiting values of $du_1/dt, \ldots, du_p/dt$, at (x); these we denote by $\Omega_1(x), \ldots, \Omega_p(x)$; similarly the limiting values of $d\Omega_1/dt, \ldots, d\Omega_p/dt$ may be denoted by $\Omega_1'(x), \ldots, \Omega_p'(x)$. We then form an array of p rows, in which the elements of the first column consist of $\Omega_1(x), \ldots, \Omega_p(x)$, the elements of the second column consist of $\Omega_1'(x), \ldots, \Omega_p'(x)$, and so on. From what has been said above, we can infer, by a limiting process, that the necessary and sufficient condition that there should exist a rational function, infinite only at the place (x), of order $q+1$, is that the $(q+1)$th column of this array should be linearly dependent upon all, or some, of the preceding columns. If this dependence does not require all the preceding columns, this will be because there exists a function, infinite only at (x), of lower than the $(q+1)$th order; the expression of the function of order $q+1$, in the neighbourhood of (x), will then not involve all the negative powers $t^{-q}, t^{-q+1}, \ldots, t^{-1}$, of order less than t^{-q-1}. If no one of the first p columns of the array depends on those preceding, in whole or in part, then the $(p+1)$th column depends on these p columns (the array having only p rows) and involves them all. In this case there is no function, of the kind considered, of any of the first p orders, but there is a function of order $p+1$; this indeed is the case for a general position of (x). We have proved that all the functions $\Omega_1(x), \ldots, \Omega_p(x)$ do not vanish, at any place (x); it may in fact be proved that all the functions $\Omega_1'(x), \ldots, \Omega_p'(x)$, equally, do not vanish at any place (x); or in general, not all the functions $\Omega_1^{(k)}(x), \ldots, \Omega_p^{(k)}(x)$. Hence we may regard the elements of any column of the array as the homogeneous coordinates of a point in space $[p-1]$. Clearly not all the indefinite number of points representing the columns of the array can lie in the same prime of this space, since this would mean the existence of an expression $C_1\Omega_1(x)+\ldots+C_p\Omega_p(x)$ which is constant. There must therefore be, in this indefinite series of points, just p points which are independent, and suffice to determine the space, while all the others depend on these p. There exists therefore a rational function, infinite only at the place (x), of every order except p orders. This result, called Weierstrass's *gap theorem*, is found in Weierstrass, *Math. Werke*, IV, p. 225. See also Band II, p. 235.

That there should exist a rational function, infinite only at (x), of less order than $p+1$, evidently involves that the determinant formed from the first p columns of the array, is zero; and clearly this is so only for a limited number of places (x). It can be shewn that this number is at most $(p-1)p(p+1)$, but the places need not be all distinct. For $p > 1$, however, there are at least $2p + 2$ distinct places; there are more unless the fundamental curve be hyperelliptic (Hurwitz, *Math. Annal.* XLI, 1893, p. 409; an exposition will be found in the writer's *Abel's Theorem*, pp. 41, 90). Geometrically, these exceptional places, for which there exists a rational function of order less than $p + 1$, are characterised by the existence of a ϕ-curve vanishing thereat to the pth order (so having only $p - 2$ other zeros, at most). For instance, for the curve of genus 2 represented by $y^2 z^4 = (x, z)_6$, it may be proved that the ϕ-curves are $z^2 (x, z)_1 = 0$, and there are 6 cases where $(x, z)_1$ has a zero of the second order, namely when it is one of the (supposed different) factors of $(x, z)_6$. Here $6 = 2p + 2 = (p - 1) p (p + 1)$. For the general plane curve of order 4 (genus 3), the ϕ-curves are the lines of the plane, and the points in question are the $24, = (p - 1) p (p + 1)$, inflexions. But, for example, for the particular quartic curve represented by $x^4 + y^4 + z^4 = 0$, the number of such points which are distinct is $12, = 2p + 6$. In general, it can be proved, for $p > 3$, that the number of distinct points of the kind is

$$\geqslant 2p + 6 + 8 (p - 3)/(p^2 - 3p + 4)$$

(see Segre, *Rend. Lincei*, VIII, 1899, p. 89; and Berzolari, *Enzykl. Math.* III$_2$, 3, p. 436). When the fundamental curve is not hyperelliptic, and is representable by a canonical curve of order $2p - 2$ in space $[p - 1]$ (see p. 81 preceding), the points in question, on this canonical curve, are those at which a prime has p coincident intersections therewith. For instance, on the complete intersection of a quadric and cubic surface, in ordinary space ($p = 4$), there are $(p - 1) p (p + 1)$, $= 60$, points, at which the osculating plane of the curve has 4 coincident intersections. More generally, upon any curve of genus p, a set of a linear series g_r^n exists having r points coincident at one point of the curve. It will be proved later that there are $(r + 1)(n + pr - r)$ positions of this point for which one of the $n - r$ remaining points of the set coincides with this point. For $n = 2p - 2$, $r = p - 1$, this number is $(p - 1) p (p + 1)$.

Ex. Prove that for a hyperelliptic curve whose genus p is odd, there exists no rational function of order p.

Note II. On the theory of special sets, and an extension of Clifford's theorem, in the case when the fundamental curve is general for its genus. By a special set of points, as has been explained, is meant a set lying on an adjoint curve of order $N - 3$, if N be the order of the fundamental curve, or, as we say, on a ϕ-curve. It is understood that the points of the set are generally not at the multiple points of the fundamental curve. When this curve is not a plane curve, a special set is one whose points form part of a canonical set thereon. We have seen that in any set of a linear series $g_r{}^n$, consisting of coresidual special sets, of freedom r, with n points in a set, there are $n-r$ points giving independent conditions for unconditioned ϕ-curves passing through them, and that all such ϕ-curves contain the remaining r points of the set. Thus, denoting the points of such a set by $(a_1), \ldots, (a_n)$, and forming the matrix of p rows and n columns, in which the jth column consists of $\phi_1(a_j), \ldots, \phi_p(a_j)$, the values of a set of independent ϕ-polynomials at (a_j), this matrix must be of rank $n-r$. The series $g_r{}^n$ being complete, and $r = n-p+r'+1$, this, with appropriate notation, is the same as saying that every one of the $r(r'+1)$ determinants, expressed, for $k = 1, 2, \ldots, r'+1$ and $\sigma = 1, \ldots, r$, by

$$\begin{vmatrix} \phi_1(a_1), & \ldots, & \phi_1(a_{n-r}), & \phi_1(a_{n-r+\sigma}) \\ \cdots\cdots\cdots\cdots\cdots\cdots\cdots\cdots\cdots\cdots\cdots\cdots\cdots\cdots\cdots \\ \phi_{n-r}(a_1), & \ldots, & \phi_{n-r}(a_{n-r}), & \phi_{n-r}(a_{n-r+\sigma}) \\ \phi_{n-r+k}(a_1), & \ldots, & \phi_{n-r+k}(a_{n-r}), & \phi_{n-r+k}(a_{n-r+\sigma}) \end{vmatrix},$$

must vanish; this is only equivalent to saying that there are $r'+1$ ϕ-curves containing $(a_1), \ldots, (a_n)$. *When these $r(r'+1)$ conditions connecting the points $(a_1), \ldots, (a_n)$ are independent*, we can take $n - r(r'+1)$ arbitrary points of the fundamental curve, and thence determine $r(r'+1)$ others, so that just $r'+1$ ϕ-curves contain the n points, where r is determined from n and r' by $r = n-p+r'+1$. When one such set of n points is found, it determines a series $g_r{}^n$, of which any particular set is fixed by r points of it. This number, defining a particular set of an established series, cannot, clearly, be greater than the number, $n-r(r'+1)$, of arbitrary points of the curve, from which the initial set of n points is found. Thus we have $n-r(r'+1) \geqslant r$, or $p \geqslant (r+1)(r'+1)$, or $n-r \geqslant rp/(r+1)$, which, unless r is as great as $p-1$, involves Clifford's equation $n-2r > 0$. Since the initial set of n points has $n-r(r'+1)$ arbitrary points, the series $g_r{}^n$, of which, in every set, r points are arbitrary, depends only on $n-r(r'+1)-r$, that is, $p-(r+1)(r'+1)$ parameters.

The preceding determination of a special series $g_r{}^n$ from $n-r(r'+1)$ arbitrary points of the curve, can be stated more geometrically in terms of the canonical curve of order $2p-2$ in space $[p-1]$, which is applicable because it is assumed that the curve is general for its genus, and in particular is not hyperelliptic. Two facts are then to be used; (i), that in a space $[s]$, of s dimensions, a space $[k]$ depends on $(k+1)(s-k)$ constants (as, for instance, a line in ordinary space depends on 4 constants); (ii), that the number of conditions for a space $[k]$, in a space $[s]$, to meet a given curve of this space, is $s-k-1$. The problem of a special series $g_r{}^n$ on the canonical curve, is that of finding a set of n points on this curve lying in $r'+1$, or $p-n+r$, primes of the space $[p-1]$, namely lying in a space $[n-r-1]$. In the space $[p-1]$, spaces $[n-r-1]$ depend on parameters whose number is $(n-r)[p-1-(n-r-1)]$, or

$$nr' + n - r(r'+1);$$

for such a space to meet the canonical curve it must satisfy $p-1-(n-r)$, or r' conditions for each meeting; to meet the curve n times, if all the meetings are independent, it must satisfy nr' conditions. Thus spaces $[n-r-1]$, meeting the canonical curve in n points, if these require independent conditions, depend on $n-r(r'+1)$ constants. As the position of a point on the canonical curve depends on one parameter, we infer that, of the n intersections, $n-r(r'+1)$ may be taken arbitrarily.

Consider for example a $g_1{}^n$, for which $r'+1=p+1-n$, and $n-r \geqslant rp/(r+1)$, that is $n \geqslant 1+\tfrac{1}{2}p$. Taking $n-r(r'+1)$, or $2n-(p+1)$ arbitrary points of the curve, the other $p+1-n$ points of a set can be determined so that $p+1-n$ ϕ-polynomials vanish in all the points of the set. The set of $p+1-n$ supplementary points is not unique; it was found by Brill indeed (*Math. Ann.* xxxvi, 1890, p. 354) that, to put with the set of $2n-(p+1)$ arbitrary points, this supplementary set of $p+1-n$ points can be taken in

$$\sum_{\lambda=0}^{\mu} (-1)^\lambda \binom{p}{\lambda} \binom{2p-1-n-2\lambda}{p-2}$$

ways, where μ is the greatest integer in $\tfrac{1}{2}(p+1-n)$. When n has its least value, which is $1+\tfrac{1}{2}p$ when p is even, the number of quite arbitrary points of the set being 1, this number is $\dfrac{2}{p}\dbinom{p}{n}$; or when n has its least value for p odd which is $1+\tfrac{1}{2}(p+1)$, the number of arbitrary points of the set being 2, this number is $\dfrac{4}{p-1}\dbinom{p}{n}$. Thus on a curve which is general for its genus, there exists a special linear series, of freedom 1, of sets of $1+\tfrac{1}{2}p$, or $1+\tfrac{1}{2}(p+1)$ points, but not less. Thus the sets of this series can have no point in common, and these numbers are the least order of existing rational function on the curve, and, of such function, respectively 1 or 2 poles may be taken arbitrarily. We may consider some simple illustrations: (*a*) On a plane quartic curve ($p=3$), there exists a rational function of 3 poles, of which 2 may be taken arbitrarily, the other being either of the two remaining intersections, with the curve, of the line joining these two; (*b*) On the curve of intersection of a quadric and a cubic surface ($p=4$), there exist two series $g_1{}^3$, a single point of a set being arbitrary, the other two points of the set being on one of the two generators of the quadric which pass through this point; (*c*) On the canonical curve of order 8 and genus 5 in space [4], which we may denote by $^5c^8[4]$, there exists a series $g_1{}^4$, of which a set is determined by taking two arbitrary points of the curve, and drawing a plane through these to meet the curve in 2 other points, which, by the formula quoted, is possible in $\dfrac{4}{p-1}\dbinom{p}{n}$ or 5 ways. And in aggregate there are ∞^τ such series g_1^4, where $\tau = p - (r+1)(r'+1)$, namely ∞^1 such series. That there are 5 planes through two arbitrary points of the curve each meeting the curve again in 2 points, is obvious directly by remarking that, by planes through the 2 arbitrary points, the curve is projected into a plane sextic curve, which, being of genus 5, has 5 double points. We may obtain a set of 4 points on this sextic lying on $(r'+1=)2$ adjoint cubic curves, of which two points A, B are arbitrary, by describing cubics through the 5 double points, the points A, B of the curve, and a further point P of the curve, and choosing P so that the further intersection of these cubics also lies on the sextic curve. Particular series g_1^4 on the sextic are obtained by the lines through any one of the double points, or by the conics through any 4 of the double points. As examples of the general theorem for series g_1^n when n has not its least value, we may instance a g_1^{p-2}; if $p-5$ arbitrary points be taken, 3 others can be associated with them in $\frac{1}{6}p(p-1)(p-5)$ ways, thus making a set of $p-2$ points through which 3 ϕ-curves pass. Or a g_1^{p-1}; if $p-3$ arbitrary points be taken, 2 others can be associated with them in $\frac{1}{2}p(p-3)$ ways, thus making a set of $p-1$ points through which there pass 2 ϕ-curves.

More generally, to find a set of n points lying on 3 ϕ-curves, defining then a series g_r^n in which $r = n - p + 3$, we may take $n - 3(n-p+3)$, or $3p - 9 - 2n$ points arbitrarily; but this number must be $\geqslant n - p + 3$, namely $n \leqslant \frac{4}{3}(p-3)$. Considering the canonical curve in space $[p-1]$, through the n points so determined, which lie in a space $[p-4]$, can be put spaces $[p-3]$, by means of which the curve can be projected birationally into a plane curve of genus p and order $2p - 2 - n$; this order is then $\geqslant \frac{2}{3}(p+3)$. Considering in turn the cases in which p is of the forms 3ϖ, $3\varpi + 1$, $3\varpi + 2$, we find that the order of the plane curve is in all cases $\geqslant p + 2 - \varpi$, namely $p + 2 - [p/3]$, where $[p/3]$ is the integer part of $p/3$. In all cases $\frac{2}{3}(p+3)$ is greater than $1 + \frac{1}{2}p$; but $p + 2 - \varpi$ is less than p when $p \geqslant 9$. The case when $n = p - 3$ may be regarded as included in this result; by projection of the canonical curve from $p-3$ arbitrary points thereon, we have a representation of the curve as a plane curve of order $p+1$, with the equivalent of $\frac{1}{2}p(p-3)$ double points (as found above, in discussing series g_1^{p-1}). A corollary may be noted: In order that a plane curve of order n should represent a curve general for its genus, it must have multiple points equivalent to at least $\frac{1}{2}(n-2)(n-4)$ double points (for instance, a plane quintic curve is incompetent to represent a general curve of genus 6, when it is without a multiple point; or a general curve of genus 5, when it has one double point. When it has two double points it may represent a curve of genus 4; it is then obtainable by projection of the canonical sextic curve).

Some further indications of general theorems find their proper place here:

(*a*) As we know the number of coefficients entering into the plane curve of order n', or $2p-2-n$, which we have just obtained by projection of the canonical curve, from the points of a special set g_r^n thereon, we can form an estimate of the number of fixed constants, or *moduli*, on which a general curve of genus p depends. The special set on the canonical curve was found from k, $=3p-9-2n$, arbitrary points, and there are ∞^r coresidual sets, where $r=n-p+3$. Thus the series g_r^n depends on $k-r$, or $4p-3n-12$ constants. The plane curve of order n' has $\frac{1}{2}n'(n'+3)$ coefficients, of which 8 are disposable by suitable linear transformation of the coordinates in the plane; this plane curve has the equivalent of $\frac{1}{2}(n'-1)(n'-2)-p$ double points, of which the existence by itself of any one would impose one condition for the coefficients of the curve. There are then *at least* $\frac{1}{2}n'(n'+3)-8-[\frac{1}{2}(n'-1)(n'-2)-p]$, or $3n'+p-9$ independent constants; of which, however, we can dispose of $4p-3n-12$ by suitable choice of the set of points of the canonical curve from which we project. There remain then, at least,

$$3n'+p-9-(4p-3n-12)$$

absolute constants, namely $3p-3$, for the general curve of genus p. This provisional reasoning, however, (1) is only for curves general for their genus, and, in particular, excludes the case of hyperelliptic curves; thus it supposes $p>2$, though it happens that 3 is the right number of constants, in this case, for a general curve represented by $y^2z^4=(x,z)_6$, being the number of independent cross ratios of the roots of the sextic $(x,z)_6$; (2) supposes that, for a general curve, the existence of d double points requires d conditions for the constants of the curve; (3) supposes that the curve has no birational transformations into itself involving an arbitrary parameter; such a transformation would involve a lessening of the number of constants upon which a series g_r^n depends. It will in fact be proved below that no such transformations exist for $p>1$. For $p=0$ there are such transformations depending on 3 parameters; for $p=1$, there is a transformation depending on one parameter. The number $3p-3$ was found by Riemann (1857, Werke, *Theorie der Abel'schen Fctnen*, § 12), who estimates the total number of constants in a rational function of order m as $2m-p+1$, since if its m poles be assigned there are further $m-p+1$ constants; and, then, regarding the curve as represented by an equation $F(\eta,\xi)=0$, proves that there are $2m+2p-2$ branchings of η as a function of ξ (as above, p. 84). There are then $2m+2p-2-(2m-p+1)$ of these which cannot be arbitrarily assigned by suitable choice of ξ. Recent criticisms of the argument for the number $3p-3$ will be found in Severi-Löffler, *Alg. Geom.* Leipzig, 1921, pp. 321, 394; and in Enriques-Chisini, *Teoria geom.* III, Bologna, 1924, pp. 112, 359.

Ex. Consider the curve $y^3z^3+y^2z^2u_2+yzu_4+u_6=0$, where u_m is homogeneous of order m in x and z. Let (a, b, c) be any point of this curve. Shew that by the transformation obtained by putting ξ, η, ζ proportional respectively to $zc(xc-za)$, $zc(yc-zb)$, $(xc-za)^2$, of which the reverse forms give x, y, z proportional respectively to $\xi(\xi a+\zeta c)$, $\xi^2 b+\eta\zeta c$, $\xi^2 c$, the curve is changed to a quintic curve with a point of self-contact, whose equation, with a further suitable change, of the form, $\eta=\eta_1+\lambda\xi$, may be taken to be $\eta_1^3\zeta^2+\eta_1^2\zeta v_2+\eta_1 v_4+\zeta w_4=0$, where v_i, w_i are homogeneous in ξ, ζ, of order i. For the sextic and quintic, respectively, the ϕ-polynomials are of the forms $z[yz+(x, z)_2]$, $\eta_1\zeta+(\xi, \zeta)_2$. A ϕ-curve which touches the fundamental curve at each of its intersections is given, in these cases respectively, by $z(x-\mu z)^2=0$, $(\xi-\sigma\zeta)^2=0$, where μ, σ are arbitrary constants; namely there are ∞^1 such ϕ-curves. Hence, by a theory which cannot be given here, the curves each depend, not on $3p-3$ or 9, but on 8 moduli (cf. the writer's *Abel's Theorem*, 1897, pp. 270, 94). There are, as in general, $2^{p-1}(2^p-1)$ or 120, other ϕ-curves having contact with the fundamental curve at all their intersections therewith. (For an elementary geometrical theory, in this case, cf. W. P. Milne, *Proc. Lond. Math. Soc.* xxv, 1926, p. 174.)

It may happen that there are ∞^2 (or more) such particular contact ϕ-curves; for instance for a general plane quintic curve, every line of the plane, repeated, is such a curve, and the aggregate is ∞^2. In this case the number $2^{p-1}(2^p-1)$, of proper contact ϕ-curves, or 2016, is reduced to 2015, in accordance with the general theory (cf. W. P. Milne, *Journ. Lond. Math. Soc.* II, 1927, p. 79, for this particular case). In the notation of the general theory referred to, the two exceptional ϕ-systems are associated respectively with an even and an odd characteristic.

(*b*) Another question is what is the number of constants upon which a general curve of order n and genus p normal in space $[r]$ depends. Thus, though all rational curves are birationally equivalent, a conic *in ordinary space* depends on 8 constants (for example, a single conic exists which meets an arbitrary line and also seven other lines which meet this, and do not meet one another); and, more generally, the rational curve of order r in space $[r]$ depends on $(r-1)(r+3)$ constants, as we shall see. While the canonical curve of order $2p-2$ in space $[p-1]$ depends on $(p-1)(p+4)$ constants (for instance, a plane quartic curve on 14; and the intersection of a quadric surface and a cubic surface on $9+(19-4)$, or 24 constants). It can be proved that the number in general is $(r+1)[n-r-rp/(r+1)]+(r+1)^2-1+3p-3$, the first term being the number τ, $=p-(r+1)(r'+1)$, of constants upon which a *complete* series g_r^n, where $r=n-p+r'+1$, depends ($\tau \geqslant 0$; p. 91); the second term, the number of constants in a general linear transformation of coordinates in space $[r]$; and the last term, the number of absolute constants of the class of all birationally equivalent curves of genus p; namely is $n(r+1) - (r-3)(p-1)$.

If the series of prime sections of the curve be of deficiency h, and specialness $r'+1$, the number $(n-r)(r+1) - rp$ must be diminished by $h(r'+1)$; e.g. the octavic curve of genus 5, intersection (save for two skew conics) of a cubic and quartic, depends on 31 constants. When the curve is not general, this number may be negative and need increase (p. 91). Examples are, the plane quintic curve $(p=6)$, depending on $3n+p-1$ constants; the space septimic $(p=6)$, partial intersection of a quadric and quartic, depending on $4n$; and the space octavic $(p=9)$, complete intersection of a quadric and quartic, depending on $4n+1$. In general, a complete intersection on a quadric depends on $2n+p+8$ constants $(n>4)$. See Noether, *Berlin. Abh.* 1882, pp. 18, 58.

As to surfaces, a Veronese quartic surface, in space $[5]$, depends on 27 constants, and can be put through 9 points of its space; a Del-Pezzo surface, of order r, in space $[r]$, depends on r^2+10. Cf. Room, *Proc. Camb. Phil. Soc.* XXVII, 1931, p. 518. In general, see Castelnuovo-Enriques, *Enzykl. Math.* III, C. 6 b, p. 713.

(*c*) It was proved by Noether that the canonical curve of order $2p-2$ in space of $p-1$ dimensions lies on $\frac{1}{2}(p-2)(p-3)$ quadrics (an exposition, of the proof, for quadrics, and manifolds of higher dimension, will be found in the writer's *Abel's Theorem*, Cambridge, 1897, p. 155). For general values of p, these quadrics have no other common intersection than the canonical curve, unless the canonical curve possesses a special series $g_1{}^3$ $(p>4)$, in which case the quadrics intersect in a rational ruled surface of order $p-2$. But, in particular for $p=6$, if the canonical curve contain a special series $g_2{}^5$, being the representative of a plane quintic curve, the quadrics intersect in a Veronese surface (Enriques, *Rend....Bologna*, XXIII, 4 May 1919).

For a curve of order n and genus p, in space $[r]$, when $r=n-p$, if also $\frac{1}{2}r(r-1)-p>0$, it can be shewn that the curve lies on $\frac{1}{2}r(r-1)-p$ quadrics *at least*.

(*d*) We have seen that a set of a special series $g_r{}^n$, on the canonical curve, lies on a space $[n-r-1]$, the sets of the complementary series $g_r{}^{n'}$, where $n'=2p-2-n$, and $r=n-p+r'+1$, lying similarly each on a $[n'-r'-1]$. Supposing $r'\geqslant r$, it can be shewn that all the spaces $[n-r-1]$ generate a manifold of dimension $n-1$, and order $\binom{r'+1}{r}$, which likewise can be defined from the spaces $[n'-r'-1]$.

If τ, given by $p-(r+1)(r'+1)$ be $\geqslant 1$, all the spaces of this manifold contain a space $[\tau-1]$; and the locus of these spaces $[\tau-1]$, for all the ∞^τ series $g_r{}^n$, is a manifold of dimension $2\tau-1$. The theorem arises very naturally from the expression of the everywhere finite integrals given in Chap. VII following, to which we refer.

Ex. 1. Suppose $p = 6$. In general, by the preceding theory, there exist, on the canonical curve of order 10, in space [5], five special series g_1^4, any such series being established by taking an arbitrary point on the curve, and thence determining three other points, so that the 4 points lie on a plane. The curve can be represented on a plane by a sextic curve, with 4 nodes, obtainable by projecting the canonical curve from a set of 4 coplanar points thereof. The ϕ-curves of the plane sextic, 6 in number, proportional to coordinates in the space [5], not only transform the sextic curve into the canonical curve, but transform the plane into a rational surface in the space [5], containing the canonical curve, which is of order 5, because two of the ϕ-curves in the plane have 5 intersections (other than the 4 nodes). This is called a Del-Pezzo surface. The five series g_1^4 on the plane sextic are obtained, either by lines through a node, or by conics through the 4 nodes; in each case the determining curves meet a ϕ-curve (which is represented on the surface by a prime section) in two variable points. Thus there are 5 systems of conics on the Del-Pezzo surface, one conic of each system through an arbitrary point of the surface, the planes of the conics meeting the canonical curve in sets g_1^4. The complementary series to the g_1^4 is a g_2^6, of which the sets lie in spaces [3], or solids; any set of g_2^6 lies with any set of the complementary g_1^4 in a prime, or space [4], so that the plane of the latter meets the solid of the former in a line. To establish a series g_2^6 we may take $n' - r'(r+1)$, with $n' = 6$, $r' = 2$, $r = 1$, that is two arbitrary points of the canonical curve, and find thence 4 points of the curve lying in a solid with the 2 arbitrary points. In the plane representation, the five series g_2^6 are given, either by a conic through 3 of the nodes of the plane sextic curve, or by a line of the plane, and in every case the determining curve meets an adjoint cubic curve in the plane (which represents a prime section of the Del-Pezzo surface) in 3 points; thus the rational cubic curve through the points of a set of a g_2^6, in the solid containing this set, is one of a system of cubic curves lying on the Del-Pezzo surface, of which one curve can be put through 2 arbitrary points of the surface. Further, in the plane any curve determining a set of a series g_1^4 meets any curve determining a set of the complementary g_2^6 in two points; thus on the Del-Pezzo surface, any conic, of either system, meets any cubic curve, of the complementary system, in two points.

Besides the g_1^4 or g_1^{p-2}, on the canonical curve, there are series g_1^5 or g_1^{p-1}, of which each set lies in a solid, a set being obtainable by taking $p - 3$, or 3, arbitrary points of the curve, and determining the other 2, so that the five lie in a solid, which is possible in $\frac{1}{2}p(p-3)$, or 9, ways. In the plane representation, we take three arbitrary points of the plane sextic curve, A, B, C, and a further point P so that the remaining intersection Q of the two adjoint cubic curves through A, B, C, P may also lie on the sextic curve. The complementary series of a g_1^5 is likewise a g_1^5. Beyond these series, and the canonical series g_1^{10}, and the g_1^4, g_2^6 discussed, there appear to be no special series on the canonical curve, when this is perfectly general.

But when the curve is particular there may be other series. Consider plane sextic curves, (a) which have 4 nodes of which 3 are in line; (b) which have a triple point and a single double node; (c) which have 4 nodes all in a line. In case (a), the line has no further intersection with the sextic curve, but the 3 nodes are not independent conditions for the line (cf. footnote, p. 72, under (b)); the six adjoint cubic curves contain 5 consisting of the line and an arbitrary conic through the fourth node; two such conics have 3 further intersections. Thus it may be proved that the Del-Pezzo surface has a node, and lies on 5 quadrics passing through this; the canonical curve lies also on another quadric not containing the node. There are now 4 special series g_1^4, two of the 5 generally existing coinciding with one another. In case (b), the adjoint cubics have one fixed double point, and one simple base point, and the Del-Pezzo surface degenerates, containing as part a rational ruled quartic surface, which lies on all quadrics containing the canonical curve. This curve has now a g_1^3 (corresponding to lines in the plane through the triple point), of which the complementary series, g_3^7, is determined by primes through any generator, containing one set of g_1^3; the sets of g_3^7 lie on rational cubic curves. The ruled surface contains ∞^1 conics, each of which contains a set of a g_1^4 lying on the canonical curve; any prime through such a conic meets the surface beside in two generators. In case (c), where the 4 nodes of the plane sextic are in line, the sextic curve now contains this line, and the adjoint cubic curves break up into this line and a variable conic. The Del-Pezzo surface is again degenerate, consisting of a Veronese surface, of order 4, and a plane meeting this in a conic. The canonical curve is obtained by the intersection of the Veronese surface with a cubic primal passing through a conic of the surface; and now contains a g_2^5, corresponding to the intersections of the plane quintic curve with a variable line. The six quadrics containing the canonical curve contain also the Veronese surface (cf. (c), p. 96 above).

Ex. 2. The reader may similarly consider the particular cases for $p = 7$, taking a plane septimic curve with 8 double points as representative curve.

Ex. 3. By remark (d) above, p. 96, the ∞^1 spaces $[\frac{1}{2}(p-1)]$ containing the sets of one series g_1^n on the canonical curve of odd genus p, where $n = 1 + \frac{1}{2}(p+1)$, meet in a point, generating a cone of order $\frac{1}{2}(p-1)$. There are ∞^1 such series, and the locus of this point is a curve; D. W. Babbage (*Proc. Camb. Phil. Soc.* xxviii, 1932, p. 426) finds for $p = 5$ that this curve is of genus 6 and order 10.

Ex. 4. On a non-singular plane curve of order n there is a special series g_1^{n-1} determined by lines through a fixed point O of the curve. The complementary series consists of sets obtained by adjoining O to the set of intersections of the fundamental curve with a general curve of order $n - 4$.

Ex. 5. A plane curve of order n with less than $n - 2$ double points $(p > \frac{1}{2}(n-2)(n-3))$ is unobtainable by projection from a curve of order n in higher space; namely is a *normal* curve. But a curve of order n with more than $\frac{1}{2}(n-2)(n-3)$ double points $(p < n-2)$ can always be so obtained. When $p = n - 2$, a complete series g_2^n is obtainable by variable adjoint curves of order $n - 2$ through $2p - 2$ (or $2n - 6$) arbitrary fixed points; this series agrees with that determined by the lines of the plane when the $2p - 2$ fixed points are a canonical set.

The enumerative problem of finding the number of curves satisfying a number of conditions equal to that of the constants on which such a curve depends arises below (Vol. vi, Chap. ii). For a geometrical discussion of this problem for conics (and quadrics) see Ursell, *Proc. Lond. Math. Soc.* xxx, 1929, p. 322. For rational cubics see Todd, *Proc. Royal Soc.* A, cxxxi, 1931. For a rational quartic curve in space [4], see White, *Journ. Lond. Math. Soc.* iv, 1929; Welchman, *Proc. Camb. Phil. Soc.* xxviii, 1932; Todd, *Proc. Camb. Phil. Soc.* xxvi, 1930, p. 332; Babbage, *Journ. Lond. Math. Soc.* viii, 1933.

Note III. On the points common to two plane curves. The Cayley-Bacharach theorem. It is a familiar fact, which attracted attention very early in the study of plane curves, that, of the mn common points of two curves of orders m and n, which have no multiple points, not all can be arbitrarily assigned. When $m \geqslant n$, the equation of the curve of order m can be so modified, with the help of the equation of the other curve, as to contain only

$$mn - \tfrac{1}{2}(n-1)(n-2)$$

assignable coefficients, so far as its intersections with the curve of less order are concerned. Thus we can assume (what is a particular case of the subsequently proved Riemann-Roch theorem) that, among the mn intersections, there are $\tfrac{1}{2}(n-1)(n-2)$ such that all curves of order $m(\geqslant n)$ drawn through the others, pass through these.

Assuming this, we can prove that, in order that mn given points should be the intersections of two curves of orders m and n, with $m > n$, the coordinates of the mn points must be connected by $mn - 3n + 1$ conditions; when $m = n$, they must be connected by $n^2 - 3n + 2$ conditions (e.g. 6 points, to be the common points of a conic and a cubic curve, must satisfy one condition; 9 points, to be common to two cubic curves, must have their coordinates subject to two conditions). To formulate the proof we mentally divide the mn points, when $m > n$, into three sets: A, of $\tfrac{1}{2}n(n+3)$ points; B, of $mn - [\tfrac{1}{2}n(n+3) + \tfrac{1}{2}(n-1)(n-2)]$ points; C, of $\tfrac{1}{2}(n-1)(n-2)$ points. We can suppose the points A taken arbitrarily, and therefore of sufficient generality to determine a curve, $f = 0$, of order n, passing through them. That the points B should lie on this curve $f = 0$ requires $mn - \tfrac{1}{2}n(n+3) - \tfrac{1}{2}(n-1)(n-2)$ conditions connecting the coordinates of the points $A + B$, whose aggregate number is $mn - \tfrac{1}{2}(n-1)(n-2)$. This number is less than the number, $\tfrac{1}{2}m(m+3)$, of arbitrary points through which a curve of order m can be described, being less than this by $\tfrac{1}{2}(m-n+1)(m-n+2)$. We put then a curve, $\phi = 0$, of order m, through the points $A + B$; that its remaining $\tfrac{1}{2}(n-1)(n-2)$ intersections with $f = 0$ should agree with the points C requires $(n-1)(n-2)$ conditions connecting the coordinates of the points $A + B + C$. The total number of conditions required is thus $mn - \tfrac{1}{2}n(n+3) - \tfrac{1}{2}(n-1)(n-2) + (n-1)(n-2)$. This is the number $mn - 3n + 1$ enunciated. The number of points in the set B is, we easily see, $(m-n)n - 1$, and the set B does not exist unless $m > n$. Indeed, when $m = n$, there remain, after taking the set A, of $\tfrac{1}{2}n(n+3)$ points, only $\tfrac{1}{2}(n-1)(n-2) - 1$ of the given n^2 points. In this case, therefore, we begin by selecting $\tfrac{1}{2}n(n+3) - 1$ arbitrary points; the $\tfrac{1}{2}(n-1)(n-2)$ remaining intersections of the two curves, of order n, which can be put through the selected points, must agree with the $\frac{1}{2}(n-1)(n-2)$ remaining given points. This requires n^2-3n+2 conditions connecting the coordinates of the given points, as was said.

Relations among the mn intersections of two curves were investigated by Jacobi (*Werke*, III, pp. 285, 610); these can be used to prove Abel's Theorem (Chap. III, above), as in Clebsch-Gordan, *Abelsche Fctnen*, 1866, p. 44. Proof of Jacobi's relations (for functions of any number of variables) is given in Netto's *Algebra*, II, 1900, p. 165, with references to Kronecker. See also Harnack, *Math. Ann.* IX, 1876, p. 371, who obtains the results in the following form (likewise capable of extension to any number of variables): let $f=0$, $\phi=0$ be two curves, of orders n and m, and $u=0$ an arbitrary line; also, the coordinates being x, y, z, let J be the Jacobian determinant $\partial(f, \phi, u)/\partial(x, y, z)$. Then, if Φ be any polynomial, homogeneously of order $m+n-3$, the sum of the values of the fraction $u\Phi/J$, at all the mn common points of the curves, is zero. For *the two given curves*, whose coefficients enter into these equations, this gives, in general, $\frac{1}{2}(m+n-1)(m+n-2)$ relations.

We now proceed to consider the conditions to be satisfied by a general curve, $\psi=0$, of order r (without multiple points), in order that it may pass through the mn points common to two given curves, $f=0$, of order n, and $\phi=0$, of order m, also without multiple points. We put $r=m+n-3-\alpha$, taking $\alpha=0$ when $r\geqslant m+n-3$, and suppose $r\geqslant m$, $r\geqslant n$. We prove that, when $r>m+n-3$, the mn common points of $f=0$, $\phi=0$ give independent conditions for the curve $\psi=0$, of order r, to pass through these points; but that, when $r\leqslant m+n-3$, these mn points give only $mn-\frac{1}{2}(\alpha+1)(\alpha+2)$ independent conditions; in this latter case there are in fact

$$mn-\tfrac{1}{2}(\alpha+1)(\alpha+2)$$

points, among the mn, such that curves $\psi=0$ through these necessarily contain the others*.

The results we reach are, *in part*, contained in a much more general theorem, which will be considered in some detail in Chap. VIII, below. For the sake of clearness we enunciate this general theorem: In space $[s]$, of s dimensions, consider the manifold which is the intersection of

---

* In particular, curves of order $m+n-3$ through all but one of the intersections of two given irreducible curves of orders n and m necessarily contain the remaining intersection. This is in accordance with the identity

$$\tfrac{1}{2}(m+n-1)(m+n-2)=mn-1+\tfrac{1}{2}(m-1)(m-2)+\tfrac{1}{2}(n-1)(n-2).$$

It may be compared with the known result that, for a non-degenerate curve of order $m+n$, having double points, these double points give independent conditions for curves of order $m+n-3$ required to pass through them. For the general case of the adjoint ϕ-curves of a degenerate curve, see Noether, *Acta Math.* VIII, 1886, p. 161.

h primals, each given by the vanishing of a single polynomial homogeneous in the $s+1$ homogeneous coordinates; suppose that this manifold of intersection has no multiple part of dimension $s-h$. Let the orders of the intersecting primals be n_1, n_2, ..., n_h. Then the number of independent linear conditions for the coefficients of a primal of order ρ, in order that it may contain the manifold of intersection, is obtainable as the coefficient of t^ρ in the ascending expansion of the function

$$(1-t^{n_1})(1-t^{n_2})...(1-t^{n_h})/(1-t)^{s+1}.$$

This number is called the *postulation* of the manifold for primals of order ρ. The proof of the statement may be found in Bertini, *Geom. d. Iperspazi*, 1907, p. 263; with which cf. F. S. Macaulay, *Modular Systems*, Cambridge Tracts, No. 19, 1916, p. 65. If we apply this general theorem to the case now under consideration, for which $s=2$, and the manifold of intersection is that of two curves of orders n, m, whose common points are simple, the postulation of these mn common points, for a curve of order r, is the coefficient of t^r in the ascending expansion of $(1-t^n)(1-t^m)(1-t)^{-3}$, namely, is $(r+2)_2 - (r-m+2)_2 - (r-n+2)_2 + (r-m-n+2)_2$, with the omission of any binomial coefficient, $(\lambda+2)_2$, of these for which $\lambda < 0$. Whence for $r \geqslant m+n$, all these terms enter, and it is easy to verify that their sum is mn; for $r \geqslant m$, $r \geqslant n$, but $r < m+n$, the postulation is, thus, $mn - \frac{1}{2}(r-m-n+1)(r-m-n+2)$, which remains equal to mn for $r=m+n-1$ or $m+n-2$; for $r=m+n-3-\alpha$, with $\alpha \geqslant 0$, the postulation is $mn - \frac{1}{2}(\alpha+1)(\alpha+2)$. These results are in accordance with the theorem, also part of the general theorem referred to, that when $\psi = 0$ passes through all the (simple) intersections of $f=0$, $\phi=0$, there exist two curves $f'=0$, $\phi'=0$, of appropriate orders, such that we have an identity $\psi = f'f + \phi'\phi$.

In the present note we prove a more detailed theorem: *For $r \geqslant n$, $r \geqslant m$, $r=m+n-3-\alpha$, with $\alpha \geqslant 0$, the curves $\psi = 0$, of order r, put through $mn - \frac{1}{2}(\alpha+1)(\alpha+2)$ of the intersections of two non-singular curves $f=0$, $\phi=0$, of orders n, m, with simple intersections, have a freedom $i+1+\frac{1}{2}(r-n)(r-n+3)+\frac{1}{2}(r-m)(r-m+3)$, that is are, in tale, $i+(r-n+2)_2+(r-m+2)_2$, where i is the number of linearly independent curves of order α containing the $\frac{1}{2}(\alpha+1)(\alpha+2)$ excepted intersections (in general no curves of order α can be put through this number of points); and further that the curves $\psi = 0$, so described, will pass through these excepted points if made to contain i of them, suitably chosen (when $i < \frac{1}{2}(\alpha+1)(\alpha+2)$).* The

$$mn - \tfrac{1}{2}(\alpha+1)(\alpha+2)$$

chosen points are thus equivalent only to $mn - \frac{1}{2}(\alpha+1)(\alpha+2)-i$ independent points as conditions for curves of order r put through them; and the choice of i further independent points, among the $\frac{1}{2}(\alpha+1)(\alpha+2)$ excepted points, makes the number of conditions, in all, equal to $mn - \frac{1}{2}(\alpha+1)(\alpha+2)$; thus, curves $\psi=0$, of order r, put through $mn - \frac{1}{2}(\alpha+1)(\alpha+2)$ points, of the intersections of $f=0$, $\phi=0$, which provide independent conditions, pass through the remaining intersections. Obviously, the $i+(r-n+2)_2+(r-m+2)_2$ curves through the $mn - \frac{1}{2}(\alpha+1)(\alpha+2)$ selected points may contain, necessarily, *some* of the $\frac{1}{2}(\alpha+1)(\alpha+2)$ remaining points, as we shall illustrate by examples. The theory we give was initiated by Cayley (*Papers*, I, 1843, p. 27), and developed by Bacharach (*Math. Ann.* XXVI, 1886, pp. 275–99); it serves as an application of the general theory of this chapter. From this theory we assume two lemmas: (1), Among the intersections of a curve $\jmath=0$, of order n, which may have multiple points, with an adjoint curve of order μ, let there be a set of p points (not at the multiple points) which lie on h adjoint curves ϕ, of order $n-3$; denote by H a set of h points suitably chosen from these p points, and by K the remaining $p-h$ points of these; also denote by M the intersections (other than the multiple points) of the curve of order μ with $f=0$ other than the p points $H+K$. Then every adjoint curve of order μ through the sets M, H contains the set K (the set $H+K$ defines a series $g_h{}^p$). (2), An adjoint curve of order μ, with $\mu>n$, put through $n\mu-p+1$ of the intersections of $f=0$ with a particular adjoint curve of order μ (other than the multiple points), contains $f=0$ entirely, if these $n\mu-p+1$ points are independent conditions for such adjoint curves. Here p as usual denotes the genus of $f=0$.

Now consider the whole set of intersections, with $f=0$, of order n, beside those at the multiple points, of an adjoint curve of order m (which we may denote by c^m), in which $m>n-3$. These points are of number $n(m-n+3)+2p-2$. Divide this set into two sets: one set A, and a remainder set B; this latter consisting of β points, where

$$\beta=p-[kn-\tfrac{1}{2}k(k+3)], \ =\tfrac{1}{2}(s+1)(s+2)-[\tfrac{1}{2}(n-1)(n-2)-p],$$

in which k is any number >0 and $<n-3$, small enough to make $\beta\geqslant 0$, and $s=n-3-k$. It is easy to see that the two forms of β are equal, and, from the second form, one more than the number of independent points through which we could put an adjoint curve, c^s, of order s, if the multiple points of $f=0$ were independent conditions for this c^s. Take, further, a set A_0, of $\frac{1}{2}k(k+3)$ points of $f=0$, so chosen as to be independent conditions for non-adjoint curves of order k put through them; let $\omega=0$ be the curve, c^k, through A_0; denote the residual intersections of $\omega=0$ with $f=0$ by B_0, so that B_0 is of $kn-\frac{1}{2}k(k+3)$, or $p-\beta$ points.

The composite set $B+B_0$ consists then of p points. The set B_0 is of number $kn-\frac{1}{2}k(k+3)$, which is greater than the number, $\frac{1}{2}k(k+3)$, of independent points through which a non-adjoint curve of order k could be drawn, because $k<n-3$; and the number of points in B_0 is $k(n-3)-\frac{1}{2}(k-1)(k-2)+1$, or $k(n-3)-q+1$, where q is the genus of the curve $\omega=0$. Thus, by lemma 2, all curves of order $n-3$ through the set B_0 contain $\omega=0$ as a part, provided these points give independent conditions on $\omega=0$ for curves of order

$n-3$ not necessarily adjoint to $f=0$. The result certainly follows for $k=1$, or $k=2$, the number of points being then $>k(n-3)$.

Now denote by $i(\geqslant 0)$ the number of curves, c^s, of order s, adjoint to $f=0$, if any, which can be drawn through the set B $(0<s<n-3)$. If $\psi=0$ be such a curve, the curve $\omega\psi=0$ is a curve, adjoint to $f=0$, of order $n-3$, through the p points $B+B_0$, and there are i such curves at least, since there are i curves $\psi=0$. Let $i+j$ be the total number of curves of order $n-3$, adjoint to $f=0$, through the set $B+B_0$. Certainly $j=0$ when $k=1$, or $k=2$, and more generally when the $p-\beta$ points B_0 are independent for curves of order $n-3$, as we have remarked. This is ensured if, but does not require* that, the $p-\beta$ points B_0 should be independent for curves of order $n-3$ adjoint to $f=0$; and this sufficient condition is satisfied if there are no other sets of $p-\beta$ points of $f=0$ which are coresidual with the set B_0; this again is true if $f=0$ have no multiple points, since the $\frac{1}{2}k(k+3)$ points A_0 are supposed independent for curves of order k.

The sets $A+A_0$, $B+B_0$, together, form the complete intersection, with $f=0$, of a composite adjoint curve of order $m+k$, consisting of the adjoint curve of order m through $A+B$, together with the non-adjoint curve $\omega=0$; and the set $B+B_0$ consists of p points, and lies on $i+j$ adjoint ϕ-curves of $f=0$. Wherefore (see lemma 1), the general adjoint curve of order $m+k$, put through the set $A+A_0$, gives on $f=0$ a series of freedom $i+j$; and this curve will contain all the points $B+B_0$ if made to contain $i+j$ of them which give independent conditions (when $i+j<\beta$). Now, the set A_0 gave independent conditions for non-adjoint curves of order k; thus it likewise gives independent conditions for curves of order $m+k$ adjoint to $f=0$. Wherefore, discarding the set A_0, we infer that a curve of order $m+k$, adjoint to $f=0$, which contains all but β of the intersections with $f=0$ of an adjoint curve of order m (namely, passes through the set A) has, on $f=0$, a freedom $i+j+\frac{1}{2}k(k+3)$. Here i is the number of curves of order s, adjoint to $f=0$, which pass through these β remaining intersections $(s=n-3-k)$, and $i+j$ is the number of adjoint curves of order $n-3$, passing through the p points constituted by these β remaining intersections, and through $p-\beta$ points of $f=0$, which are residual, by a non-adjoint curve $\omega=0$ of order k, to $\frac{1}{2}k(k+3)$ arbitrary general points of $f=0$. If we denote $m+k$ by r, so that $s=m+n-3-r$, and recognise that the absolute

---

* For a plane sextic curve $f=0$, with $p=4$, having its six double points on a conic, taking $k=1$, and hence $\beta=0$, a set A_0, of 2 points, gives 4 collinear points B_0, which are independent for general cubic curves, but equivalent only to 3 for cubic curves adjoint to $f=0$. Thus, here also, $j=0$, the adjoint cubics through B_0 reducing to the line and the conic containing the six double points $(i=1)$.

freedom, in the plane, of curves of order r, $> n-3$, is greater than the freedom on $f=0$ by $\frac{1}{2}(r-n)(r-n+3)+1$, we reach the conclusion that the tale of the curves of order r, adjoint to $f=0$, under consideration, is $i+j+(r-m+2)_2+(r-n+2)_2$. We have stated the proof subject to $n-3<m<r<m+n-3$. When $f=0$ has no multiple points, we have proved that $j=0$; in this case also $\beta=\frac{1}{2}(s+1)(s+2)$. We can apparently allow the possibility $s=0$ ($k=n-3$, $\beta=1$), and hence $r=m+n-3$; and also the possibility $k=0$, with $r=m$; and remove the condition $m>n-3$. Then the conditions are $n\leqslant r$, $m\leqslant r$, $r\leqslant m+n-3$, and we obtain the result as stated above, when $f=0$ has no multiple points.

But though, when $i>0$, it does not follow that non-singular curves of order r, put through any $mn-\beta$ intersections of the non-singular curves of orders n and m, necessarily pass through all the other β points, unless put also through i independent points of these, it may well be that there are *some* of the β points through which they do all pass. Namely, going back to the argument, the linear series of sets of p points determined on $f=0$ by the set $B+B_0$, which is special since $i>0$, may have points common to all sets of the series. It may be, in particular, that, among the β points B there are β_1 points, such that curves of order $n-3$ through these β_1 points (otherwise unrestricted, $f=0$ being now without multiple points), and through the set B_0, do not necessarily pass through the remaining $\beta-\beta_1$ points of B, these latter furnishing exactly $\beta-\beta_1$ conditions for curves of order $n-3$ through B_0 and the set of β_1 points. We have remarked ($f=0$ being non-singular) that curves of order $n-3$ through the set B_0 contain the curve $\omega=0$. Hence, the possibility is that, among the β points, there are β_1 points such that curves of order s through these do not necessarily pass through the remaining $\beta-\beta_1$ points of B. If this is so, the series determined by curves of order $m+k$ through the set $A+A_0$ has $\beta-\beta_1$ points of B common to all sets of the series (p. 87 preceding); and this would be so if curves of order $m+k$, through the point A only, had this character. From the arbitrariness of the set A_0, we shall assume that the converse is true, namely that, under the assumed property of the set B, with regard to curves of order s through them, all curves of order $m+k$, through the set A only, necessarily contain the $\beta-\beta_1$ points of the set B.

For example (Bacharach, *loc. cit.* p. 284) let $n=6$, $m=7$, $k=1$; consider curves of order 8, through a set A of 36 points, common to curves of orders 6 and 7, when these curves are such that the remaining 6 intersections, forming the set B, have 5 points lying in a line. The conics through these 5 points are 3 in number, and do not necessarily contain the sixth point. Hence: Curves of order 8, through the 36 points A, and through two arbitrary points, A_0, of the sextic curve, will determine, on this curve, the linear series defined by the 10 points which consist of B and the 4 remaining intersections, B_0, with the sextic, of the line through A_0. These octavic curves will all pass through the sixth point of the set B (other than the 5 collinear points); we assume that octavic curves, through the set A only, have the same property. More generally, if, among the $\frac{1}{2}(s+1)(s+2)$ points B, there be $s+i+1$ in line ($i>0$), curves of order s through these collinear points consist of this line, and an arbitrary curve of order $s-1$, not necessarily containing the $\frac{1}{2}s(s+1)-i$ points remaining from the set B. Or, if, of the

$$\tfrac{1}{2}(m+n-r-1)(m+n-r-2)$$

excepted intersections of the given curves $f=0$, $\phi=0$, there be $i+m+n-r-2$ which are collinear ($i>0$), then curves of order r through the $mn-\frac{1}{2}(m+n-r-1)(m+n-r-2)$ specified intersections, all pass through the $\frac{1}{2}(m+n-r-2)(m+n-r-3)-i$ noncollinear remaining points of the excepted intersections. These points lie on, at least, i curves* of order $m+n-r-4$.

We also include the following example of the application of the principles developed in this chapter. *Let n, h, ρ be positive integers, such that $\rho<n-3$, $1<h<\frac{1}{2}n(n+3)$, $\frac{1}{2}\rho(\rho-3)<h-1\leqslant\rho(\rho+3)$. Thus the number q, given by $q=h+n\rho-\frac{1}{2}(\rho+1)(\rho+2)$, is $>\frac{1}{2}\rho(\rho+3)$ and $>\rho(n-3)$, while also $n^2-q>\frac{1}{2}n(n+3)-h$; for these involve, respectively, only $h-1+\rho(n-3-\rho)>0$, $h>\frac{1}{2}(\rho-1)(\rho-2)$, and*

$$(n-\rho-1)(n-\rho-2)>0.$$

Now suppose that, of the n^2 intersections of two non-singular curves of order n, there are q points, forming a set Q, which lie on a curve of order ρ, itself not passing through any of the n^2-q other common points. If, from these other n^2-q points, we choose a set B, of $\frac{1}{2}n(n+3)-h$ points, which give independent conditions for curves of order n passing through them, then, all curves of order n through these points B, pass of themselves through the set, R, of

$$\tfrac{1}{2}(n-\rho-1)(n-\rho-2)$$

points which, with B, make up the set of n^2-q points; and these points R do not lie on a curve of order $n-\rho-3$. Also, the remaining q intersections of any two of the ∞^h curves, of order n, which pass through the points $B+R$, lie upon a curve of order ρ.

For, by the theorem just proved, the curves of order $n+\rho$ put through all but a set R of $\frac{1}{2}(n-\rho-1)(n-\rho-2)$ points, among the intersections of two curves of order n, have absolute freedom

---

* The general question is: In what way must $\frac{1}{2}(s+1)(s+2)$ points of a curve of order s be situated, in order to contain among them the maximum number which is possible, of points which do not lie necessarily on curves of order s put through the others? Cf. also Bacharach, *loc. cit.* p. 287.

$i+1+\rho(\rho+3)$, where i is the number of curves of order $n-\rho-3$ which pass through the set R. Suppose that a set, Q, of q, or

$$h+n\rho-\tfrac{1}{2}(\rho+1)(\rho+2)$$

points, among the points through which the curves of order $n+\rho$ are put, lie upon a curve of order ρ, which does not contain any of the other n^2-q intersections of the two curves of order n. Consider, on one of the two curves of order n, say $f=0$, the series of sets of points coresidual with the $q+\tfrac{1}{2}(n-\rho-1)(n-\rho-2)$ points $Q+R$; the number of points in the set $Q+R$ is also $h+\tfrac{1}{2}n(n-3)$, and the freedom of the series in question is thence

$$h+\tfrac{1}{2}n(n-3)-\tfrac{1}{2}(n-1)(n-2)+j,$$

where j is the number of curves of order $n-3$ which pass through the set $Q+R$; in virtue of $q>(n-3)\rho$, remarked above*, such curves of order $n-3$, containing the q points Q, must contain entirely the curve of order ρ; thus, j is the number, say i, of curves of order $n-3-\rho$ which contain the set R. Wherefore, the freedom, on $f=0$, of the series defined by the set $Q+R$ is $h-1+i$; this series is given, however, by curves of order n through the remaining points, B, of intersection of the two curves of order n, whose number is $\tfrac{1}{2}n(n+3)-h$. These points are therefore equivalent only to $\tfrac{1}{2}n(n+3)-h-i$ independent points for curves of order n, such curves through them having an equation of the form

$$\lambda f+\lambda_1 f_1+\ldots+\lambda_h f_h+\mu_1\psi_1+\ldots+\mu_i\psi_i=0.$$

We have, however, supposed these points to be independent for curves of order n; thus $i=0$, there are no curves of order $n-\rho-3$ through the points R, and all curves of order $n+\rho$ through the points B, and through the q points Q, pass through the points R. The freedom of these curves is $1+\rho(\rho+3)$, which we have supposed $\geqslant h$; among such curves will therefore be the composite curves consisting of the curve of order ρ through the points Q, taken with the $h+1$ curves of order n through the $\tfrac{1}{2}n(n+3)-h$ independent points B. Thus, all these curves of order n pass through the points R.

Now, again on $f=0$, consider the freedom of the linear series defined by the set, Q, of q points. We have proved that a curve of order $n-3$ through the points of this set contains, as part, the curve of order ρ on which the set lies; the number of such curves of order $n-3$ is thus $\tfrac{1}{2}(n-\rho-1)(n-\rho-2)$, and the freedom of the series on $f=0$ defined by these points is thus

$$h+n\rho-\tfrac{1}{2}(\rho+1)(\rho+2)-\tfrac{1}{2}(n-1)(n-2)+\tfrac{1}{2}(n-\rho-1)(n-\rho-2),$$

which is $h-1$, the same as the freedom on $f=0$ of the series deter-

* The condition $q>(n-3)\rho-\tfrac{1}{2}(\rho-1)(\rho-2)$ is equivalent only to $h>0$; so that the condition $h>\tfrac{1}{2}(\rho-1)(\rho-2)$ is unnecessary when the points Q are independent for curves of order $n-3$ through the set R.

mined by curves of order n through the points B; by what has been proved, this should be so. But, by the theorem of coresiduation, the series can also be obtained by drawing curves of order ρ through the further intersections, with $f = 0$, of the curve of order ρ which contains the defining set Q. If these residual points give independent conditions for curves of order ρ, the freedom of this system of curves of order ρ is

$$\tfrac{1}{2}\rho(\rho+3) - \{n\rho - [h + n\rho - \tfrac{1}{2}(\rho+1)(\rho+2)]\},$$

which again is $h - 1$. Wherefore, any curve of order n through the set B (and, therefore, through the set $B + R$) meets $f = 0$ in points lying on a curve of order ρ. This proves the result enunciated, $f = 0$ being any one of the ∞^h curves in question.

Ex. 1. If $n - 1$ of the n^2 intersections of two non-singular plane curves of order n lie in line, then all curves of order n through $\tfrac{1}{2}(n-1)(n+4)$ independent points, chosen from the other intersections, pass through the remaining $\tfrac{1}{2}(n-2)(n-3)$ of these $n^2 - n + 1$ points; and these latter do not lie on a curve of order $n - 4$. Further any two curves of order n through the $n^2 - n + 1$ points have their remaining intersections in line. (This is the case when $\rho = 1$, $h = 2$. Cf. Zeuthen, *Lehrbuch d. abzähl. Geom.* 1914, p. 241; White, *Proc. Camb. Phil. Soc.* xxii, 1924, p. 5.)

Ex. 2. Prove, for $r = 1$, and $r = 2$, the following result (Bacharach, *loc. cit.* p. 292; after Olivier, *Crelle*, lxx, 1869, p. 159): If curves $U = 0$, $V = 0$, $W = 0$, of order n, have in common a set, O, of $n^2 - \tfrac{1}{2}(n-r)(n-r+3)$ points, and $U' = 0$ be the curve of order $n - r$ through the set, A, of remaining intersections of $V = 0$, $W = 0$, supposed to give $\tfrac{1}{2}(n-r)(n-r+3)$ independent conditions for $U' = 0$; with a similar statement for $V' = 0$ through the remaining common points B of $W = 0$, $U = 0$, and for $W' = 0$ from $U = 0$, $V = 0$; then prove that the curves $U' = 0$, $V' = 0$, $W' = 0$ have in common a set, O', of $\tfrac{1}{2}(n-r)(n-3r+3)$ points; and, that the remaining intersections, A', of $V' = 0$, $W' = 0$, in number $\tfrac{1}{2}(n-r)(n+r-3)$, lie on the curve $U = 0$, with a similar statement for $W' = 0$ and $U' = 0$, and for $U' = 0$, $V' = 0$. There are thus 8 sets of points, indicated by O (U, V, W), A (U', V, W), ..., A' (U, V', W'), ..., O' (U', V', W'); all intersections of every two of the curves enter in this enumeration, except those forming the three sets $U = U' = 0$, $V = V' = 0$, $W = W' = 0$. The theorem may be considered in connexion with an identity of the form $UU' + VV' + WW' = 0$, which can be set up by remarking that the composite curve $UU' = 0$ contains all the common points of $V = 0$, $W = 0$.

Note IV. On canonical forms for the equation of a manifold, or an aggregate of manifolds. It was proved in Chap. ii that the equation of a general plane cubic curve can be supposed to be of the form $X^3 + Y^3 + Z^3 + 3mXYZ = 0$, where X, Y, Z are linear in the original coordinates; this is in accord with the fact that X, Y, Z together involve 9 coefficients, which, with m, are of the same number, 10, as in a general cubic curve. The properties of the curve are investigated very easily with this equation. Similarly many properties of a cubic surface become clear when it is known that its equation may be expressed by the vanishing of the sum of the cubes of five linear functions of the coordinates (which involve together the same number, 20, of constants as in the general equation). Conversely, however, it is not certain that a specified form of equation which appears to involve the right number of coefficients, is as general as the equation which it is sought to represent; for instance, the sum of 5 fourth powers of linear functions of the three coordinates in a plane involves 15 coefficients, which is the same as the number in the equation of a general plane quartic curve; in fact, however, if the sum of 5 fourth powers be arranged in powers of the coordinates, the fifteen coefficients are not independent functions, and the quartic curve so obtained is not the general plane quartic curve. Or again, in space of three dimensions, 5 linear functions $u_1, \ldots, u_5$ involve 15 *ratios of* their coefficients, and three sums of squares of these, say

$$a_1 u_1{}^2 + \ldots + a_5 u_5{}^2, \quad b_1 u_1{}^2 + \ldots + b_5 u_5{}^2, \quad c_1 u_1{}^2 + \ldots + c_5 u_5{}^2$$

thus involve in all 27 *ratios of* the coefficients involved, which is the same number as enter in the equations of 3 quadric surfaces. But it is not the case that 3 given quadric surfaces can have their equations put into these forms. On the other hand, four given quadric surfaces can have their equations arranged as sums of squares of the same *six* linear functions (involving 38 coefficients, instead of the necessary 36) (cf. Terracini, *Ann. d. Mat.* xxiv, 1915, pp. 1 ff.).

Some reference to the literature dealing with these questions may be added here. Let two homogeneous polynomials of order n in $r+1$ variables $x_0, \ldots, x_r$, say $\phi(x_0, \ldots, x_r), f(x_0, \ldots, x_r)$, be called conjugate when one (and therefore the other) of the two constants $\phi(\xi_0, \xi_1, \ldots, \xi_r) f(x_0, \ldots, x_r), \quad f(\xi_0, \ldots, \xi_r) \phi(x_0, \ldots, x_r)$ vanishes, where ξ_i is the operator $\partial/\partial x_i$. Each form, if general, involves $(n+r, r)$ coefficients. Let F be a specified form of order n in the same variables involving this number of constants, $c_0, c_1, \ldots$. Let $\partial F/\partial c_i$ be denoted by $F_i(x_0, \ldots, x_r)$. Then it can be shewn, by examining the determinantal condition for the $(n+r, r)$ coefficients in F to be independent, that the necessary and sufficient condition that this form F should be possible for f, is that it should be impossible for a form ϕ of order n to exist which is conjugate to all the $(n+r, r)$ forms F_i. In particular when, as is often the case, $r+1$ of the parameters c_i enter in F only in a linear function $y, = c_0 x_0 + \ldots + c_r x_r$, which enters in F, part of the condition expressed is that $\partial F/\partial y$ be conjugate to all the first polars $\partial \phi/\partial x_0, \ldots, \partial \phi/\partial x_r$, of the form ϕ. And a particular case of this arises when F is of the form $y^n + F_1$, where F_1 does not contain $c_0, \ldots, c_r$; then, this part of the condition, exhausting the reference to $c_0, \ldots, c_r$, is that $\phi = 0$ should have a double point at the point whose coordinates are

$(c_0, c_1, \ldots, c_r)$. In particular, if $(r+1)^{-1}(n+r, r)$ be an integer, say h, then f can be expressed as a sum of h nth powers of linear forms provided h points $(c_0, \ldots, c_r)$ exist for which there is no manifold $\phi = 0$, of order n, having these as double points. For example, considering the converse theorem, however 5 points be taken in a plane, a (degenerate) quartic curve exists having these as double points; thus a ternary quartic cannot be expressed as a sum of five fourth powers of linear forms. Again, a cubic primal in space of four dimensions contains 35 terms, and the sum of the cubes of seven linear forms contains the same number; but if seven points be taken, the cubic primal defined by the chords of the rational quartic curve through these points has a double point at all of them. Thus a cubic primal in space [4] cannot be expressed as a sum of seven cubes. Reversely, a binary form of odd order $2k-1$ can be expressed as a sum of k powers, there being no binary form of order $2k-1$ with k double factors. Likewise a plane quintic curve can be expressed as a sum of seven fifth powers; for seven points can be taken (three in a line) which cannot be the double points of a quintic curve; and a cubic surface can be expressed by the sum of five cubics, since a cubic surface cannot have five general nodes. For these considerations see Lasker, *Math. Ann.* LVIII, 1904, p. 434, and Wakeford, *Proc. Lond. Math. Soc.* XVIII, 1920, p. 403. More recently, J. Bronowski (*Proc. Camb. Phil. Soc.* XXIX, 1933, pp. 69, 245) has remarked that the possibility of the expression of a primal of order n by h powers of linear forms is that r primals of order n having nodes at $h-1$ given points should have further free intersections; for instance, two plane curves of order n with $\frac{1}{6}(n+1)(n+2)-1$ nodes (when this is integral) have $\frac{1}{3}(n-2)(n-4)$ further intersections; this vanishes only when $n=2$ or $n=4$; thus a curve of order $3k \pm 1$, unless $k=1$, is expressible by a sum of $\frac{1}{2}[3k^2+3k+1 \pm (2k+1)]$ powers. The question of the expression as a sum of powers of linear forms has been developed from another point of view; for instance, the impossibility of expressing a conic as a sum of two squares is equivalent with the impossibility of drawing a chord to a Veronese surface in space of 5 dimensions (given by coordinates $\xi^2, \eta^2, \zeta^2, \eta\zeta, \zeta\xi, \xi\eta$, where ξ, η, ζ are arbitrary) from an arbitrary point of this space.

Cf. Palatini, *Atti...Torino*, XXXVIII, 1902, p. 43; *Rend. Lincei*, XII, 1903, p. 378; *Atti...Torino*, XLIV, 1909, p. 362; and Terracini, *Rend. Palermo*, XXXI, 1911, p. 392; and Scorza, *Rend. Palermo*, XXV, 1908, p. 193, and XXVII, 1909, p. 148.

Ex. 1. Let ξ, η, ζ, τ and ω be linear functions of the coordinates (x, y, z, t) in ordinary space, $\phi(\xi, \eta, \zeta, \tau)$ a cubic polynomial of any specified form such that the conditions that the first polars of this, regarded as a polynomial in x, y, z, t, should be conjugate to the first polars of a given cubic polynomial $f(x, y, z, t)$, give sixteen independent conditions for the coefficients in ξ, η, ζ, τ; and let $\psi(x, y, z, t)$ be a given quadric polynomial. Then $f(x, y, z, t)$ can be written in the form $\phi(\xi, \eta, \zeta, \tau) + \omega\psi(x, y, z, t)$. For instance $\phi(\xi, \eta, \zeta, \tau)$ may be

$$\eta\zeta\tau + \zeta\xi\tau + \xi\eta\tau + \xi\eta\zeta;$$

but not $\xi^2\eta + \eta^2\zeta + \zeta^2\tau$.

Ex. 2. A general plane-curve of order n can have its equation expressed by the vanishing of a symmetrical determinant of n rows and columns, in which every element is linear in the coordinates. If ξ, η, ζ be linear in the coordinates, a conic and a plane cubic curve are respectively given by the vanishing of the discriminants of the two quadric forms, in u, v, and in u, v, w, $\xi u^2 + 2\zeta uv + \eta v^2$, $m(\xi u^2 + \eta v^2 + \zeta w^2) + 2(\xi vw + \eta wu + \zeta uv)$; and the result for a plane quartic was obtained by Hesse, *Crelle*, XLIX, 1855, p. 243.

A direct proof is suggested by A. C. Dixon, *Proc. Camb. Phil. Soc.* XI, 1902, pp. 350, 351; and XII, 1904, pp. 449–53. Let $v_{11} = 0$ be one of the ∞^{n-1} curves of order $n-1$ which touch the given curve, $f = 0$, in $\frac{1}{2}n(n-1)$ points; denote the most general curve of order $n-1$ which passes through the points of contact by $\lambda_1 v_{11} + \lambda_2 v_{12} + \ldots + \lambda_n v_{1n} = 0$. There exist then identities of the form $v_{1r}v_{1s} = v_{11}v_{rs} + fw_{rs}$, from which the polynomials v_{rs} $(r, s = 2, \ldots, n)$ may be determined. It can then be proved that the minor of v_{rs} in the determinant of n rows and columns $|v_{rs}|$ is of the form $f^{n-2}\beta_{rs}$, wherein β_{rs} will be linear in the coordinates; and then that the given curve $f = 0$ is given by $|\beta_{rs}| = 0$.

Ex. 3. Using v_i, or w_i, for a homogeneous polynomial of order i in x and z, shew that a ternary quartic in x, y, z, is capable of the form $f = (y + v_1)\phi + v_2{}^2$, where ϕ is a cubic. Hence shew that the quartic surface $z^2 t^2 = f$ is birationally equivalent to a cubic surface. Shew also that the sextic curve discussed on p. 95 above is capable of the form $\psi = (yz + w_2)(y^2 z^2 + yzv_2 + v_4) + (yzv_1 + v_3)^2$; thence prove that the sextic surface $z^4 t^2 = \psi$ is birationally reducible to the quartic surface $z^2 t^2 = f$ cited.

CHAPTER V

THE PERIODS OF ALGEBRAIC INTEGRALS.
LOOPS IN A PLANE. RIEMANN SURFACES

As has been indicated, at the present stage of the theory of algebraic geometry, many results find their clearest statement in connexion with the theory of algebraic integrals; just as the first exhaustive investigation of the genus of an algebraic curve was in fact by Abel, for the theory of algebraic integrals (*Oeuvres*, 1881, p. 145). And, notwithstanding the initial feeling which may arise, of the incongruity of the two theories, the geometrical reader will find it desirable to have an understanding of the main results of the latter theory. We therefore give now an account, with the objects, (i) of shewing how to determine the number of periods of an algebraic integral; (ii) of explaining the conception and use of the so-called Riemann surface; (iii) of making connexion with an arithmetic theory of plane curves. All these are helpful towards an extension of the theory of curves to the theory of surfaces; in particular, the geometrical aspect of the theory of Riemann surfaces is intimately related with a wide theory, Analysis Situs, or Topology, which promises to give a descriptive alternative to much involved computation. It must be understood that what we give is, in the space we allot to it, necessarily very incomplete. It will be sufficient if it indicate clearly the nature and bearings of the ideas involved; detailed developments will be found in many other places.

Meaning and number of periods. We suppose the independent complex variable x to be represented, in the known way, upon a Euclidean plane, regarded as closed at infinity by the single point $x = \infty$, like a sphere. Then, taking three simple examples, it is a familiar fact that the integrals

$$\text{(i)}, \ \int_c^x \frac{dx}{x}; \quad \text{(ii)}, \ \int_0^x \frac{dx}{1+x^2}; \quad \text{(iii)}, \ \int_0^x \frac{dx}{(1-x^2)^{\frac{1}{2}}}, \quad (c \neq 0),$$

are not determined by the assignment of their limits of integration. Each is indeterminate by the addition of a certain constant, and therefore by any integer multiple of this constant. Denoting any value of either integral by I, and any integer by k, the general values of the integrals are, respectively, $I + 2\pi i k$, $I + k\pi$, $I + 2\pi k$. The respective additive constants, $2\pi i$, π, 2π, are obtainable, in the several cases, by a circuit of the upper limit (x), respectively round $x = 0$, round one of the points $x = \pm i$, and round both the points $x = \pm 1$. There is an obvious difference between the case (iii), and the cases (i), (ii), in that the integral (iii) does not become infinite as x approaches to $+1$ or -1, while the integral (i) becomes infinite when x approaches to $x=0$, and (ii) when x approaches to $+i$ or $-i$; another difference is that the subject of integration in (iii) is ambiguous in sign. Now consider an integral

$$\text{(iv)}, \quad \int_0^x \frac{dx}{[(1-x^2)(1-e^2x^2)]^{\frac{1}{2}}},$$

where e is a constant, which we may take to be real and positive and < 1; this is a generalisation of (iii), but, unlike (iii), does not become infinite for $x = \infty$. Let (x, y) denote any point of the curve whose equation is $y^2 = (1-x^2)(1-e^2x^2)$; then we may understand (iv) to denote the integral

$$\int_{(0,\,1)}^{(x,\,y)} \frac{dx}{y},$$

taken by a path of pairs of associated complex values of x and y, all representing points of the quartic curve, from the point $x=0$, $y=1$, to the general point given by the upper limit. Two such paths of integration, extended over a continuum of possibly complex points belonging to the curve, from the same initial point to the same final point, may not lead to the same value of the integral. But the difference between the values obtained by two such paths is evidently equal to what is found by taking the integral over a certain closed path; that, namely, which begins from the lower limit, passes by one of these paths to the upper limit, and then returns to the lower limit by a path which is the reverse of the second of the two paths referred to. The determination of the complete aggregate of values of which the integral is capable, for given positions of the lower and upper limits, thus requires an examination of the values obtainable by all possible closed paths of integration, of which every point (possibly complex) belongs to the curve.

More generally, let $f(x, y) = 0$ represent any curve; let $R(x, y)$ be any rational function of x and y, supposed to be associated so as to satisfy the equation of the curve; let (a), (x) denote a fixed and a variable point of the curve, respectively. We may then similarly consider the integral $\int_{(a)}^{(x)} R(x, y)\, dx$, of which the path of integration is by points of the curve. If this integral become logarithmically infinite for some positions of the upper limit (x), a closed path about such a point leads to a value for the integral which does not vanish, as in the simple cases (i), (ii), considered at first, wherein the subject of integration was a single-valued function of x only. Suppose that $R(x, y)$ is such that there are no such points of logarithmic infinity; there may still be closed paths, as in the case (iv) considered above, for which the value of the integral is not zero, arising from the circumstance that $R(x, y)$ is not a single-valued function of x only. We prove in fact that, if p be the genus of the curve $f(x, y) = 0$, the general value of the integral, when there are no logarithmic infinities, is of the form $I + m_1\Omega_1 + \ldots + m_{2p}\Omega_{2p}$, where I is a particular value, $\Omega_1, \ldots, \Omega_{2p}$ are certain constants, *the periods*, depending upon $f(x, y)$ and $R(x, y)$, and $m_1, \ldots, m_{2p}$ are integers; by proper choice of the path these integers may be chosen at will. When the function $R(x, y)$ is such that the integral has logarithmic infinities, its general value will involve also a sum of integral multiples of constants obtained by circuit of these infinities.

Of this fundamental theorem we consider two methods of proof, the former of a natural and elementary kind, by means of the theory of *loops*, the latter with the help of a so-called *Riemann surface*, constructed to represent the (complex and real) points of the curve $f(x, y) = 0$. Although the latter is finally of much greater importance, it may at first seem artificial; and our exposition, making appeal, for the sake of brevity, to intuitive notions of topology, is not final.

The method of loops in a plane. For the first method, we suppose (what is shewn in Chap. II not to be an essential limitation) that the curve represented by the equation $f(x, y) = 0$ has no multiple points other than double points at which the tangents are distinct; further, the equation being of order m in x and y, we suppose that the terms of order m have m distinct factors, and that the tangents of the curve having equations of the form $x = $ constant, all touch the curve at simple points, and each only once. The values of x for which two values of y are equal are then to be found from $f(x, y) = 0$, $\partial f/\partial y = 0$. Among these are the values for the double points, say of number δ; the other solutions correspond to points of contact of tangents of the curve with equation of the form $x = $ constant. The number of these latter, say w, is given by $w = m(m-1) - 2\delta$, which is $w = 2m + 2p - 2$, since, as we have seen (Chap. IV, p. 77), $p = \frac{1}{2}(m-1)(m-2) - \delta$. Regarding the complex value x as represented upon a plane, for a general value $x = a$ there will be m values of y, say $b_1, \ldots, b_m$; and the values of y, satisfying the equation $f(x, y) = 0$, which are in the neighbourhood of the value $y = b_i$, will be given, when x is sufficiently near to a, by an expression of $y - b_i$ as a power series in $x - a$. By continuation of the power series the complete range of values of y satisfying the equation, as x varies, can be computed from the equation itself, unless we come to a value of x for which two of the m values of y are equal. If this be a double point, (x_0, y_0), at which, then, by what has been supposed, neither tangent of the curve has an equation of the form $x = $ constant, there will be two expansions, corresponding to the two branches of the curve, both of $y - y_0$ in powers of $x - x_0$; the values of y to be associated with x, as this passes through $x = x_0$, are thus determined by the values with which the path approaches this point. But two values of y become equal also at a point $x = x_0$, $y = y_0$, when there is a tangent of the curve given by $x = x_0$; the two values of y for a value of x in the neighbourhood of x_0, are then given by two power series expressing $y - y_0$ in powers of $(x - x_0)^{\frac{1}{2}}$, wherein the coefficients of even powers of $(x - x_0)^{\frac{1}{2}}$ are the same, but the coefficients of odd powers of $(x - x_0)^{\frac{1}{2}}$ differ in sign. Either series is then obtainable from the other by a circuit of the complex variable x round the point x_0 of the representing plane. It is the possibility of this which explains, when we are considering an algebraic integral $\int R(x, y)\, dx$, the existence of closed circuits of integration which give non-vanishing values of the integral, other than those arising from points of logarithmic infinity of the integral. In what follows immediately, we exclude the consideration of points of logarithmic infinity of the integral; and we assume that a closed path of integration for x gives a zero value for the integral unless the path surrounds points, by circuit of which a value of y is changed into another value of y (of the kind considered above).

To illustrate the ideas consider first the integral $\int_{(a)}^{(x)} x^{i-1}\, dx / y$, wherein $y^2 = (x - c_1) \dots (x - c_{2p+2})$, the roots $c_1, \dots, c_{2p+2}$ being all different, and the value of x at the lower limit (a) not being one of these, while i is one of $1, 2, \dots, p$; thus $(p \geqslant 1)$ the integral does not become infinite for $x = \infty$. The relation between x and y does not satisfy all the limitations which we have supposed above to hold in general; but this is immaterial for our purpose. A simple closed circuit by x round any one of the so-called *branch points* given by $x = c_1, \dots, x = c_{2p+2}$, starting with a pair of values x_0, y_0 for x, y, evidently leads to the values $x_0, -y_0$. Suppose the value $x = a$ to be arbitrary; consider closed paths of integration for a, each of which begins at $x = a$, with one of the two values of y corresponding thereto from the equation of the curve, either $y = b$ or $y = -b$; then proceeds to the neighbourhood of one of the $2p + 2$ branch points; makes a circuit about this, which we think of as lying in the immediate neighbourhood of the branch point, but described in either direction of circulation about this point; and then comes back to $x = a$ along the path of approach. Such a path we call a *loop*. It is supposed that to the value of x belonging to any point of the path there is associated the value of y arising continuously, from the initial value, by means of the equation of the curve. The $2p + 2$ possible loops will be considered in a definite order; and it is supposed that no two of them have a point of intersection other than the initial point $x = a$. If we describe first the loop (c_1), and then the loop (c_2), the result of the first is to lead from the values (a, b), for x and y, to the values $(a, -b)$; and the result of the second, beginning with $(a, -b)$, is to lead back to (a, b). The value obtained for the integral $\int x^{i-1}\,dx/y$ round this pair of loops, by what we have assumed, is independent of the position of $x = a$, and the same as would be obtained by any closed path passing once round the points (c_1), (c_2) but including no other branch points, with a proper initial value for y. The contribution to the value of the integral which arises by the part of a single loop which is in the immediate neighbourhood of a branch point is zero; the effect of this small part is to change the sign of y. The value obtained for the integral by a closed path passing once round the two branch points is not zero, being, as we easily see, the same save for sign, as twice the integral taken by a simple path from $(c_1, 0)$ to $(c_2, 0)$. Now consider any closed circuit which, beginning at $x = a$ with the value $y = b$, comes back to $x = a$ *with the same value $y = b$*. Such a circuit, in virtue of the principle we have assumed, must give the same value for the integral as a composite circuit, of the kind we may denote by $m_1(c_1) + \ldots + m_{2p+2}(c_{2p+2})$, wherein (c_i) denotes the circuit of the loop defined by c_i, and $m_1, \ldots, m_{2p+2}$ are integers; but, in order that, as we have postulated, the final value of y in the circuit may agree with the initial value, the sum of these integers must be even (including zero). It can then be easily seen, if (c_1, c_2) denote such a simple composite circuit about the branch points c_1, c_2 as we have described, in a specified order, beginning and ending with the values a, b for x and y, that the general composite circuit spoken of is equivalent to a circuit which may be expressed in the form $\mu_1(c_1, c_2) + \mu_2(c_2, c_3) + \ldots + \mu_{2p+1}(c_{2p+1}, c_{2p+2})$, wherein $\mu_1, \ldots, \mu_{2p+1}$ are also integers (not necessarily of zero or even sum). In particular, a circuit of this form is that in which every one of $\mu_1, \ldots, \mu_{2p+1}$ is 1; this circuit is evidently equivalent, so far as the value which it gives for the integral is concerned, to a single circuit round $x = \infty$, which would give zero for the value of the integral (there being no branch point at $x = \infty$). Thus, so far as the value of the integral is concerned, any composite circuit which begins and ends with the same pair of associated values for x, y, may be supposed to have the form $\nu_1(c_1, c_2) + \nu_2(c_2, c_3) + \ldots + \nu_{2p}(c_{2p}, c_{2p+1})$, where $\nu_i = \mu_i - \mu_{2p+1}$; and this involves only $2p$ circuits (c_r, c_{r+1}). We thus reach the conclusion that the general value of the integral $\int_{(a)}^{(x)} x^{i-1}\,dx/y$ is of the form $I + \nu_1\Omega_1 + \ldots + \nu_{2p}\Omega_{2p}$, where I is the value of the integral for a particular path arbitrarily chosen, $\Omega_1, \ldots, \Omega_{2p}$ are definite constants, and $\nu_1, \ldots, \nu_{2p}$ are integers. Conversely, by proper choice of path from (a) to (x), we can suppose $\nu_1, \ldots, \nu_{2p}$ to have any assigned integer values. The number of periods of an algebraic integral was known to Galois, who refers to Jacobi and

Abel; see his letter to Auguste Chevalier, 29 May 1832 (*Œuvres mathématiques d'Évariste Galois...*, with an introduction by M. Émile Picard, Paris, 1897, pp. ix, 30).

Consider now more generally the integral $\int R(x, y) dx$, supposed to have no logarithmic infinities, the fundamental equation $f(x, y) = 0$ being simplified as above, so that there are $2m + 2p - 2$ simple branch points, all arising for finite values of x. Every one of these has the property that, when x makes a circuit about it in its immediate neighbourhood, a particular pair of the m roots y of the equation $f(x, y) = 0$ are changed either into the other, the other $m - 2$ roots being unaffected. Suppose that these branch points are considered in a definite order, and a loop, as explained, put about each, all these loops having the same arbitrary initial point, $x = a$, and no two intersecting except at $x = a$. Denote the values of y satisfying the equation $f(a, y) = 0$, supposed different, by $y_1, ..., y_m$, in any definite order. Every one of the loops must then be associable with a symbol of two numbers chosen from $1, 2, ..., m$, say (α, β), or (β, α); the meaning being that, if we start from $x = a$ with one of the two values y_α, y_β for y, and follow the value of y, with the help of the equation $f(x, y) = 0$, round the loop, we shall arrive back at $x = a$ with the other of these two values; the direction in which the loop is described is in fact indifferent. If we follow the same loop with an initial value of y other than y_α or y_β, the final value of y will be the same as the initial value. There may be several branch points associated with the same symbol (α, β); but the whole set of numbers $1, 2, ..., m$ must occur if we consider the binary symbols for all the loops; and, more generally, it must be possible, starting at $x = a$ with any assigned value of y, say y_i, to make a selection from the loops, so that the circuit of these selected loops, in a proper order, will lead, from the initial value $y = y_i$, to any other assigned value, $y = y_j$, as final value. For, otherwise, there will be a set of values of y, chosen from $y_1, ..., y_m$, but not including all these, which are interchanged among themselves by every possible combination of the loops; and this will involve, it is easy to see, that the polynomial in y, $f(x, y)$, can be written as the product of two, or more, polynomials in y all with coefficients rational in x. We assume that this is not so, the curve $f(x, y) = 0$ being irreducible.

Not every one of the possible $\frac{1}{2}m(m-1)$ binary symbols, obtainable by taking two numbers from $1, 2, ..., m$, will occur, in general, among the binary symbols associated with the loops as originally drawn. But, by suitable combinations of the loops as drawn, we can obtain interchanges among $y_1, ..., y_m$ other than those arising for these single loops; and it will be convenient to denote such a combination of single loops which, for instance, interchanges y_β and y_γ, by (β, γ), or (γ, β). Consider in particular the combination

$(\alpha, \gamma)(\alpha, \beta)(\alpha, \gamma)$, where (α, β), (α, γ) are among the original single loops. By the loop (α, γ), the roots $(y_\alpha, y_\beta, y_\gamma)$ are changed, respectively, to $(y_\gamma, y_\beta, y_\alpha)$; by the loop (α, β) these are then changed, respectively, to $(y_\gamma, y_\alpha, y_\beta)$; and then, finally, these are changed, respec- 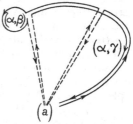 tively, by the loop (α, γ), to $(y_\alpha, y_\gamma, y_\beta)$; the composite loop thus interchanges y_β and y_γ, but leaves y_α unaltered; it may then be denoted by (β, γ), or (γ, β). As indicated by the diagram, the composite loop, deformed so as to include self de- structive portions (indicated by the dotted lines), may be regarded as a loop to enclose the branch point (α, β) which approaches this by a detour about the branch point (α, γ). If we interchange β and γ we may prove in the same way that $(\alpha, \beta)(\alpha, \gamma)(\alpha, \beta)=(\beta, \gamma)$; and the two results (since the repetition of a loop gives no change in the initial value of y) may both be expressed by

$$(\alpha, \beta)(\alpha, \gamma)=(\alpha, \gamma)(\beta, \gamma)=(\beta, \gamma)(\alpha, \beta);$$

or, in words, by saying that the order of two succeeding symbols $(\alpha, \beta)(\alpha, \gamma)$, which have a common number, α, may be altered, by carrying either symbol, unchanged, over the other; if the symbol, over which this transference is made, be modified, by changing the number in it which is common to the two symbols, into the other number which enters in the transferred symbol; thus in the first equation which is written, (α, γ) is carried over (α, β), being put to the left, with the simultaneous change of (α, β) to (γ, β). In applying this, we may for clearness refer to it as the *rule of trans- ference*. It is clear, on the other hand, that a succession of two symbols (α, β), (γ, δ), which have no number in common, has the same effect as the same two symbols in the other order (γ, δ), (α, β), the values of y interchanged by one symbol not being affected by the other; we may suppose this remark included in the rule.

Now write down the whole aggregate of the original w binary symbols, one for each of the w branch points, in order, as they arise from the loops as originally drawn, the symbols written to the right referring to operations carried out first. We desire to shew that, by the rule of transference, the aggregate of operations given thereby can be so modified as to be expressible by the aggregate

$$(m-1, m)(m-1, m)\ldots.(3, 4)(3, 4)(2, 3)(2, 3)(1, 2)\ldots(1, 2),$$

where there are; first, $2p+2$ symbols $(1, 2)$; then, two symbols $(2, 3)$; then, two symbols $(3, 4)$, and so on; and, finally, two symbols $(m-1, m)$; in all $2p+2+2(m-2)$ symbols. For this, we use the principles: (a), the effect of the whole sequence of w loops, upon any one of the values of y, must be to change this into itself, the whole sequence being clearly equivalent to a circuit about $x = \infty$, at which, by hypothesis, there is no branch point; (*b*), by a suitable selection from the loops it must be possible to change any root y_i to any other root y_j, as already remarked. And we use certain rules, which are consequences of the rule of transference; these, for clearness, may be stated: (i), a pair of consecutive symbols (representative of different loops but) containing the same two numbers, may be simultaneously carried over another symbol immediately consecutive to, or preceding, this pair, without change of the numbers of this symbol; for example $(\alpha, \gamma)(\alpha, \beta)(\alpha, \beta)$ may be replaced by $(\alpha, \beta)(\alpha, \beta)(\alpha, \gamma)$, transferring (α, β) over (α, γ), and then (α, β) over the changed value of (α, γ), which is (β, γ); (ii), in a set of three consecutive symbols such as those just considered, where two consecutive symbols involving the same numbers are preceded (or followed) by a symbol containing a number which occurs in the two like symbols, this number may be replaced, in both the like symbols, by the other number of the single (preceding, or following) symbol, without change in the order of the three symbols; for instance $(\alpha, \gamma)(\alpha, \beta)(\alpha, \beta)$ is the same as $(\alpha, \gamma)(\beta, \gamma)(\beta, \gamma)$; for the former has been seen to be the same as $(\alpha, \beta)(\alpha, \beta)(\alpha, \gamma)$, and in this (α, γ) may be carried over both the symbols (α, β) in turn; we may call this the *rule of attraction*; (iii), in a sequence of three pairs of symbols, such as $(\gamma, \delta)(\gamma, \delta) (\beta, \gamma) (\beta, \gamma) (\alpha, \beta) (\alpha, \beta)$, in which the first pair of symbols involve the same numbers α, β, the second pair involve the same numbers β, γ, of which one (β) occurs in the symbols of the first pair, and the third pair also involve the same numbers γ, δ, of which one (γ) has occurred in the second but not in the first pair, we may replace the numbers α, β, γ, δ by α, γ, β, δ, namely the composite symbol is the same as

$$(\beta, \delta)(\beta, \delta)(\beta, \gamma)(\beta, \gamma)(\alpha, \gamma)(\alpha, \gamma).$$

This may be proved by two applications of the rule of attraction; it may be called the rule of *permutation of numbers*.

With these rules, suppose, in the sequence of w binary symbols to be considered, that the first symbol, that written on the right, is (α, β). Regarding the sequence of symbols from right to left, we may come to other symbols (α, β), and these may all be carried over intervening symbols, in turn, and placed to follow the first symbol, at its left. We thus have a batch of symbols (α, β) at the right. We may suppose the number of these symbols is *even*. For suppose it is odd; then the effect, upon y_α, of this batch of symbols, is to change it into y_β; but, as, by principle (*a*), the complete sequence of the w symbols must change y_α into itself, as we survey the whole sequence to the left of the batch of symbols (α, β), we must come to a symbol involving the number β (and not (α, β)), say the symbol (β, γ); let this be carried over, so as to follow immediately to the left of the batch of symbols (α, β); so far, then, y_α would be changed to y_γ; thus, as we survey the whole sequence further, from right to left, we must find a symbol involving γ, say (γ, δ); let this be carried over any intervening symbols to follow immediately on (β, γ). Proceeding thus, the sequence must be replaceable by one of the form $\ldots(\alpha, \lambda)(\lambda, \kappa)\ldots(\delta, \gamma)(\gamma, \beta)(\beta, \alpha)\ldots(\beta, \alpha)$, with other of the w symbols to the left of (α, λ). But now we can carry the batch $(\lambda, \kappa)\ldots$ (γ, β) over the symbol (α, λ), until this follows the last (α, β), where it will appear in the form (α, β). We then have, for the whole sequence, a form $\ldots(\lambda, \kappa)\ldots(\delta, \gamma)(\gamma, \beta)(\alpha, \beta)(\beta, \alpha)\ldots(\beta, \alpha)$, with an *even* number of symbols (α, β) on the right. Arguing in the manner thus indicated, we can finally reach the conclusion that the whole sequence of w symbols can be replaced by $[\rho, \sigma][\mu, \rho]\ldots[\beta,\gamma][\alpha,\beta]$, wherein each such parenthesis as $[\alpha, \beta]$ means a sequence of an even number of symbols (α, β), and one number of this parenthesis occurs in the following parenthesis, until we come to the penultimate parenthesis $[\mu, \rho]$, of which the number ρ occurs in the last parenthesis $[\rho, \sigma]$.

But herein, by repeated use of the rule of attraction (ii), we can modify the parentheses until each of them, except the first (written at the right) contains only two binary symbols. For instance, $(\rho, \sigma)(\rho, \sigma)(\rho, \sigma)(\mu, \rho)$ is first, by the application of this rule to the three symbols $(\rho, \sigma)(\rho, \sigma)(\mu, \rho)$, the same as $(\rho, \sigma)(\mu, \sigma)(\mu, \sigma)(\mu, \rho)$, and then, by the application of this rule to the three symbols $(\rho, \sigma)(\mu, \sigma)(\mu, \sigma)$, the same as $(\rho, \sigma)(\mu, \rho)(\mu, \rho)(\mu, \rho)$; whereby, if $[\rho, \sigma]$ contains more than two symbols, two of them are transformed to symbols (μ, ρ), which may be included in the parenthesis $[\mu, \rho]$. Thereby, the sequence of symbols is reduced to consist, first, of a parenthesis of $2p+2$ symbols all of the same form, followed by $m-2$ parentheses each of two like symbols, each parenthesis (except the last) being described by two numbers of which one occurs in the preceding. The numbers occurring must, by principle (b), consist of the numbers $1, 2, \ldots, m$, in some order. We can then use the rule of permutation of numbers, (iii), or, more simply, can rename the roots y_i, so as finally to obtain the sequence stated $[m-1, m]\ldots[3, 4][2, 3][1, 2]$.

For this theory, and more detail, see Severi, *Algebraische Geometrie*, Leipzig, 1921, pp. 205–210; Lüroth, *Math. Ann.* IV, 1871, p. 181; Clebsch, *Math. Ann.* VI, 1873, p. 216; Bertini, *Rend. Lincei*, III, 1894, p. 106; Clebsch u. Gordan, *Abelsche Functnen*, Leipzig, 1866, pp. 80–90; Weierstrass, *Werke*, IV, 1902, p. 329.

Now consider the application of this sequence of loops to the proof of the number of periods of an algebraic integral. First take the case when the sequence consists only of the $2p+2$ symbols

$(12) (12) \dots (12) (12)$. A form of the equation $f(x, y) = 0$ for which this is the case is $y^2 = (x, 1)_{2p+2}$, where $(x, 1)_{2p+2}$, a polynomial of $2p + 2$ different linear factors, can be denoted by $u_0 u_1 \dots u_p$, where $u_i = (x - a_i)(x - c_i)$. We have already remarked that every circuit about the branch points which leads to the same value of y as that with which we start out, may be regarded as equivalent to an aggregate, with integer coefficients, of $2p$ circuits, each definable by a pair of loops, which we may denote by $\alpha_1 = (a_1, c_1)$, $\beta_1 = (c_1, a_2)$, ..., $\alpha_i = (a_i, c_i)$, $\beta_i = (c_i, a_{i+1})$, ..., $\alpha_p = (a_p, c_p)$, $\beta_p = (c_p, c)$. The circuit is thus also equivalent to an aggregate, with integer coefficients, of the following $2p$ aggregates, built up from the preceding pairs of loops,

$$\alpha_1, B_1 = \beta_1 \beta_2 \dots \beta_p; \quad \dots\dots; \quad \alpha_i, B_i = \beta_i \dots \beta_p; \quad \dots\dots;$$
$$\alpha_{p-1}, B_{p-1} = \beta_{p-1} \beta_p; \quad \alpha_p, B_p = \beta_p;$$

these $2p$ circuits are those represented in the diagram.

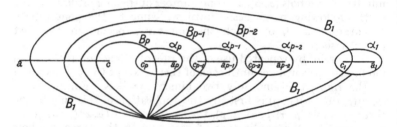

The periods of any integral $\int R(x, y)\,dx$, which is without logarithmic infinities, may be taken to be the values obtained by integration round these circuits.

Next, take the case when there are three values of y, and the loops enclosing the branch places have been reduced to the $2p + 4$ represented by $(23)(23)(12) \dots (12)$. Then an integral, whose closed path of integration starts with the value y_3 for y, and finishes with the same, can only give a value which is not zero if the path enclose both the branch points defining the two loops (23); but even such a path is ineffective, since it can clearly be deformed, over the two branch points (12), until it is a circuit round $x = \infty$, which is not a branch point. An integral path, beginning with y_2, which encloses both the branch points defining the two loops (23), gives in fact the same value as an integral path beginning with y_3 which encloses both these points, described in the opposite direction; this is clear if we replace the closed path by two open paths, passing from one of the two branch places to the other, remembering that the contribution to the integral due to the portion of a loop immediately near a branch place is zero. Thus a closed integration with the initial value y_2, about the two branch points defining the two loops

(23), can be deformed to be round $x = \infty$, and gives rise to no period. For paths which give non-vanishing periods, there remain then only paths beginning with the values y_1 or y_2, enclosing the whole or part of the $2p + 2$ branch places (12). Of these, a path enclosing all the $2p + 2$ points can effectively, we have seen, be deformed, over the branch points (23), to a path round $x = \infty$. We are thus finally reduced to a path such as was considered when only the branch points (12) were present; so that the number of periods is still $2p$.

Consider next the case of $2p + 6$ branch places, given by the scheme (34) (34) (23) (23) (12) ... (12). As before, a path enclosing the two branch points (34), which begins with either of the values y_4 or y_3, leads to no period; and effectively such a path needs consideration only when the initial value is y_4. A path, with initial value y_3, which encloses the four branch points (34) (34) (23) (23) may be deformed to be about $x = \infty$, and is also ineffective; a path, with initial value y_3, enclosing only (23) (23) is thus reducible to one enclosing only (34) (34), and is equally ineffective. Similar remarks hold, for circuit of the branch points (34) (34) (23) (23), when the initial value is y_2. We are thus reduced as before to $2p$ circuits about the branch points (12).

And the general case can be treated in the same way, beginning always with the branch points denoted by the symbols at the left of the scheme. The general conclusion is that there are $2p$ periods, obtainable by circuits of the branch points (12), with y_1 or y_2 as initial values.

Riemann surfaces. We pass now to a method of representing the corresponding values of x and y which satisfy the (irreducible) equation $f(x, y) = 0$, which adds greatly to clearness of thought. It is in intimate connexion with what has been said, but differs from it by employing, not one plane for the values of x only, but superposed planes with the points of which are associated the values of y as well as of x.

Consider first, as a simple example, the function $y = x^{\frac{1}{2}}$, or the equation $y^2 = x$, in particular for values of x near to $x = 0$. The values of x being represented upon a plane, there are two values of y for any small value of x; but a circuit by x about $x = 0$ changes one of these continuously into the other. Now suppose, in order to convey the idea, we have two equal circular pieces of paper, one laid over the other, both with their centres at $x = 0$, say at the point O; let them both be cut, from the centre O, along overlying radii, so that the radius, OA, of the upper sheet, is replaced by two edges of the cut, say OA and OA', while the underlying radius, OB, of the lower sheet, is replaced by the edges OB, OB' of the cut in the lower sheet, OB lying beneath OA, and OB' beneath OA'; next, let the edge OA, of the upper sheet, be joined to the edge OB', of the lower sheet, say by sticking paper; and suppose that it is possible, simultaneously, to join the edge OB, which lies underneath OA, to the edge OA' of the upper sheet, by sticking paper passing through the other sticking paper without any connexion therewith (except at O). Then a circuit of the whole circumference of the upper sheet, beginning at A' and ending at A, may be followed without break of continuity by a circuit of the whole circumference of the lower

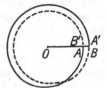

sheet, beginning at B' and ending at B, from which we can pass without break to A'. Two sheets, so supposed to cross, without connexion (save at one point O), evidently furnish a representation of the values of $x^{\frac{1}{2}}$ about $x=0$, there being a complete (1, 1) correspondence between the points of the two sheets, and the values assumed by y in the neighbourhood of $x=0$.

Consider next, for example, the function $y=[x(x-1)]^{\frac{1}{2}}$, or the equation $y^2=x(x-1)$, there being now *two* branch points, one at $x=0$, the other at $x=1$; the factor $(x-1)^{\frac{1}{2}}$ is single valued for values of x in the immediate neighbourhood of $x=0$, and the factor $x^{\frac{1}{2}}$ is single valued for values of x in the immediate neighbourhood of $x=1$. This time, imagine two pieces of paper, one laid over the other, to cover a portion of the plane of x which contains $x=0$ and $x=1$; let a slit be cut in the upper sheet from $x=0$ to $x=1$, and an underlying slit be cut in the lower sheet between the same two points; then, let one edge of the slit in the upper sheet be supposed joined to the edge, of the lower sheet, which lies underneath the other edge of the slit in the upper sheet; and, likewise, the other edge, of the slit in the upper sheet, be joined to the still free edge of the slit in the lower sheet, with the supposition, however, that the two joins can maintain their several identities though interpenetrating. Two such sheets will obviously serve to represent the values of $y=x^{\frac{1}{2}}(x-1)^{\frac{1}{2}}$ in the portion of the plane of x above which the sheets are laid. In this example, we may suppose the two sheets to be each continued to infinity without further intersection, each being closed as if laid over the surface of a sphere. Then we shall have a representation, for all finite and infinite values of x, of the values of y satisfying the equation $y^2=x(x-1)$, by a two-sheeted surface, having two branch points, between which is a *cross line*. In the previous example, $y^2=x$, such a representation is also obtained by supposing a branch point at $x=\infty$, in addition to that at $x=0$, there being a cross line, at which the two sheets cross, extending from $x=0$ to $x=\infty$.

The preceding ideas have a simple application to the representation of the values of y given by the equation $y^2=u_0u_1\dots u_p$, where $u_i=(x-a_i)(x-c_i)$, all the $2p+2$ values a_i, c_i being different. For this, it is evidently only necessary to think of a two-sheeted surface, with a cross line between the two branch points a, c, a cross line between the two branch points a_1, c_1, and, in general, a cross line between a_i and c_i; to every pair of values (x, y), satisfying the equation, corresponds a point of this surface, and conversely, the pairs (x, y), $(x, -y)$ being represented by the points in the two sheets which lie over the point representing x in the plane below. Instead, then, of a closed curve of integration on the plane of x, which so passes round the critical values of x as to finish with the same value of y as the initial value, we shall consider, on the two-sheeted surface, a closed curve which finishes, *in the same sheet*, at its initial point. A period of an integral $\int R(x, y) dx$, which has no logarithmic infinities, will be the result of integration along such a closed curve, unless this gives a zero value. This value will be zero for a small closed curve, lying wholly in one sheet of the surface, in the immediate neighbourhood of a point (as follows from Cauchy's theorem for a single-valued function of a complex variable); it will also be zero for a curve enclosing a region of the double surface which is capable of being divided into *cells* bounded by such small closed curves. Thus the value of the integral, taken round any closed curve on the double surface, may be the sum of the values obtained by integration round several other closed curves. The problem then immediately arises, of finding, if possible, a funda-mental set of closed curves on the double surface, such that the integral, taken round any closed curve whatever, is expressible in terms of the values obtained by integration round the fundamental curves. If this can be done, the values of the integral taken round these fundamental curves will be the periods of the integral. The introduction of the multiple-sheeted surface, though simplifying very much the representation of the values of y belonging to a given x, thus raises a new problem, of the topology, or connectivity, of this surface, to be solved independently of the particular integral under consideration. And it is convenient to regard this problem as different from that of the number of independent loops in the plane x, which we have dealt with, though this may furnish sug-gestions. Consider for example, first, the simple case of the two-sheeted surface associated with the equation

$$y^2 = (x-a)(x-c)(x-a_1)(x-c_1).$$

Let an oval cut be made in the upper sheet, enclosing the branch points c_1, a_1; and let another oval cut be made, from a point on one side of the first oval cut, to the point immediately opposite on the other side, this second cut passing over both the cross lines, so as to lie partly in the lower sheet (as indicated by the dotted lines in the diagram). Then any closed path of integration on the two-sheeted

surface may be shewn to be equivalent to the repetition, a certain
number of times, of integration
along one side of the former oval
cut (along APB, in the diagram),
taken with the repetition, a
certain number of times, of integration along one side of the

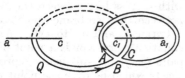

latter cut (along CQB, in the diagram). Thus, as was found by the
method of loops, there are two periods. The double-sheeted surface,
in this case, so far as its topology is concerned, is like the surface of
an anchor ring; on this, two independent closed circuits can be made,
one through the hole, the other round the hole. Or, it is like the
surface of a solid constituted by a sphere with a handle attached.
More generally, it may be shewn that the two-sheeted surface
associated with the equation $y^2 = u_0 u_1 \dots u_p$, where $u_i = (x - a_i)(x - c_i)$
is, in its topology, like the surface of the solid formed by a sphere
with p independent handles.

A general equation $f(x, y) = 0$, with m values for y, simplified as
explained above, so that the loops to the critical points on the plane
of x may be given by a scheme (p. 117, above)

$$(m-1, m)(m-1, m) \dots (2, 3)(2, 3)(1, 2) \dots (1, 2),$$

in which there are $2p+2$ symbols $(1, 2)$, may similarly be represented by a surface of m sheets, consisting of, (i), two sheets, (1) and
(2), connected by $p+1$ cross lines, with a branch point at both ends
of each cross line; (ii), a sheet (3), connected by a single cross line
with the sheet (2); (iii), a sheet (4), connected by a single cross line
with the sheet (3); and so on, until we come to the last sheet (m),
connected only, by a single cross line, with the preceding sheet
$(m-1)$. Analogously then to the reduction of the loops in the plane
of x, above given, it may be shewn, first, that the presence of the
sheet (m) adds nothing to the possibility of drawing independent
circuits of integration on the surface, circuits thereon being reducible to circuits on the sheet $(m-1)$; then, that circuits on the
sheet $(m-1)$ may similarly be reduced to circuits on the sheet
$(m-2)$, and so on. Till, finally, it appears that effectively all period
circuits are obtainable by consideration of the sheets (1) and (2); if
any periods exist, these sheets are connected by at least two cross
lines $(p > 0)$. The system of independent period circuits is thus of
the same character as for the two-sheeted surface considered above.

Topology of a general Riemann surface. It appears
desirable to sketch the relations of what precedes, which is based
directly on the fundamental algebraic equation, with a very interesting wider theory; it will be necessary, however, to assume as intuitive many properties that require detailed consideration.

We consider, then, a surface which we suppose to lie entirely in the finite part of space. It consists of real points, some called interior points, and others not interior points; the aggregate of the latter points consitutes the boundary; we shall suppose here that the boundary consists of one or more closed curves, of finite number. An interior point is one which is an interior point of a single small region, of which every point is a point of the surface, having the property that the points of this small region are in (1, 1) continuous correspondence with the points of a simple closed area in a plane; this may, for instance, be a triangle, and in any case has a definite boundary curve to which there corresponds (continuously) a single boundary curve of the region of the surface. The correspondence is to be such that, to any sequence of points of either region which has a limiting point belonging to the region, there corresponds a sequence of points of the other region, with a limiting point corresponding to the former. Later on it is assumed that the correspondence is such that the points of the region considered on the surface correspond to the values of one branch of an algebraic function of the complex variable which is represented by the points of the plane region. The definition does not, for example, allow the vertex of a double cone to be an interior point of a small region of the surface of the cone, but does allow a branch point of a two-sheeted surface, such as we have described, to be such an interior point; though, in this latter case, the boundary curve of the small region on the surface winds twice round the point. The small region of the surface may be called a *cell*; and, with a definite direction of circulation assigned round its boundary curve, it may be called an *oriented cell*. Two oriented cells on the surface may be said to be *adjacent* when they have no interior point in common, but a portion of the boundary curve of one coincides with a portion of the boundary curve of the other, and these two portions are described in opposite directions by the orientations of the two cells. Two oriented cells may be called adjacent, however, also when the common portion of their boundary curves consists only of discrete points (the cells having no interior points in common). We may then have on the surface an aggregate of oriented cells of which each is adjacent to one or more of the cells of the aggregate, and in such a way that the portions of boundary curves, of cells of the aggregate, which belong each only to one cell, form together one or more continuous closed curves, each with a definite direction of description. The simplest case is that of a single simple closed curve forming the boundary of a region entirely filled with adjacent cells, the single closed curve having a definite direction of description which, over every portion, agrees with that of the cell whose boundary curve contains this portion. If C be such a single simple closed curve, we denote its bounding character, which we have described, by writing $C \sim 0$. And, if the boundary of an aggregate of adjacent cells consist of two simple closed curves, each with a definite direction of description, in consonance with that of the portions of boundary of cells from which the closed curve is made up (as explained), these curves being A and B, then we write $A + B \sim 0$; or, if we assign to B the opposite direction, denoting it then by B_1, we write $A \sim B_1$. We shall assume that the whole surface under consideration may be regarded as an aggregate of adjacent cells, and that the portions of boundary curves which are not common to two adjacent cells, form (one or more) boundary curves, each with a definite direction of description, which, for every portion of such a boundary curve, is that derived from the orientation of the cell to which this portion belongs; and further, that such a boundary curve does not cross itself or any other boundary curve. This assumption involves not only that the surface is of one piece, but also that it is consistently orientable (sometimes described as one-sided). A *closed* surface is one for which no boundary curve exists. It will be understood that the relation expressed by $C \sim 0$ may be established by infinitely many different ways of dividing the surface into cells. This relation holds when the closed curve C can be continuously deformed (remaining on the surface) into a small closed curve lying entirely in the immediate neighbourhood of a single point of the surface, but this is not a necessary condition for $C \sim 0$.

It may be worth while to make this last remark clear by considering a simple concrete case. Consider a conical hill, of which the whole surface is accessible, including the summit. Evidently a circuit round the base of the hill is ~ 0, for it can be continuously deformed, up the surface of the hill, to vanish about a point at the summit. But now suppose a tunnel, of circular section, is cut through the hill (from one side of the hill to the other), and the surface of this tunnel (including its roof) is added to the outside surface of the hill. The circuit round the base of the hill (below the tunnel level) cannot now be continuously deformed (remaining on the surface of the hill) to a closed curve at the summit. But it is still ~ 0. For let A, B be points of the circuit at the base, on opposite sides of the hill, and let ASB be a path from A up one side of the hill, to the summit S, and down by the other side to B; denote the same path described in the opposite direction by BSA. The points A, B, of the circuit at the base, divide this into two portions, which we may distinguish by taking a point P in one and a point Q in the other, so that $APBSA$ is a closed circuit, and $BQASB$ is also a closed circuit; the circuit $APBSA$ is deformable, on the surface of the hill, into a circuit λ about one entrance to the tunnel; and the circuit $BQASB$ is deformable into a circuit μ, about the other entrance, in both cases with retention of the direction of description of the circuits. It is easy to see that, if λ, μ be deformed to coincide along a section of the surface of the tunnel in its interior, they will, along this section, be described in opposite directions. Thus the surface of the hill bounded by the circuit $APBQA$ at its base, can be divided into adjacent regions, such that the common portion of the boundary of any two adjacent regions is, as belonging to the two regions, oppositely directed. So that $APBQA$ is ~ 0.

Consider now such a surface as we have described, with or without boundary curves. It may be that every closed curve drawn thereon is ~ 0; this is the case for instance for the complete surface of a sphere, as also for a portion of the surface of a sphere bounded by any simple closed curve drawn thereon. But, on the curved surface of a right circular cylinder, bounded by the circumferences of the two flat ends, closed curves can be drawn, encircling the cylinder, which, not utilising the two bounding circumferences, are not ~ 0. In this case we remark, (i), that if one such closed curve be drawn, say C, any other such curve is $\sim C$; (ii), that, if the surface be cut along a generator, from the circumference at one end to the circumference at the other end (an operation which does not separate the surface into two pieces), then, on the mutilated surface, every closed curve is ~ 0. In what follows we shall assume that, on the surface which we consider, it is possible to find a finite number of simple closed curves $C_1, C_2, \ldots, C_r$, such that every other closed curve is $\sim$ a linear aggregate of these curves, each described a certain number of times; or, in symbols, that, for every closed curve C on the surface, there is a set of positive or negative integers, $m_1, \ldots, m_r$, such that $C \sim m_1 C_1 + \ldots + m_r C_r$; the number r, taken as small as may be possible by proper choice of $C_1, \ldots, C_r$, will be called, for the present, the *rank* of the surface. We shall consider how the rank may be modified by the introduction of cuts in the surface, this being, in many cases, a simple way of computing r. The final object is, for the case of a surface which represents the pairs (x, y) satisfying an algebraic equation, $f(x, y) = 0$, to make a connexion between the number r, which gives the number of periods of an algebraic integral associated with this equation, and the genus p of the equation.

The operations of cutting the surface which we employ may be: (i), the making of an infinitesimally small hole, say a *pinhole*, whereby the surface acquires a pinhole boundary besides those it may have had; the hole may be extended to a slit, or be of finite size, so that it does not interfere with other cuts made in the surface; (ii), the making of a cut from a point of a boundary curve to another point of the same, or a different, boundary curve; when the cut, beginning at a point of a boundary curve, has been partially made, the two edges may be counted as new boundary curves, and the cut, instead of ending on a new boundary curve, may end at one edge of the portion of the cut already made. Such a cut is often called a cross-cut (Querschnitt); we shall call it a *traverse*; (iii), the making of a (infinitesimal) pinhole, followed by a traverse which begins and ends at the boundary of the pinhole; the total cut so obtained will be called an *oval cut*; it is sometimes called a *loop cut*.

We now develop in order a set of twelve results, for which, throughout, we make appeal to intuitive conceptions:

I. A surface with two, or more, simple closed boundary curves, which do not meet, must have $r > 0$. For, make a traverse from a point of one boundary curve to a point of another boundary curve. We can evidently pass, on the surface, from a point on one side of this cut to the opposite point on the other side; for instance by following one of the two boundary curves; it appears clear, therefore, that on the original surface, before the cut is made, a closed curve can be drawn which does not constitute by itself the complete boundary of a portion of the surface.

II. If a surface having only one simple closed boundary curve have $r > 0$, it is possible to make thereon an oval cut, and it is also possible thereon to make a traverse, without dissevering the surface into two or more separate pieces. For suppose an oval cut to be possible, and to be made, which does dissever the surface; the number of pieces must then be two, one bounded by the original boundary curve of the surface taken with one edge of the oval cut, the other bounded by the other edge of the oval cut; this latter then shews how to draw on the original surface a closed curve forming by itself the complete boundary of a portion of the surface. Thus, if every oval cut, made on the original surface, dissevered the surface, every closed curve on the original surface would be ~ 0; and this is contrary to $r > 0$. We can thus make, on the original surface, an oval cut which does not dissever the surface. And this oval cut, with a traverse joining one edge of it to the original boundary line of the surface, constitutes a (looped) traverse which does not dissever the surface.

III. It follows from I and II that a surface with a boundary curve (or curves) which is dissevered into separate pieces by every traverse must have $r = 0$. This is a consequence of I for a surface with two or more boundary curves, and of II for a surface with only one boundary curve. It is also true that a surface without boundary which is dissevered by every oval cut must have $r = 0$.

IV. For a surface without boundary, the introduction of a pinhole boundary does not alter the value of r. We shall therefore suppose all the surfaces which we consider to have such a boundary, if they have no other. For a surface with one or more boundary curves, the making of a new pinhole boundary increases the value of r by unity.

V. If a surface have $r = 0$, it follows from I that it cannot have more than one simple boundary curve; and it follows also from

$r = 0$ that every oval cut, as also every traverse, dissevers the surface. For a surface with $r > 0$, having one or more simple boundary curves, a traverse, which does not dissever the surface, diminishes r by unity. Such a traverse, in fact, destroys the possibility of completing, into a closed curve, a path which passes, on the surface, from a point on one edge of the traverse into the opposite point of the other edge.

VI. It follows from IV and V that, on a surface with one or more boundary curves, an oval cut, which does not dissever the surface, does not alter the value of r. For such an oval cut may be obtained by making a pinhole boundary, and then a traverse beginning and ending thereat. As said in V, the possibility of such an oval cut involves $r > 0$. The boundary curve assumed (when there is but one) may be only a pinhole boundary, of which the oval cut must then be independent. For a surface without boundary the theorem is not true, as we see for instance by considering the surface of an anchor ring.

VII. If a traverse dissever a surface of rank r into two portions, of respective ranks r_1 and r_2, then $r = r_1 + r_2$. The possibility of a traverse involves that the given surface has at least one boundary curve. To prove this theorem, let r_1 closed curves be taken on one portion of the dissevered surface, and r_2 closed curves on the other portion, in terms of which all closed curves, of the two portions respectively, can be represented with the connexion $\sim$, as in the definition of the rank. Then we assert that the aggregate of these $r_1 + r_2$ curves would, on the original surface before the traverse was made, be a system in terms of which every closed curve of the surface could be represented; and that there is no aggregate of curves consisting of multiples of curves from the system of r_1 curves, taken with multiples of curves from the system of r_2 curves, which, on the original surface, is ~ 0. The former assertion is equivalent with $r \leqslant r_1 + r_2$. In regard to this, it is evidently true that closed curves lying entirely in one of the two portions of the surface are representable by aggregates of curves from the $r_1 + r_2$ curves; but also, a closed curve of the original surface, which crosses the track along which the traverse is subsequently made, can be replaced by an aggregate of curves each lying entirely in one of the two portions of the surface, namely by adding, thereto, parts lying along the track of the traverse, to be described twice over, in opposite directions. The second assertion, equivalent to $r \geqslant r_1 + r_2$, is that the $r_1 + r_2$ curves cannot be replaced by a smaller system after removal of the traverse. Now, by hypothesis, the traverse dissevers the surface, and thus renders impossible any path on the surface connecting a point of one of the r_1 curves with a point of any one of the r_2 curves; if then a certain aggregate of curves from the r_1 curves, with a certain aggregate of curves from the r_2 curves, forms the complete boundary of a portion, α, of the original surface, there must be points of the traverse lying in this portion α. But, from such a point, a traverse cannot pass to the boundary of the original surface, without passing over the boundary of the portion α; this boundary is made up, however, of curves from the $r_1 + r_2$ curves, and these are supposed so drawn that none of them meets the traverse. There is thus no aggregate of curves, from the $r_1 + r_2$ curves, which is ~ 0; which is the second assertion. If there were a less number than $r_1 + r_2$ curves, on the original surface, in terms of which the $r_1 + r_2$ specified curves could be expressed, with the connexion $\sim$, these specified curves would be connected in the way now proved impossible.

VIII. If an oval cut dissever a surface of rank r, with one or more boundary curves, into two pieces of ranks r_1 and r_2, then $r + 1 = r_1 + r_2$. For, to make the oval cut, we first introduce a pinhole boundary, thereby increasing the rank to $r + 1$ (by IV), and afterwards make a traverse beginning and ending at this point, in this case a dissevering traverse.

IX. From VII and VIII we infer that, if a surface of rank r, with one or more boundary curves, be dissevered into two portions, of respective ranks r_1 and r_2, by means either of a traverse, or of an oval cut, then $r - 1 = q + (r_1 - 1) + (r_2 - 1)$, where $q = 1$ in the case of a traverse, but $q = 0$ in the case of an oval cut. If, for a surface with a boundary, for which $r > 0$, a traverse be made which does not dissever the surface, the rank, r_1, of the surface with this traverse, is given (V) by $r_1 = r - 1$; this result is included in the formula we have given if we put $q = 1$ and omit the term $r_2 - 1$. If an oval cut be made which does not dissever the surface, we have (by VI) $r_1 = r$, likewise obtainable from the formula by putting $q = 0$, and omitting $r_2 - 1$.

X. Hence we can infer, by induction, that if, on a surface of rank r, with one or more boundary curves, there be made q traverses, and also oval cuts of any number (including none), and the result be an aggregate of distinct pieces, with respective ranks $r_1, r_2, \ldots$, then

$$r - 1 = q + \sum_i (r_i - 1).$$

This is a capital result.

XI. A particular theorem, useful to us, is that, for a surface with only a single pinhole boundary, the rank r is even. Suppose $r > 0$. We can then (by II) make a non-dissevering traverse, beginning and ending at the point boundary, so obtaining (by V) a surface of rank $r - 1$. As this has two boundary curves, the edges of the cut, it follows (by I) that $r - 1 > 0$. Hence (by III) we can further make a non-dissevering traverse, passing from one edge to the other of the first traverse, thereby obtaining (by V) a surface of rank $r-2$, with only one boundary curve, constituted by the edges of the two traverses. If $r-2>0$, we can proceed as before, using the single boundary curve as we did the point boundary; then, with two additional traverses, we obtain a surface of one boundary with rank $r-4$. The process can be continued if $r-4>0$. If r were odd, we 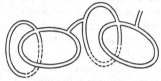 should reach a single surface of rank unity with one boundary curve, reducible, by a single traverse, to a single surface of zero rank, with two boundary curves; and (by I) such a surface does not exist. Thus r is even; and the process stops with a surface of zero rank, with one boundary curve, after $\frac{1}{2}r$ pairs of traverses have been made. The operation is illustrated by the diagram, where the dotted lines are to indicate the absence of intersection of the traverses.

XII. Now consider a surface constructed to represent the pairs of values of x, y which satisfy an irreducible algebraic equation $f(x, y)=0$, consisting of m sheets laid over the surface of a sphere, connected together at branch points. We need not limit ourselves to the hypothesis of merely simple branch points, in which only two sheets of the surface wind into one another, such as we have described in detail; it is clearly possible, in a similar way, to represent the k values of y, $=x^h$, where $h=1/k$, when x is small, by means of k sheets, winding into one another about $x=0$, having $(k-1)$ cross lines through this point, over which every sheet preserves its identity. We may then suppose the general branch point, of the surface under consideration, to involve k_i sheets, passing into one another over k_i-1 passage lines, each of which is a passage line, or cross line, at some other branch point. We assume that the surface can be constructed, to represent the m values of y satisfying $f(x, y)=0$, and satisfies the conditions we have explained. It will be closed; but we take an arbitrary pinhole boundary. Putting $w_i=k_i-1$ for each branch point, and $w=\Sigma w_i$, we prove that the rank of the surface is given by $r=w-2m+2$. We can suppose that there are two polar regions, within which no two of the m sheets are connected by branch points. By cutting through all the m sheets, say along arctic and antarctic circles, we can separate 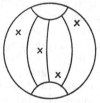 these polar regions from the rest of the surface, leaving m pieces about each pole. Supposing the pinhole boundary to be, in one of the m sheets, in one of the (in all) $2m$ closed cuts thus made, these cuts constitute one traverse and $2m-1$ oval cuts. We can suppose, by slight deformation if necessary, that no branch point of the surface lies immediately under another branch point; and then, if s be the number of branch points, we can make s cuttings, through all the sheets, along meridian curves, from arctic circle to antarctic circle, so separating the branch points, any one of which now lies, in one of the m sheets, between two meridian curves. This is equivalent to making ms further traverses on the surface. Between two such cuttings along meridian curves, enclosing a branch point connecting say k_i sheets, there will be, beside $m-k_i$ simple pieces, all separate, and each of zero rank, a single piece consisting of k_i sheets connected at a branch point; and this single piece also has $r=0$. The total number of pieces, after the cutting, is

$$\alpha, \ =2m+\Sigma(m-k_i+1), \ \text{or} \ 2m+sm-w;$$

the total number of traverses is $q, \ =1+ms$; and every piece is of zero rank. Whence, by the formula $r-1=q+\Sigma(r_i-1)$, we have $r-1=1+ms-\alpha, \ =w-2m+1$; as was said.

This simple proof of the value of r is due to Carl Neumann (*Riemann's Theorie*, Leipzig, 1884, p. 169). Riemann's own proof (*Ges. Werke*, 1876, p. 107) is by a contour integration round the boundary of the surface, after it has been dissected, as explained above, by r traverses, into a single surface with zero rank. The result is of importance to us, as establishing the connexion between r, and the genus, p, of the algebraic equation $f(x,y)=0$. For we have previously shewn (Chap. IV, p. 83) that $w=2m+2p-2$. Wherefore $r=2p$. This shews that the number of periods of an algebraic integral, associated with $f(x,y)=0$, which has no logarithmic infinities (which, by the definition, is r), is equal to $2p$, as was previously proved in a different way, under the hypothesis of simple branch points.

Curve systems on a surface. We proceed now to indicate how the preceding theory of the topology of a Riemann surface is connected with another theory which goes back to Euler and Moebius. Consider in space a set of points, α_0 in number, and a set of lines, α_1 in number, each joining one of the points to another, no two of the lines having an intersection which is not one of the points. We may, in many ways, consider the lines, each used only once, as forming, properly grouped, a set of closed circuits, each with a certain number of *diagonals*, joining pairs of (the given) points on the circuit, this being so done that any of the circuits is connected, to other of the circuits, by one or more of the given lines. The given lines are thus distributed into, portions of circuits, diagonals, and *joins of circuits*. If any of the points is then common to, say, s circuits, we suppose it replaced by s points, of which one is connected to the $s-1$ others by newly introduced lines; this introduces changes in the numbers α_0 and α_1, as well as new lines of the third of the three categories spoken of. After this modification, denote $1-\alpha_0+\alpha_1$ by k. Then the sum of the numbers of lines which are either diagonals or joins of circuits is $k-1$; for the contribution to $\alpha_1-\alpha_0$ by each polygonal circuit is zero, there being as many points as lines entering in the circuit, and each diagonal and each join means a contribution of unity to α_1, with no addition to α_0. Next consider a closed surface, with one pinhole boundary, dissected, as we have described, by q traverses and any number of oval cuts, into pieces of respective ranks r_i. Suppose that, on the undissected surface, in place of these traverses and oval cuts, we draw lines. The points of intersection of these lines, and the lines themselves, then form such a system as we have described; in this system the oval cuts, and the first traverse which was possible (beginning and ending at the point boundary) are represented by polygonal circuits, but the other traverses are represented either by diagonal lines or by joins of circuits. Thus the numbers q and k are the same. If we denote by α_2 the number of pieces into which the surface was dissected, there will be α_2 regions on the undissected surface on which the lines are drawn, with the property that we cannot pass, on this surface, from the interior of one of these regions to the interior of any other, without crossing one of the lines; we call this the number of *faces*. The formula $r-1=q+\Sigma(r_i-1)$ thus gives, for the rank of the closed surface with one point boundary, $r=1+1-\alpha_0+\alpha_1+\Sigma r_i-\alpha_2$, which is $r=2-(\alpha_0-\alpha_1+\alpha_2)+\Sigma r_i$. Of this result a very simple example is given by Euler's well-known formula for the surface of a polyhedron, which has α_0 vertices, α_1 edges, and α_2 faces, where every face has zero rank, and the whole surface of the polyhedron has zero rank ($\Sigma r_i=0$, $r=0$); the formula then becomes $\alpha_0-\alpha_1+\alpha_2=2$.

We may illustrate the results of this chapter by some simple examples.

Ex. 1. If, in the curved portion of the surface of a right circular cylinder, we cut a hole, the remaining part of this curved portion may be dissected into a single piece by two traverses joining the hole to the circumferences of the two ends. For this piece we have $r_1=0$. Putting $q=2$ in the formula $r-1=q+\Sigma(r_i-1)$, we have $r-1=2-1$; thus $r=2$, for the curved portion in which the hole exists. It is indeed easy to see that, if we draw two closed curves on this surface, one encircling the hole, the other encircling the cylinder (not intersecting the former), then any other closed curve on the surface forms by itself, or with one, or both, of these, the complete boundary of a part of the surface.

Ex. 2. The part of a plane which is interior to a circle but exterior to k other circles which lie inside the first (not intersecting this or one another) has, likewise, $r=k$.

Ex. 3. The surface, of one piece, with one boundary line, and one portion overlying another, illustrated in the diagram, is reduced to a

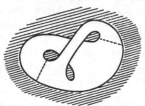

single piece of rank zero by two traverses (such as those dotted in the diagram). Thus $r=2$.

Ex. 4. The surface of the convex solid formed by cementing one face ABC, of a pyramid $ABCD$, to the larger face PQR of another pyramid, has evidently the connectivity of the complete surface of a circular cylinder, including its ends (or of the surface of a sphere), namely $r=0$.

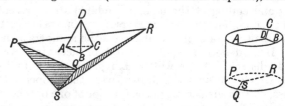

If the lines be drawn on the surface of the cylinder which correspond to the edges in the duplicate pyramid, we may employ the formula $r=2-(\alpha_0-\alpha_1+\alpha_2)+\Sigma r_i$, with $\alpha_0=8$, $\alpha_1=12$, $\alpha_2=7$, $r_i=0$ except for the curved portion of the surface of the cylinder which has rank unity; and this also gives $r=0$.

Ex. 5. For the surface of the solid which is formed by joining each of n spheres (lying externally to one another) to the other $n-1$, by thin circular cylinders (each cylinder meeting in closed curves the two spheres which it joins), we may find the rank by dissection. If we make a cutting along each of the $n(n-1)$ closed curves in which the cylinders meet the spheres, there will result n spheres, each with $n-1$ holes, together with the curved surfaces of $\frac{1}{2}n(n-1)$ cylinders. For each such sphere we have $r_i=n-2$, and for each such curved cylindrical surface we have $r_i=1$. The cuttings (if the pinhole boundary be properly taken) consist of one traverse, and $n(n-1)-1$ oval cuts. Thus, by $r-1=q+\Sigma(r_i-1)$, we have $r-1=1+n(n-2-1)+\frac{1}{2}n(n-1)(1-1)$, or $r=(n-1)(n-2)$. In particular, for $n=2$ the surface is similar to the surface of a sphere; for $n=3$ it is similar to the surface of an anchor ring.

Ex. 6. We have shewn how a Riemann surface can be dissected to a single piece of zero rank. Hence, on the undissected surface a system of

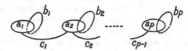

lines can be drawn consisting of p couplets linked in order by $p-1$ other lines, a couplet consisting of two simple closed curves with one intersection. For this system of lines we have $\alpha_0 = p + 2(p-1)$, $= 3p - 2$; $\alpha_1 = [3 + 4(p-2) + 3] + p - 1$, $= 5p - 3$, since the first and last couplets have 3 edges, the intervening couplets have each 4 edges, and the linking lines are $p - 1$ in number; and passage remains possible on the surface between any two points, or $\alpha_2 = 1$. Thus we have, in agreement with $r = 2 - (\alpha_0 - \alpha_1 + \alpha_2) + \Sigma r_i$, the result $\alpha_0 - \alpha_1 + \alpha_2 = 2 - r = 2 - 2p$.

Ex. 7. Similarly a system of lines on the Neumann model, considered in XII above, is indicated by the cuttings, for which $\alpha_0 = 2ms$, $\alpha_1 = 2ms + ms$, $\alpha_2 = 2m + sm - w$; and these give $\alpha_0 - \alpha_1 + \alpha_2 = 2m - w$, which is also $2 - 2p$.

As is briefly noticed in Vol. VI (in connexion with the so-called Zeuthen-Segre invariant *I*) the expression $\alpha_0 - \alpha_1 + \alpha_2$ is capable of wide generalisation. It is clear from what is said here that, if the value of this be proved to be independent of the system of lines on the Riemann surface by which it is defined, it can be made the basis of the determination of the number of periods of the algebraic integrals.

CHAPTER VI

THE VARIOUS KINDS OF ALGEBRAIC INTEGRALS. RELATIONS AMONG PERIODS

THE present chapter is in the nature of a Note, collecting together briefly details in regard to the characters of the various kinds of algebraic integrals which present themselves for a fundamental algebraic equation $f(x, y) = 0$.

From the point of view in which the equation $f(x, y) = 0$ is regarded as representing a plane curve, in the case in which this curve has no multiple points other than double points with distinct tangents, and the lines $x = 0$, $y = 0$ are in general positions with respect to this curve, we have shewn that, if p be the number of everywhere finite algebraic integrals belonging to the curve, then p is also the number of adjoint polynomials of order $m - 3$ (if m be the order of the curve); and further that these polynomials have each $2p - 2$ zeros on the curve (other than at the multiple points), no one of which is common to all of them. Also that the number of tangents which can be drawn to the curve from an arbitrary point is $2m + 2p - 2$. We have also shewn that the theory, for a curve whose multiple points have any complexity, can be reduced to this case, by birational transformation, which will leave the value of p, as defined, unaffected. We have then shewn, for the simple case, by means of the formula $w = 2m + 2p - 2$, using loops of integration, that a general algebraic integral without logarithmic infinities, associated with the curve, has $2p$ linearly independent periods; and we have obtained this same result by use of the Riemann surface. It is very desirable that the notions associated with the theory of the Riemann surface should be familiar; and we regard this point of view as always in the background. But the algebraic theory of the algebraic integrals owes very much to Weierstrass, whose lectures (*Ges. Werke*, IV) should be consulted. When the Riemann surface has been dissected by the loop pairs $a_1, b_1, \ldots, a_p, b_p$, and the linking cuts $c_1, \ldots, c_{p-1}$, as described above (Chap. V), it is simply connected, any closed curve theorem being reducible to a point boundary. On this surface then (as follows from Cauchy's theorem for the contour integral of a single-valued function of a complex variable) every algebraic integral, associated with it, which is without logarithmic infinities, is determined by its initial and final points of integration alone. An algebraic integral having also logarithmic infinities, say at $(z_1), \ldots, (z_k)$, may also be rendered dependent only on the limits of its integration, namely by introducing further cuttings of the surface along curves joining $(z_1), \ldots, (z_k)$ to an arbitrary point of the surface (or in simpler ways, as will be seen).

Consider any algebraic integral $\int R\,dx$, where R is rational in the x, y, which are throughout connected by the rational (irreducible) equation $f(x, y) = 0$. For the neighbourhood of a place at which the local parameter is t, substitution for x and y gives to the expression $R\,dx/dt$ a form

$$R\frac{dx}{dt} = \frac{M}{t^{\mu}} + \frac{N}{t^{\nu}} + \ldots + \frac{A}{t} + B + Ct^{\rho} + Dt^{\sigma} + \ldots,$$

where $\mu, \nu, \ldots$, in finite number, are positive integers, for which $\mu > \nu > \ldots > 1$, and $\rho, \sigma, \ldots$, in indefinite number, are likewise positive numbers $(\rho < \sigma < \ldots)$. The integral $\int R\,dx$, regarded as a function of its upper limit (x, y), is thus infinite at the place in question like $-M/(\mu-1)t^{\mu-1} - N/(\nu-1)t^{\nu-1} - \ldots + A \log t$, in which $A \log t$ is the logarithmic part, and the other parts may be described as algebraic. We can, however, shew how to construct integrals of the following kinds: (1), An integral which has no place of algebraic infinity (so that the coefficients $M, N, \ldots$ are absent in the expansion about every place), but has a logarithmic infinity at each of two arbitrary places (and not elsewhere). The coefficients A which arise for these two places are then (Chap. III, p. 47) equal and of opposite sign. If the two places be denoted by (ζ) and (γ), and, by division by the constant A, the integral be taken so that the multiplier of the logarithmic term at (ζ) is $+1$, and at (γ) is -1, such integral, integrated from the place (a) to the place (x), will be denoted by $P_{\zeta, \gamma}^{x, a}$, and called an *elementary integral of the third kind*. (2), An integral which has no place of logarithmic infinity, but is algebraically infinite at one arbitrary place, the infinite term in the expansion of the integral, in terms of the parameter, for the neighbourhood of this place, being $-t^{-1}$, there being no negative powers of t of higher than the first order. If this place be (ζ), and the integral be taken from (a) to (x), such integral will be denoted by $E_{\zeta}^{x, a}$, and called an *elementary integral of the second kind*. (3), We have shewn (Chap. IV, p. 64) how to construct everywhere finite integrals. If p such integrals, linearly independent of one another, taken from (a) to (x), be denoted by $u_1^{x, a}, \ldots, u_p^{x, a}$, the definitions of $P_{\zeta, \gamma}^{x, a}$ and $E_{\zeta}^{x, a}$ are equally satisfied by $P_{\zeta, \gamma}^{x, a} + U$, $E_{\zeta}^{x, a} + V$, where U, V denote linear aggregates of $u_1^{x, a}, \ldots, u_p^{x, a}$ with constant coefficients. Conversely, if P, P' be any two functions satisfying the definition of $P_{\zeta, \gamma}^{x, a}$, with the same limits, and infinities, and multipliers, the difference $P' - P$ is necessarily an everywhere finite algebraic integral, and is thus a linear aggregate such as U or V;

and the similar remark is possible for the difference of two elementary integrals $E_\zeta^{x,a}$, with the same pole.

From the integral $E_\zeta^{x,a}$ can be formed, by differentiation in regard to ζ, an integral which is algebraically infinite, at the place (ζ) only, to any integral order. For, let (x) and (ζ') be places, both in the neighbourhood of (ζ)—on a branch of the curve, of which (ζ) is the centre—in the range for which the local parameter for (ζ) is valid; let the passage from (ζ) to (x) be by the value t, and from (ζ) to (ζ') by the parameter τ.

We can then suppose (x) so near to (ζ') that we pass from the latter to the former by a parameter t' equal to $t - \tau$. Because (x) is in the neighbourhood of both (ζ) and (ζ'), we have expressions of the forms

$$E_\zeta^{x,a} = -t^{-1} + M + Nt + \ldots, \quad E_{\zeta'}^{x,a} = -t'^{-1} + M' + N't' + \ldots,$$

leading to

$$E_{\zeta'}^{x,a} - E_\zeta^{x,a} = -(t - \tau)^{-1} + t^{-1} + M' + N't' + \ldots - (M + Nt + \ldots);$$

if we now suppose (ζ') to approach to (ζ), so that t remains constant but τ diminishes in absolute value, this difference is expressible as a power series in τ, of the form $-\tau t^{-2} - \tau^2 t^{-3} + \ldots + \omega$, where ω will ultimately remain finite when (x), later, approaches to (ζ). It follows then that, as τ vanishes, the limit of $(E_{\zeta'}^{x,a} - E_\zeta^{x,a})/\tau$ is an algebraic integral, with (x) as variable place, which becomes infinite, when (x) approaches to (ζ), only like $-t^{-2}$. This limit we denote by $D_\zeta E_\zeta^{x,a}$. We shall see (in the next chapter) how to obtain an explicit expression of this, from the expression of $E_\zeta^{x,a}$, by differentiation under an integral sign. Precisely similar reasoning shews that we can obtain, as $D_\zeta^{k-1} E_\zeta^{x,a}$, an algebraic integral, in (x), which becomes infinite only when (x) approaches to (ζ), like $-(k-1)!/t^k$. And similar reasoning shews that we may obtain $E_\zeta^{x,a}$, by differentiation in regard to ζ, from $P_{\zeta,\gamma}^{x,a}$, namely that $E_\zeta^{x,a} = D_\zeta P_{\zeta,\gamma}^{x,a}$, where the differentiation is in regard to the parameter at (ζ).

It is easy to form an explicit expression for the integral $P_{\zeta,\gamma}^{x,a}$, and convenient to explain the process by regarding the fundamental equation as representing a curve. First, suppose (ζ) and (γ) to be ordinary simple points of this curve, not branch places; let the line joining these points meet the curve again in the (generally) simple points $A_1, \ldots, A_{m-2}$. The general adjoint curve of order $m-2$ drawn through the points $A_1, \ldots, A_{m-2}$ has other intersections with the curve, beside those at the multiple points, whose number is $2p - 2 + m - (m - 2)$, or $2p$; and this curve contains, homogeneously, $2p - p + 1$, or $p + 1$ arbitrary coefficients; this follows from Chap. IV,

p. 78. If $L=0$ be the equation of the line joining the points (ζ) and (γ), and ϕ be the general adjoint polynomial of order $m-3$ (containing p arbitrary homogeneous coefficients), it follows that the adjoint curve in question, through $A_1, \ldots, A_{m-2}$, has the equation $L\phi+cU=0$, where c is an arbitrary constant, and $U=0$ is a particular curve satisfying the conditions. Consider now the algebraic integral $\int(L\phi+cU)\,dx/Lf'(y)$; by the conditions of adjointness satisfied by $L\phi+cU=0$, this integral is everywhere finite save at the points (ζ) and (γ), the other zeros of $L=0$ being compensated by zeros of $L\phi+cU$. At (ζ) and (γ), the integrand having a pole of the first order, the integral is logarithmically infinite; and the constant c can be chosen so that the multiplier of the logarithm is $+1$ at (ζ); it is therefore -1 at (γ), as we have shewn. We see that the integral is of the form $\int cU\,dx/Lf'(y)+\int\phi\,dx/f'(x)$, of which the second part gives the addition we have found above, of p everywhere finite integrals, to a general integral $P_{\zeta,\gamma}^{x,a}$. When (ζ), (γ) are any two distinct places of the curve, a precisely similar construction can be made, if an appropriate birational transformation of the curve be first introduced. In the simple case (and therefore in general) the integral $E_\zeta^{x,a}$ can be constructed in like manner, if for $L=0$ we take the tangent of the curve at the point (ζ).

Now consider any algebraic integral $\int R(x,y)\,dx$ whatever. It is clear, if the places at which this integral has algebraic infinities be places (ξ), and the places where it has logarithmic infinities be places (ζ)—both kinds of infinity may enter at the same place—and, if (γ) be an arbitrary place, that we can form an algebraic integral of the form

$$-\int_a^x R(x,y)\,dx+\sum_\xi[AE_\xi^{x,a}+A_1D_\xi E_\xi^{x,a}+\ldots+A_{k-1}D_\xi^{k-1}E_\xi^{x,a}]$$
$$+\sum_\zeta BP_{\zeta,\gamma}^{x,a},$$

wherein $A, A_1, \ldots, A_{k-1}, B$ are appropriate constants, which becomes infinite, if at all, only at the single place (γ). This, however, it does not do, since the sum of the values of the constants B, at the places where $\int R(x,y)\,dx$ has logarithmic infinities, is necessarily zero (Chap. III, p. 47). This integral then must be an aggregate of everywhere finite integrals. In other words, the given integral $\int R(x,y)\,dx$ is expressible as an aggregate, (i), of integrals $P_{\zeta,\gamma}^{x,a}$, as many as the number of places (ζ) at which $\int R(x,y)\,dx$ has logarithmic infinities, the place (γ) being arbitrary; (ii), of integrals $E_\xi^{x,a}$ and their derived integrals $D_\xi^{i-1}E_\xi^{x,a}$; (iii), of integrals which are everywhere finite. The aggregate (ii), however, can be much simplified; it can be shewn that, if $(c_1), (c_2), \ldots, (c_p)$ be quite arbitrary general places of the curve, the aggregate (ii) can be replaced by the sum of an appropriate rational function and a linear aggregate

$$\lambda_1 E_{c_1}{}^{x,a} + \ldots + \lambda_p E_{c_p}{}^{x,a},$$

wherein $\lambda_1, \ldots, \lambda_p$ do not depend on (x); or, alternatively, if (c) be a quite arbitrary place, the aggregate (ii) can be replaced by the sum of an appropriate rational function and a linear aggregate

$$\lambda_1 E_c{}^{x,a} + \lambda_2 D_c E_c{}^{x,a} + \ldots + \lambda_p D_c{}^{p-1} E_c{}^{x,a},$$

where $\lambda_1, \ldots, \lambda_p$ are independent of (x). This is easy to prove; it will be sufficient to take the former statement: Let (z) be any place; there exists (Chap. IV, p. 78) an unique rational function of (x), having poles of the first order at $(c_1), \ldots, (c_p)$ and (z), with coefficient of t^{-1}, in the expansion of this function in terms of the parameter t for (z), when (x) is near to (z), equal to -1, which also vanishes at a further arbitrary place (a). Denote this rational function by $\Psi(x, a; z, c_1, \ldots, c_p)$. From this function, by differentiation with respect to the parameter t at (z), we can form a rational function of (x), with poles of the first order at $(c_1), \ldots, (c_p)$, and a pole at (z) of order k; denote this function by $D_z{}^{k-1}\Psi$; its single infinite term in its expansion near (z) will be $-(k-1)!/t^k$. With the integral $\int R(x, y)\, dx$ we can then form an aggregate of the form

$$- \int_a^x R(x, y)\, dx + \sum_\xi \left[A\Psi(x, a; \xi, c_1, \ldots, c_p) \right.$$
$$\left. + A_1 D_\xi \Psi(x, a; \xi, c_1, \ldots, c_p) + \ldots + A_{k-1} D_\xi{}^{k-1}\Psi(x, a; \xi, c_1, \ldots, c_p) \right],$$

wherein $A, \ldots, A_{k-1}$ are appropriate constants, which has poles only at $(c_1), \ldots, (c_p)$, beside possessing the logarithmic infinities which $\int R(x, y)\, dx$ may have. This aggregate is then of the form

$$\lambda_1 E_{c_1}{}^{x,a} + \ldots + \lambda_p E_{c_p}{}^{x,a} + \Sigma B P_{\zeta,\gamma}^{x,a} + \mu_1 u_1{}^{x,a} + \ldots + \mu_p u_p{}^{x,a},$$

where $\lambda_1, \ldots, \lambda_p, B, \mu_1, \ldots, \mu_p$ are certain constants, the summation extends to all the logarithmic infinities of $\int R(x, y)\, dx$, and $u_1{}^{x,a}, \ldots, u_p{}^{x,a}$ are independent everywhere finite integrals. And this is equivalent to the statement we have made. The second statement follows similarly, the places $(c_1), (c_2), \ldots, (c_p)$ being all at one place (c).

It appears from the results we have found that the nature of the infinities and the periods, of any algebraic integral, can be studied by considering integrals $P_{\zeta,\gamma}^{x,a}$, and a sum of integrals of the form $\lambda_1 E_{c_1}{}^{x,a} + \ldots + \lambda_p E_{c_p}{}^{x,a} + \mu_1 u_1{}^{x,a} + \ldots + \mu_p u_p{}^{x,a}$, where (γ), and $(c_1), \ldots, (c_p)$ are arbitrary places of the fundamental curve. The number $2p$, of constant multipliers, in the last integral sum, is the same as the number of periods previously found for an algebraic integral without logarithmic infinities. Some detailed formulae for an integral sum of this form are given in the next chapter. In addition now to the reductions thus found by algebraic methods, we consider further simplifications which can be made if the integrals are considered on a Riemann surface upon which the dissection, which reduces the surface to a simply connected surface, *is supposed to be given*. This dissection is along p pairs of oval loops, $a_1, b_1, \ldots, a_p, b_p$, linked by cuts $c_1, c_2, \ldots, c_{p-1}$. We suppose a definite direction attached to each of $a_1, b_1, \ldots, a_p, b_p$, and speak, correspondingly, of the positive or left edge, and of the negative or right edge, of each cut; *and we suppose that the positive direction of the loop b_i issues from the left edge of the cut a_i.*

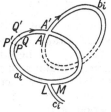

If P, P' be points, opposite to one another, respectively on the right and left edges of any cut a_i, the element of an algebraic integral $\int R(x, y)\, dx$, in passing from P' to a consecutive point Q' of the same edge, is the same as in passing from P to the consecutive point Q, opposite to Q', of the other edge; hence it can be shewn that the value at P', of the integral $\int R(x, y)\, dx$, which is a single-valued function of its upper limit on the dissected surface, exceeds the value of this integral at P by a quantity which is independent of the position of P on the cut. This quantity may then be defined as that obtained for the integral by a circuit of the positive edge of the cut b_i, *in the negative direction* (from A to A' in the diagram). This quantity will be called a period, and denoted by Ω_i. Similarly a *positive* circuit of the positive edge of the cut a_i gives a period Ω_i'; this is the quantity by which the value of the integral at any point on the positive edge of the cut b_i exceeds its value at the opposite point, on the right edge of this cut. From any point L, on the positive edge of the cut a_i, we can make a complete circuit of the edges of the cuts a_i, b_i, describing in all a path which (reading from left to right) may be represented by $a_i b_i a_i^{-1} b_i^{-1}$, until we reach a point M, just before L. Thus, though the cut c_i is necessary to render the surface simply connected, the values of the algebraic integral at two points opposite to one another on the two edges of this cut are the same.

Consider, now, p linearly independent everywhere finite integrals $u_1^{x,a}, \ldots, u_p^{x,a}$; then, at first without reference to the precise dissection of the Riemann surface we have explained, take $2p$ independent closed circuits on the surface, and denote the periods of the integral $u_i^{x,a}$ for these by $\Omega_{i1}, \ldots, \Omega_{ip}, \Omega'_{i1}, \ldots, \Omega'_{ip}$. Then in the array, of p rows and $2p$ columns $u_i |\Omega_{i1}, \ldots, \Omega_{ip}, \Omega'_{i1}, \ldots, \Omega'_{ip}|$, there is certainly one determinant of p rows and columns which is not zero. Otherwise the rows, in the array, would not be linearly independent, and there would exist an integral $\lambda_1 u_1^{x,a} + \ldots + \lambda_p u_p^{x,a}$, in which $\lambda_1, \ldots, \lambda_p$ are constants, which is single valued on the undissected Riemann surface, as originally constructed to represent the equation $f(x, y) = 0$. This function, of (x), would have the analytical character, of expression by powers of the parameter t belonging to the neighbourhood of any point of the surface; and would hence, by the argument previously given (Chap. III, p. 57), be expressible rationally by x and y. There exists, however, no rational function without infinities; thus this function is a constant; as it is zero when (x) is at (a), it is always zero. This, however, is contrary to the fact that the p integrals $u_i^{x,a}$ are linearly independent. Suppose now that $\Omega_{i,k}$ is the constant by which the value of $u_i^{x,a}$ at a point on the left edge of the oval cut a_k, used for dissecting the Riemann surface, exceeds the value of $u_i^{x,a}$ at the opposite point of the right edge of this oval cut a_k (equal to the value obtained for $u_i^{x,a}$ by negative circuit of (x) about the loop b_k). We can then shew in particular that the determinant formed by the first p columns, in the period-array set down, is not zero. For let $U + iV$ be any everywhere finite integral, in which U, V are *real* functions (of the ξ, ξ_1, η, η_1, when $x = \xi + i\xi_1$, $y = \eta + i\eta_1$); and let the period of $U + iV$ at the oval cut a_σ (by which the left-edge value exceeds the right-edge value) be $A_\sigma + iB_\sigma$, where A_σ, B_σ are real; and let the period of $U + iV$ at the oval cut b_σ (by which the left-edge value exceeds the right-edge value) be $A_\sigma' + iB_\sigma'$, where A_σ', B_σ' are real. If we take the integral $\int U dV$ along the complete closed circuit $a_\sigma b_\sigma a_\sigma^{-1} b_\sigma^{-1}$ before described, the value obtained is $A_\sigma B_\sigma' - A_\sigma' B_\sigma$; and the complete circuit of the single boundary of the simply connected Riemann surface obtained by the dissection $a_1, b_1, ..., a_p, b_p, c_1, ..., c_{p-1}$ gives the value $\Sigma(A_\sigma B_\sigma' - A_\sigma' B_\sigma)$, the summation being for σ from 1 to p. By a well-known theorem, however, if $x = \xi + i\xi_1$, this contour integral, which is

$$\int \left(U \frac{\partial V}{\partial \xi} d\xi + U \frac{\partial V}{\partial \xi_1} d\xi_1 \right),$$

is equal to the double integral

$$\iint \left[\frac{\partial}{\partial \xi} \left(U \frac{\partial V}{\partial \xi_1} \right) - \frac{\partial}{\partial \xi_1} \left(U \frac{\partial V}{\partial \xi} \right) \right] d\xi \, d\xi_1,$$

which is $\iint (U_\xi V_{\xi_1} - U_{\xi_1} V_\xi) d\xi \, d\xi_1$, where $U_\xi = \partial U / \partial \xi$, etc., extended over the whole Riemann surface bounded by the single contour. In virtue of the well-known equations $U_\xi = V_{\xi_1}$, $U_{\xi_1} = -V_\xi$, this is the same as $\iint (U_\xi^2 + U_{\xi_1}^2) d\xi \, d\xi_1$. This, however, is definitely greater than zero unless U_ξ, U_{ξ_1} are both everywhere zero, which would involve $V_\xi = 0$, $V_{\xi_1} = 0$, and hence that the integral $U + iV$ was a constant. A first, important, conclusion, when $U + iV$ is not constant, is, therefore, that $\Sigma(A_\sigma B_\sigma' - A_\sigma' B_\sigma) > 0$. This inequality shews that there is no everywhere finite integral for which the periods $A_1 + iB_1, ..., A_p + iB_p$, at the oval cuts $a_1, ..., a_p$ are all zero; nor, indeed, any such integral for which the periods $A_1' + iB_1'$,

$\dots, A_p' + iB_p'$, at the associated oval cuts $b_1, \dots, b_p$ are all zero. And a consequence is that, in the period-array of p rows and $2p$ columns which we have considered, the determinant of the first p columns, referring to the cuts a_σ, is other than zero, as also that of the last p columns; for otherwise, we could form an everywhere finite integral for which every one of A_σ, B_σ was zero. Hence it is possible, by taking suitable linear functions of $u_1^{x,a}, \dots, u_p^{x,a}$, to find p everywhere finite integrals, $v_1^{x,a}, \dots, v_p^{x,a}$, for which the period-array, respectively at the oval cuts a_σ, and the oval cuts b_σ, has the form (with p rows and $2p$ columns) $v_i | \epsilon_{i1}, \dots, \epsilon_{ip}, \tau_{i1}, \dots, \tau_{ip} |$, where $\epsilon_{ii} = 1$, $\epsilon_{ij}(j \neq i) = 0$. When this is done, the determinant of the last p columns is symmetrical in form, namely $\tau_{ij} = \tau_{ji}$, as we now proceed to prove. The integrals $v_1^{x,a}, \dots, v_p^{x,a}$, evidently unique when the dissection of the Riemann surface is given, are called the *normal elementary integrals of the first kind.* For the proof, consider first any two algebraic integrals whatever, say $I^{x,a}, J^{x,a}$, and denote the periods of the former at the oval cuts $a_1, \dots, a_p, b_1, \dots, b_p$ respectively by $\Omega_1, \dots, \Omega_p, \Omega_1', \dots, \Omega_p'$, the similar periods of $J^{x,a}$ being $\Phi_1, \dots, \Phi_p, \Phi_1', \dots, \Phi_p'$. Then take the integral $\int I^{x,a} dJ^{x,a}$ round the complete contour of the simply connected Riemann surface obtained by the $2p + p - 1$ cuttings $a_1, \dots, a_p, b_1, \dots, b_p, c_1, \dots, c_{p-1}$. It is easy to prove as before that the value obtained is $\Sigma(\Omega_\sigma \Phi_\sigma' - \Omega_\sigma' \Phi_\sigma)$, where $\sigma = 1, \dots, p$. As $I^{x,a} dJ^{x,a}/dx$ is an analytical single-valued function of the complex variable x, on the dissected Riemann surface, it follows, by Cauchy's theorem, that the value of the contour integral is equal to the sum of the values obtained by taking the same integral round the points of the Riemann surface, if any exist, for which the integral is infinite. If, in particular, $I^{x,a}, J^{x,a}$ are the two normal elementary integrals of the first kind, $v_i^{x,a}$ and $v_j^{x,a}$, so that $\Omega_\sigma = 0$ except for $\Omega_i = 1$, and $\Phi_\sigma = 0$ save for $\Phi_j = 1$, while $\Omega_\sigma' = \tau_{i\sigma}$, $\Phi_\sigma' = \tau_{j\sigma}$, then, as $\Sigma(\Omega_\sigma \Phi_\sigma' - \Omega_\sigma' \Phi_\sigma) = 0$, we have $\tau_{ji} - \tau_{ij} = 0$.

The same argument establishes another important result: As before, let $P_{\zeta, \gamma}^{x, a}$ denote the elementary integral of the third kind, with logarithmic infinities at (ζ) and (γ); denote the negative of the period of this integral at the oval cut a_ρ by λ_ρ. Then the integral $P_{\zeta, \gamma}^{x, a} + \lambda_1 v_1^{x,a} + \dots + \lambda_p v_p^{x,a}$ has zero periods at every one of the p oval cuts $a_1, \dots, a_p$. There cannot be two algebraic integrals, both logarithmically infinite, as described, at (ζ) and (γ), without periods at the oval cuts $a_1, \dots, a_p$, and both vanishing for $(x)=(a)$, since their difference would be an everywhere finite integral with zero periods at these cuts, and hence would be a constant, equal then to zero. Thus this integral is unique when the oval cuts are given. We shall constantly denote it by $\Pi_{\zeta, \gamma}^{x, a}$, and call it an *elementary normal integral of the third kind*. By the argument above, we can prove that the periods of this integral at the oval cuts $b_1, \ldots, b_p$ are $2\pi i\, v_1^{x,a}, \ldots, 2\pi i\, v_p^{x,a}$. For, in the integral considered above, take $I^{x,a} = v_\alpha^{x,a}$, and $J^{x,a} = \Pi_{\zeta,\gamma}^{x,a}$; then every one of the periods Ω_σ of $v_\alpha^{x,a}$ is zero except Ω_α, which is 1; and all the periods Φ_σ of $\Pi_{\zeta,\gamma}^{x,a}$ are zero; the contour integral thus gives only the result Φ_α', the period of $\Pi_{\zeta,\gamma}^{x,a}$ at the oval cut b_α. We consider then the values obtained by taking the integral round its two points of infinity, which are at (ζ) and (γ). If t be the parameter for the neighbourhood of (ζ), the essential part of the integral $\int v_\alpha^{x,a} d\Pi_{\zeta,\gamma}^{x,a}$ for this neighbourhood is $\int v_\alpha^{x,a} dt/t$, and, with a change of sign, the essential part at (γ) is of similar form. The value obtained for the two infinities is thus $2\pi i\, v_\alpha^{\zeta,\gamma}$. Thus $\Phi_\alpha' = 2\pi i\, v_\alpha^{\zeta,\gamma}$, as was said.

Ex. By applying the contour integration $\int I\,dJ$, considered above, to the case when I is the elementary normal integral of the third kind with its infinities at (x) and (a), and J is the elementary normal integral of the third kind with its infinities at (z) and (c), prove the identity

$$\Pi_{z,c}^{x,a} = \Pi_{x,a}^{z,c}.$$

The integral $\Pi_{\zeta,\gamma}^{x,a}$ may be used to give a brief proof of a theorem, which we may call *the converse of Abel's theorem*. By the direct theorem, if $(z_1), \ldots, (z_k)$ be the zeros, and $(c_1), \ldots, (c_k)$ be the poles of any definite rational function associated with the fundamental equation $f(x, y)$, and $u_1^{x,a}, \ldots, u_p^{x,a}$ be linearly independent everywhere finite integrals, each of the p expressions $u_\alpha^{z_1,c_1} + \ldots + u_\alpha^{z_k,c_k}$ is a sum of integral multiples of the periods of the integral $u_\alpha^{x,a}$, wherein the $2p$ integer multipliers are the same for all the p values of α. On the Riemann surface, dissected by the $2p$ oval cuts a_σ, b_σ, each term $u_\alpha^{z_i,c_i}$ is quite definite, independently of the path of integration from (c_i) to (z_i). Thus, taking the elementary normal integrals of the first kind, we may suppose that $v_\alpha^{z_1,c_1} + \ldots + v_\alpha^{z_k,c_k}$ is equal to $m_\alpha + m_1'\tau_{\alpha 1} + \ldots + m_p'\tau_{\alpha p}$, for $\alpha = 1, \ldots, p$, wherein $m_1, \ldots, m_p, m_1', \ldots, m_p'$ are definite integers. Conversely, if $(z_1), \ldots, (z_k)$ and $(c_1), \ldots, (c_k)$ be two sets of each k places, for which, on the dissected Riemann surface, these p equations hold, then there exists a rational function having $(z_1), \ldots, (z_k)$ as zeros, and $(c_1), \ldots, (c_k)$ as poles. For consider the function $\exp(\phi)$, or e^ϕ, where $\phi = \Pi_{z_1,c_1}^{x,a} + \ldots + \Pi_{z_k,c_k}^{x,a} - 2\pi i (m_1' v_1^{x,a} + \ldots + m_p' v_p^{x,a})$. This function is analytical on the Riemann surface, being expressible in the neighbourhood of any point for which the parameter is t, by a series of integral powers of t. It is single valued on the dissected surface. It is in fact single valued if the dissection is not made; for its value at any point of the left edge of the oval cut a_ρ is obtained from its value at the opposite point of the right edge by multiplication with e^μ, where $\mu = -2\pi i m_\rho'$, so that $e^\mu = 1$; and, for two points opposite to one another on the edges of the oval cut b_ρ, the corresponding multiplier is $e^{\mu'}$, where $\mu' = 2\pi i (v_\rho^{z_1, c_1} + \ldots + v_\rho^{z_k, c_k} - m_1' \tau_{1\rho} - \ldots - m_p' \tau_{p\rho})$; namely, by the assumption made at starting, $\mu' = 2\pi i m_\rho$ (since $\tau_{\beta\rho} = \tau_{\rho\beta}$); thus also $e^{\mu'} = 1$. Except at one of the places (z_1), (c_1), ..., (z_k), (c_k), the expression of the function is by a power series in the parameter t; at (z_1), for example, beside a power series, the function ϕ contains a term $\log t$, and the function e^ϕ vanishes to the first order. Likewise it vanishes to the first order at $(z_2), \ldots, (z_k)$, and has poles at $(c_1), \ldots, (c_k)$. Thus, by a proposition proved (p. 57) above, e^ϕ is a rational function of (x), with the specified zeros, and poles; as was to be proved.

Ex. 1. By applying the contour theorem for $\int U dV$, as in the text, where $U + iV$ denotes an integral of the first kind $n_1 v_1^{x, a} + \ldots + n_p v_p^{x, a}$, in which $n_1, \ldots, n_p$ are real, prove that, if $\tau_{\alpha, \beta} = \rho_{\alpha, \beta} + i\sigma_{\alpha, \beta}$, where $\rho_{\alpha, \beta}$ and $\sigma_{\alpha, \beta}$ are real, then the quadratic form in $n_1, \ldots, n_p$, $\Sigma\Sigma \sigma_{\alpha, \beta} n_\alpha n_\beta$, is definite and positive, for all real values of $n_1, \ldots, n_p$.

Ex. 2. Denoting the quadratic form $\Sigma\Sigma \tau_{\alpha, \beta} n_\alpha n_\beta$ by τn^2, and the linear form $\Sigma u_\alpha n_\alpha$ by un, where α, $\beta = 1, \ldots, p$, and making a summation with only integer values of $n_1, \ldots, n_p$, each extending from $-\infty$ to $+\infty$, it can be proved from Ex. 1, that the series $\Sigma \exp(\omega)$, where $\omega = 2\pi i un + i\pi \tau n^2$, is absolutely convergent, and thus expresses an integral function of the independent variables $u_1, \ldots, u_p$.

Ex. 3. If in the function $\Theta(u_1, \ldots, u_p)$, of Ex. 2, we replace u_α by $v_\alpha^{x, a} + k_\alpha$, where $v_\alpha^{x, a}$ is the normal elementary integral, it can be shewn, for general values of $k_1, \ldots, k_p$, that the function of (x) so obtained vanishes for p positions of (x), depending on $k_1, k_2, \ldots, k_p$.

Ex. 4. If $u_1^{x, a}, \ldots, u_p^{x, a}$ denote any p linearly independent everywhere finite integrals, it can be shewn, for specified places $(c_1), \ldots, (c_p)$ and specified values of $k_1, \ldots, k_p$, that the p equations

$$u_\alpha^{z_1, c_1} + \ldots + u_\alpha^{z_p, c_p} = k_\alpha, \qquad (\alpha = 1, \ldots, p),$$

can be satisfied by proper choice of the places $(z_1), \ldots, (z_p)$. This is a familiar result when $p = 1$, the coordinates of the place (z_1) being elliptic functions of k_1. In general, the rational symmetric functions of the coordinates of the places $(z_1), \ldots, (z_p)$ are single-valued analytical functions of $k_1, \ldots, k_p$. In the elliptic case, in accordance with the two-period ambiguity of the integral $u^{z, c}$, the elliptic functions have two periods; in the general case the functions are similarly unaltered when each of $k_1, \ldots, k_p$ is increased by an appropriate constant, and there are $2p$ sets of such simultaneous increments which leave the functions unaltered. For particular values of $k_1, \ldots, k_p$, the functions are indeterminate, and the places $(z_1), \ldots, (z_p)$ have not a definite set of positions.

Ex. 5. The functions solving the so-called inversion problem, expressed by the equations in the previous example, can be expressed in terms of such functions as the theta functions referred to in Exx. 2, 3.

Ex. 6. The general theory of (meromorphic) functions of p variables which are unaltered when the variables are increased simultaneously by the elements of any one of $2p$ (or fewer) columns of periods, is of great importance. A preliminary study of the possible reduction of the array of periods adds to the simplicity of the theory.

For the content of Exx. 1–4 here the reader may be referred to the author's *Abel's Theorem* (Cambridge, 1897); and for Exx. 5, 6 to the author's *Multiply-periodic Functions* (Cambridge, 1907); for the reduction of the array of periods, referred to in Ex. 6, a somewhat different formulation is found in Severi, *Rend. Palermo*, xxi, 1906, "Intorno al teorema d' Abel sulle superficie...". Krazer, *Lehrbuch der Thetafunctionen* (Leipzig, 1903), should also be consulted.

CHAPTER VII

THE MODULAR EXPRESSION OF RATIONAL FUNCTIONS AND INTEGRALS

THE present chapter is concerned with a theory of the rational functions and integrals, associated with an (irreducible) algebraic curve $f(x, y) = 0$, which began with the arithmetical work of Kronecker and Dedekind* for the theory of integer numbers. Developed in detail the theory gives an alternative to much of what has preceded; what is given here however seems a desirable, if not a necessary, accompaniment of what has already been proved.

We have explained what is meant by a rational function belonging to the curve $f(x, y) = 0$; to fix the ideas we recall certain properties of a general kind for such functions. It was seen that, as a rule, there exists no rational function of the first, nor of the second order; there is thus, in general, a least order for which there exists a rational function, associated with the curve. But we cannot expect to be able to construct a function of this least order with its poles taken arbitrarily on the curve; for instance, when the equation $f(x, y) = 0$ is of the form $y^2 - u = 0$, where u is a polynomial in x, though there exists a function, of the form $(x - a)^{-1}$, with two poles, these must be at places for which x has the same value. There is thus another least number, say k, such that, *whatever k places be taken*, a rational function exists with poles, all of the first order, at these k places, or at places chosen from these. Such function will be actually infinite, indeed, at every one of the k places taken; for if, for all sets of k places taken, the function had not k poles from among them, then k has not been taken as small as possible. From these considerations we conclude that, *if $k_1, k_2, ..., k_m$ be m positive integers, whose sum is large enough, a rational function can be constructed whose poles are multiple, and at m given arbitrary places, the function being infinite to order k_1, or less, at the first place; to order k_2, or less, at the second place; and so on.* Such a function will not always be actually of the assigned orders of infinity $k_1, k_2, ..., k_m$, at these places; for example, for the equation $y^2 - u = 0$, where u is a polynomial in x of order $2p + 2$, with unrepeated linear factors, though, as a rule, a function can be constructed with $p + 1$ arbitrary poles, we cannot construct a function with a pole of order p at a place $x = a$, $y = b$, on the curve, and a pole of the order 1 at the conjugate place $x = a$, $y = -b$; as the reader may prove $(p > 1)$.

Modular expression of a rational function. Now let z denote a rational function which has a pole of the first order at

---

* Kronecker, *Crelle*, xci, xcii, 1882, *Werke*, Bd. ii. Dedekind u. Weber, *Crelle*, xcii.

each of m places, $A_1, ..., A_m$, and is not else infinite. Consider the aggregate of rational functions existing, which have poles only at these m places, of any possible orders. Denote by $(r_1, r_2, ..., r_m)$ such a function, actually infinite at these places respectively to orders $r_1, ..., r_m$; these numbers we call, for the present, the indices of the function. It is evidently possible to subtract from this function such a rational polynomial in z, of order r_m, that the difference is not infinite at A_m; for by hypothesis z is infinite at A_m only to the first order. Thus, save for polynomials in z, it is sufficient to consider, among the aggregate of functions $(r_1, r_2, ..., r_m)$, only those for which the m-th index, r_m, is zero. Considering then the other indices $r_1, r_2, ..., r_{m-1}$, in order, there will be one (or more) of these which is greater than the others, and, if there are several with this greatest value, there will be one, with this value, which occurs first; say this is r_i; thus we have $r_1 < r_i, r_2 < r_i, ..., r_{i-1} < r_i$, but $r_i \geqslant r_{i+1}, r_i \geqslant r_{i+2}, ..., r_i \geqslant r_{m-1}$, $(1 \leqslant i \leqslant m-1)$; when this is so, let the function $(r_1, r_2; ..., r_{m-1}, 0)$ be said to be *of the i-th class*, and to be *of dimension* r_i.

Since it is only when the aggregate order of infinity of a function is great enough, that the function can be constructed with pre-scribed orders of infinity at prescribed places, it is clear that, for all functions of the i-th class (not mere constants), there is a least dimension occurring among such functions of the i-th class as actually exist. Take now any function of the i-th class which has this least dimension; say this function is

$$(s_1, s_2, ..., s_{i-1}, r_i, s_{i+1}, ..., s_{m-1}, 0),$$

where

$$s_1 < r_i, s_2 < r_i, ..., s_{i-1} < r_i; \quad r_i \geqslant s_{i+1}, ..., r_i \geqslant s_{m-1}; \quad r_i > 0,$$
$$(i = 1, 2, ..., m-1);$$

we may speak of this function as *the chosen reduced function* of the i-th class. If we denote any other existing function (with m-th index equal to zero) which is of the i-th class by

$$(S_1, S_2, ..., S_{i-1}, R_i, S_{i+1}, ..., S_{m-1}, 0),$$

where

$$S_1 < R_i, S_2 < R_i, ..., S_{i-1} < R_i; \quad R_i \geqslant S_{i+1}, ..., R_i \geqslant S_{m-1}, R_i \geqslant r_i,$$
$$(i = 1, ..., m-1),$$

then, by choosing the constant λ suitably, in the difference

$$(S_1, S_2, ..., S_{i-1}, R_i, S_{i+1}, ..., S_{m-1}, 0)$$
$$- \lambda z^{R_i - r_i}(s_1, s_2, ..., s_{i-1}, r_i, s_{i+1}, ..., s_{m-1}, 0),$$

we can secure that, in the function given by this difference, say

$$(T_1, T_2, ..., T_{i-1}, R_i', T_{i+1}, ..., T_{m-1}, R_i - r_i),$$

we have $R_i' < R_i$, and also that any of the earlier indices T_k, where $k < i$, which certainly does not exceed the greater of S_k and $R_i - (r_i - s_k)$, is less than R_i; while any one of the later indices T_l, where $l > i$, which certainly does not exceed the greater of S_l and $R_i - (r_i - s_l)$, is at most as great as R_i. Thus, in this difference, all the first i indices are less than R_i, and the succeeding indices do not exceed R_i. Then further, forming the difference

$$(T_1, T_2, ..., T_{i-1}, R_i', T_{i+1}, ..., T_{m-1}, R_i - r_i) - (z, 1)_{R_i - r_i},$$

where the polynomial in z, denoted by $(z, 1)_{R_i - r_i}$, of order $R_i - r_i$, is to be chosen so that this is a function,

$$(S_1', S_2', ..., S'_{i-1}, R_i'', S'_{i+1}, ..., S'_{m-1}, 0),$$

with last index zero, we see that the index R_i'', which does not exceed the greater of R_i' and $R_i - r_i$, is $< R_i$; further that S_k', when $k < i$, not exceeding the greater of T_k and $R_i - r_i$, is $< R_i$; and that S_l', for $l > i$, not exceeding the greater of T_l and $R_i - r_i$, does not exceed R_i. The formation of this last difference is unnecessary if $R_i = r_i$. In any case, employing the process which is expressed by

$$(S_1, S_2, ..., S_{i-1}, R_i, S_{i+1}, ..., S_{m-1}, 0)$$
$$= \lambda z^{R_i - r_i} (s_1, ..., s_{i-1}, r_i, s_{i+1}, ..., s_{m-1}, 0) + (z, 1)_{R_i - r_i}$$
$$+ (S_1', ..., S'_{i-1}, R_i'', S'_{i+1}, ..., S'_{m-1}, 0),$$

we obtain a function, $(S_1', S_2', ..., S'_{i-1}, R_i'', S'_{i+1}, ..., S'_{m-1}, 0)$, of one of two kinds; (i), in which the greatest index is R_i, but, if so, this does not occur in the first i positions, or (ii), in which the greatest index is $< R_i$. In other words, this function obtained, is either of higher class, without change of dimension, or is of less dimension, than the function $(S_1, ..., S_{i-1}, R_i, S_{i+1}, ..., S_{m-1}, 0)$ from which it was obtained. There is, however, an upper bound for the class, namely $m - 1$; and there is a lower bound for the dimension of a function of specified class. Therefore, if we denote by h_i the chosen reduced function of class i, $(s_1, ..., s_{i-1}, r_i, s_{i+1}, ..., s_{m-1}, 0)$, for $i = 1, ..., (m - 1)$, and apply the above process of reduction continually, we shall be able to express any rational function whose poles are all at the places $A_1, ..., A_m$, in the form

$$(z, 1)_0 + (z, 1)_1 h_1 + ... + (z, 1)_{m-1} h_{m-1},$$

wherein $(z, 1)_i$ is a polynomial in z; and this in such a way that no term arises in this expression, of the form $z^\alpha h_\beta$, whose dimension is higher than that of the function given by this expression.

Fundamental integral functions. Now suppose that, in the fundamental irreducible algebraic equation $f(x, y) = 0$, the highest aggregate order in x and y is n, and that the term in y^n is actually present. As has been remarked, this can always be secured by a preliminary substitution for x, of the form $x + hy$. Then y becomes infinite only for infinite values of x, and may be spoken of as an *integral* function of x. Suppose further that, for $x = \infty$, there are n distinct places, there being no cycle of two, or more, of the n

values of y (or, in another description, no branch place for $x = \infty$). This also is easily secured, as we see by considering the homogeneous equation $F(x_0, x_1, x_2) = 0$ corresponding to $f(x, y) = 0$, and changing, if necessary, the line denoted by $x_2 = 0$; for the equation $f(x, y) = 0$ this comes to using functions of the forms $x(ax + by + c)^{-1}$, $y(ax + by + c)^{-1}$ in place of x and y.

Then, we apply the preceding theory, of rational functions with poles at m given places, to the case when the function z, there spoken of, is the function x (and thus $m = n$), and the chosen fundamental reduced functions are *integral* functions, becoming infinite only for $x = \infty$. These we denote by $g_1, \dots, g_{n-1}$; and the theorem proved is that any *integral* rational function, that is one having poles for $x = \infty$ only, is expressible in the form

$$u_0 + u_1 g_1 + \dots + u_{n-1} g_{n-1},$$

where $u_0, \dots, u_{n-1}$ are polynomials in x. The dimension of the function g_i will be denoted by $\sigma_i + 1$; this is the least positive integer such that $x^{-(\sigma_i+1)} g_i$ is not infinite at any of the n places for which $x = \infty$. This number is not zero, or the function g_i, having then no infinities, would be a constant; thus $\sigma_i \geqslant 0$. The sum of the $n-1$ numbers σ_i will be denoted by p (it being proved below that this notation is consistent with previous uses of p); the sum of the dimensions of $g_1, \dots, g_{n-1}$ is thus $p + n - 1$.

From these integral functions $g_1, \dots, g_{n-1}$, we can obtain such a set of functions $h_1, \dots, h_{n-1}$ as is given by the general theorem above for the case when $z = (x - a)^{-1}$, where a is any general constant. For let $h_i = (x - a)^{-(\sigma_i+1)} g_i$, so that h_i is infinite only when $x = a$, or $z = \infty$, using z for $(x - a)^{-1}$. Let H be any rational function which is infinite only when $x = a$, or $z = \infty$, and let $\sigma + 1$ be its dimension, so that $(x - a)^{\sigma+1} H$ is infinite only when $x = \infty$, and $\sigma + 1$ is the least positive integer for which this is so. Then we have an expression $(x - a)^{\sigma+1} H = u_0 + u_1 g_1 + \dots + u_{n-1} g_{n-1}$, where $u_0, \dots, u_{n-1}$ are polynomials in x. Let the orders of these polynomials be, respectively, $r_0, r_1, \dots, r_{n-1}$; then the fact that there occurs no term, in this expression, on the right, which is of higher dimension than that of the function $(x - a)^{\sigma+1} H$, involves that $r_i + \sigma_i + 1 \leqslant \sigma + 1$. The expression is, however, the same as

$$H = z^{\sigma+1} u_0 + \dots + z^{\sigma+1} u_i z^{-(\sigma_i+1)} h_i + \dots,$$

while u_i can be written as $z^{-r_i} v_i$, where v_i is a polynomial in z of order r_i. As $\sigma + 1 - r_i - (\sigma_i + 1)$, which we may denote by t_i, is not negative, the product $z^{t_i} v_i$, or w_i, say, is a polynomial in z; thus we have an expression $H = w_0 + \dots + w_i h_i + \dots$, for any rational function H which is infinite only when $z = \infty$, in terms of the functions h_i; and it is easy to see that the dimension of h_i, in terms of z, is $\sigma_i + 1$; and that no term occurs, in this expression, on the right, having dimension higher than that of H. Conversely, given a set of functions, $h_1, \ldots, h_{n-1}$, appropriate, in accordance with the general theorem, for the expression of rational functions which are infinite only when $x = a$, we can similarly deduce a set, $g_1, \ldots, g_{n-1}$, appropriate for the expression of all integral functions.

Important properties of the fundamental integral functions. There are certain important properties of the integral functions $g_1, \ldots, g_{n-1}$ which we now develop:

(a) There can be no identity of the form

$$U_0 + U_1 g_1 + \ldots + U_{n-1} g_{n-1} = 0,$$

where $U_0, \ldots, U_{n-1}$ are rational functions of x only. For as $y, y^2, \ldots, y^{n-1}$ are integral functions of x, each is expressible linearly by $g_1, \ldots, g_{n-1}$, with coefficients which are polynomials in x; if there were such a linear equation connecting $g_1, \ldots, g_{n-1}$ as we have spoken of, then all of $y, y^2, \ldots, y^{n-1}$ would be expressible linearly by at most $n-2$ of the functions $g_1, \ldots, g_{n-1}$, with coefficients rational in x only. There would thus be a linear relation connecting $y, \ldots, y^{n-1}$ with coefficients rational in x. This is contrary to the hypothesis that the original equation, $f(x, y) = 0$, is irreducible.

(b) Consider a value $x = c$, for which all the n places of $f(x, y) = 0$ are distinct; at one of these n places let $g_1 = a_1, g_2 = a_2, \ldots, g_{n-1} = a_{n-1}$; at another of these let $g_1 = b_1, g_2 = b_2, \ldots, g_{n-1} = b_{n-1}$. We prove that not all the $n-1$ differences $a_1 - b_1, a_2 - b_2, \ldots, a_{n-1} - b_{n-1}$ are zero. For choose the constants $\lambda_1, \lambda_2, \ldots, \lambda_{n-1}$ so that the function $\lambda_1(g_1 - a_1) + \ldots + \lambda_{n-1}(g_{n-1} - a_{n-1})$, which vanishes at one of the two places spoken of, vanishes also at the $n-2$ places for which $x = c$, other than these two places. Then, if all the differences $a_1 - b_1, \ldots, a_{n-1} - b_{n-1}$ were zero, the function

$$[\lambda_1(g_1 - a_1) + \ldots + \lambda_{n-1}(g_{n-1} - a_{n-1})]/(x - c)$$

would not become infinite at any of the n places for which $x = c$; it would thus be an integral function, and, therefore, expressible linearly in terms of $g_1, \ldots, g_{n-1}$ with coefficients which are polynomials in x. We should then have such an identity connecting $g_1, \ldots, g_{n-1}$ as was shewn impossible in (a).

(c) The result (b) is a particular case of a much more general proposition, to which we now proceed. Consider a finite value of x, $x = c$, for which the n corresponding values of y, to take the most general possibility, form several cycles, there being one cycle of r values, $y^{(1)}, \ldots, y^{(r)}$, then another cycle of s values, and so on. There will then be one place expressed by a pair of equations of the forms $x - c = t^r$, $y = A + A_1 t + \ldots$, in terms of the local parameter t; then another place expressed by a pair of equations of the forms $x - c = t_1^s$, $y = B + B_1 t_1 + B_2 t_1^2 + \ldots$; and so on. For the first place, let the corresponding values, for any one, g_k, of the functions $g_1, \ldots, g_{n-1}$, in the neighbourhood of the place, be given by

$$g_k^{(1)} = a_k + a_k't + \tfrac{1}{2}a_k''t^2 + \ldots, \quad g_k^{(\alpha)} = a_k + a_k't_\alpha + \tfrac{1}{2}a_k''t_\alpha^2 + \ldots,$$

where $k = 1, \ldots, (n-1)$, $\alpha = 2, 3, \ldots, r$, and t_α is $\epsilon^{\alpha-1}t$, with

$$\epsilon = \exp(2\pi i/r).$$

For the second place there will be corresponding expressions, with $b_k, b_k', \ldots$ in place of $a_k, a_k', \ldots$, and s in place of r; and so on for all the places occurring for $x = c$. The theorem to be proved is that a certain determinant A_0, of n rows and columns, formed from the coefficients $a_k, a_k', \ldots, b_k, \ldots$ in these expansions, is not zero. This determinant is

$$A_0 = \begin{vmatrix} 1, & a_1, & a_2, & \ldots, & a_{n-1} \\ 0, & a_1', & a_2', & \ldots, & a'_{n-1} \\ 0, & \tfrac{1}{2}a_1'', & \tfrac{1}{2}a_2'', & \ldots, & \tfrac{1}{2}a''_{n-1} \\ \hdotsfor{5} \\ 1, & b_1, & b_2, & \ldots, & b_{n-1} \\ 0, & b_1', & b_2', & \ldots, & b'_{n-1} \\ 0, & \tfrac{1}{2}b_1'', & \tfrac{1}{2}b_2'', & \ldots, & \tfrac{1}{2}b''_{n-1} \\ \hdotsfor{5} \\ \hdotsfor{5} \end{vmatrix},$$

whose construction is quite simple. The second, third, ... columns are formed from coefficients in the expansions, respectively, of $g_1, g_2, \ldots, g_{n-1}$; the first r rows arise from these expansions at the first place considered for $x = c$, containing the coefficients of the powers $t^0, t^1, \ldots, t^{r-1}$ of the parameter, at this place; the following s rows similarly arise for the second place; and so on. It is in the nature of the case that $r + s + \ldots = n$. If this determinant were zero it would be possible to find constants, $\lambda, \lambda_1, \ldots, \lambda_{n-1}$, so that the n expressions, formed from the rows of this determinant,

$$\lambda + \lambda_1 a_1 + \ldots + \lambda_{n-1}a_{n-1}, \quad \lambda_1 a_1' + \ldots + \lambda_{n-1}a'_{n-1}, \quad \ldots,$$
$$\lambda_1 a_1^{(r-1)} + \ldots + \lambda_{n-1}a_{n-1}^{(r-1)},$$
$$\lambda + \lambda_1 b_1 + \ldots + \lambda_{n-1}b_{n-1}, \quad \lambda_1 b_1' + \ldots + \lambda_{n-1}b'_{n-1}, \quad \ldots,$$
$$\lambda_1 b_1^{(s-1)} + \ldots + \lambda_{n-1}b_{n-1}^{(s-1)},$$

$$\ldots\ldots\ldots\ldots\ldots\ldots\ldots$$

were all zero. This would mean, however, that the function $\lambda + \lambda_1 g_1 + \ldots + \lambda_{n-1}g_{n-1}$ vanishes to order r at the first place, to order s at the second place, and so on; and thence that the function $(\lambda + \lambda_1 g_1 + \ldots + \lambda_{n-1}g_{n-1})/(x-c)$ is not infinite for $x = c$; this function would then be an integral function, expressible linearly by $g_1, \ldots, g_{n-1}$, with coefficients which are polynomials in x. There would then exist an identity such as is proved impossible in (a). Particular consequences of this theorem are, first, the theorem (b), that if there be no branch place for $x=c$, not every one of the $(n-1)$ differences $a_1-b_1, a_2-b_2, \ldots, a_{n-1}-b_{n-1}$ vanishes; second, that in the expansions of the $(n-1)$ integral functions g_k $(k=1, \ldots, n-1)$ at the first branch place considered at $x=c$, the coefficients of t^m cannot vanish in all these functions, for any m such that $0<m<r$; since otherwise the determinant A_0 would have a row of vanishing elements $0, a_1^{(m)}, a_2^{(m)}, \ldots, a_{n-1}^{(m)}$. Thus, for general values of $\lambda, \lambda_1, \ldots, \lambda_{n-1}$, the expansion of the function $\lambda+\lambda_1 g_1+\ldots+\lambda_{n-1}g_{n-1}$, in terms of the parameter t, at a branch place of $f(x, y)=0$ (occurring for a finite value of x, in whose neighbourhood $x=c+t^r$, with y of the form $d+d_1 t+\ldots$) contains all powers of t from t to t^{r-1}.

Employing the same notation, we can deduce, from the non-vanishing of the determinant of constants, A_0, a general functional theorem. It is convenient to formulate this as an identity connecting matrices. Consider a general (finite) value of x for which the corresponding n places of $f(x, y)=0$ are distinct, let $g_k^{(1)}, \ldots, g_k^{(n)}$ denote the values of the integral function g_k at these n places. Denote by G the matrix of n rows and columns whose i-th row consists of the elements $1, g_1^{(i)}, g_2^{(i)}, \ldots, g_{n-1}^{(i)}$; it is a consequence of what we have shewn that the determinant of this matrix would only vanish if there were branch places for this value of x. Denote also by $\bar{G}$ the matrix formed from G by transposition of rows and columns. Then the product matrix $\bar{G}G$ has for elements in its first row, $n, s_1, \ldots, s_{n-1}$, where $s_k=g_k^{(1)}+\ldots+g_k^{(n)}$, and for elements in its i-th row, $s_{i-1}, s_{i-1,1}, \ldots, s_{i-1,n-1}$, where

$$s_{h,k}=g_h^{(1)}g_k^{(1)}+\ldots+g_h^{(n)}g_k^{(n)},$$

so that $s_{h,k}=s_{k,h}$. Each of these elements is the sum of the n conjugate values, for the same value of x, of a function which only becomes infinite when $x=\infty$; thus each element is expressible as a polynomial in x only. The determinant of the matrix $\bar{G}G$ is thus also a polynomial in x. One consequence of the theorem we shall obtain is that, corresponding to a place where there is a branching of index r (about which the expressions for x and y are of the forms $x-c=t^r$, $y=B+B_1 t+\ldots$, in terms of the parameter at the place), there is a factor of the determinant of $\bar{G}G$ equal to $(x-c)^{r-1}$. By what was proved above there is no such vanishing factor in this determinant unless there is a branch place. It will follow, then, that the order of the polynomial in x given by the determinant of $\bar{G}G$, is the sum of the various values of $r-1$ at all the existing branch places occurring for finite values of x. Consider such a branch place, of index r; put $\epsilon=\exp(2\pi i/r)$; let M_s denote the determinant, of s rows and columns, whose (λ, μ)-th element is ϵ^σ, where $\sigma = (\lambda-1)(\mu-1)$, namely

$$M_s = \begin{vmatrix} 1, & 1 & , & ..., & 1 \\ 1, & \epsilon & , & ..., & \epsilon^{s-1} \\ 1, & \epsilon^2 & , & ..., & \epsilon^{2(s-1)} \\ \multicolumn{5}{c}{\dotfill} \\ 1, & \epsilon^{s-1}, & & ..., & \epsilon^{(s-1)^2} \end{vmatrix} ;$$

denote the cofactor of the ρ-th element of the last column of this determinant by $H_{s,\rho}$, and put $K_{s,\rho} = H_{s,\rho}/M_s$. We evidently have $H_{s,s} = M_{s-1}$, and, hence, $K_{s,s} = M_{s-1}/M_s$. By these definitions, the following sum, of s terms,

$$K_{s,1} + \epsilon^{\rho-1} K_{s,2} + \epsilon^{2(\rho-1)} K_{s,3} + ... + \epsilon^{(s-1)(\rho-1)} K_{s,s},$$

vanishes for $\rho < s$, and is unity when $\rho = s$. Consider the expansion of the function g_k at the branch place of index r, namely

$$g_k^{(1)} = a_k + a_k' t + \tfrac{1}{2} a_k'' t^2 + ...,$$

and the other expansions when t is replaced by ϵt, $\epsilon^2 t$, ..., $\epsilon^{s-1} t$, of the forms

$$g_k^{(\lambda)} = a_k + a_k' \epsilon^{\lambda-1} t + \frac{a_k''}{2!} \epsilon^{2(\lambda-1)} t^2 + ..., \qquad (\lambda = 2, ..., s, \ s \leqslant r);$$

we infer that

$$K_{s,1} g_k^{(1)} + K_{s,2} g_k^{(2)} + ... + K_{s,s} g_k^{(s)} = \frac{1}{(s-1)!} a_k^{(s-1)} t^{s-1} \\ + A_k^{(s)} t^s + A_k^{(s+1)} t^{s+1} + ...,$$

where $A_k^{(s)}, A_k^{(s+1)}, ...,$ are proper constants. Now, recall the form of the matrix G, and the first r rows of this, which, in a general column, contain $g_k^{(1)}, g_k^{(2)}, ..., g_k^{(r)}$. By what we have remarked, if we multiply the first two rows of this matrix G respectively by $K_{2,1}$ and $K_{2,2}$, and add the results to form a new second row, this (taking $s = 2$ in what was remarked) will consist of elements

$$0, \ \frac{1}{1!} a_1' t + (t^2), \ \frac{1}{1!} a_2' t + (t^2), \ ..., \ \frac{1}{1!} a'_{n-1} t + (t^2),$$

where (t^2) denotes, in every case, a power series beginning with a term in t^2. Similarly, taking $s = 3$, if we multiply the first three rows of G respectively by $K_{3,1}, K_{3,2}, K_{3,3}$, and add the results to form a new third row, the elements of this row will be

$$0, \ \frac{1}{2!} a_1'' t^2 + (t^3), \ \frac{1}{2!} a_2'' t^2 + (t^3), \ ..., \ \frac{1}{2!} a''_{n-1} t^2 + (t^3),$$

where (t^3) denotes, in every case, a power series beginning with a term in t^3. The process may be continued, the last operation being for $s = r$, and we consider all the first r rows of the matrix G. We may repeat these statements in matrix form by writing

$$
\begin{pmatrix}
1 & , & \cdot & \cdot & \cdots, & \cdot \\
K_{2,1}, & K_{2,2}, & \cdot & , & \cdots, & \cdot \\
K_{3,1}, & K_{3,2}, & K_{3,3}, & \cdots, & & \cdot \\
\hdotsfor{6} \\
K_{r,1}, & K_{r,2}, & K_{r,3}, & \cdots, & K_{r,r}
\end{pmatrix}
\begin{pmatrix}
1, & g_1^{(1)}, & \cdots, & g_{n-1}^{(1)} \\
1, & g_1^{(2)}, & \cdots, & g_{n-1}^{(2)} \\
\hdotsfor{4} \\
1, & g_1^{(r)}, & \cdots, & g_{n-1}^{(r)}
\end{pmatrix}
$$

$$
= \begin{pmatrix}
1, & a_1 + (t) & , & \cdots, & a_{n-1} + (t) \\
0, & a_1' t + (t^2) & , & \cdots, & a'_{n-1} t + (t^2) \\
\hdotsfor{5} \\
0, & \dfrac{a_1^{(r-1)}}{(r-1)!} t^{r-1} + (t^r), & & \cdots, & \dfrac{a_{n-1}^{(r-1)}}{(r-1)!} t^{r-1} + (t^r)
\end{pmatrix},
$$

where the first matrix on the left has r rows and columns, with zeros to the right of the diagonal, the second matrix on the left consists of the first r rows of the matrix G, having r rows and n columns, as has also the matrix on the right. This latter matrix can be written as the product of the diagonal matrix of r rows and columns whose diagonal elements are $1, t, t^2, \ldots, t^{r-1}$, multiplied into a matrix of r rows and n columns, which does not vanish when $t = 0$, but reduces to the first r rows occurring in the determinant A_0 above, namely as

$$
\begin{pmatrix}
1 & & & \\
& t & & \\
& & t^2 & \\
& & & \ddots \\
& & & & t^{r-1}
\end{pmatrix}
\begin{pmatrix}
1, & a_1 + (t) & , & \cdots, & a_{n-1} + (t) \\
0, & a_1' + (t) & , & \cdots, & a'_{n-1} + (t) \\
\hdotsfor{5} \\
0, & \dfrac{a_1^{(r-1)}}{(r-1)!} + (t), & & \cdots, & \dfrac{a_{n-1}^{(r-1)}}{(r-1)!} + (t)
\end{pmatrix},
$$

where the unmarked elements of the multiplying matrix are zeros. If there be, for the same value of x, another branch place, with a cycle of r_1 values of y, an exactly similar reduction can be carried out for the r_1 rows of G which follow the first r rows; therein, the appropriate local parameter being t_1, we should use $\epsilon_1 = \exp(2\pi i/r_1)$ instead of ϵ, and in place of coefficients $a_k, a_k', a_k'', \ldots$, we should have coefficients $b_k, b_k', b_k'', \ldots$. And so for all other branch places which occur for the same value of x. The final result will be an equation in matrices, which we write $KG = \delta A$, and now explain: Every one of K, δ, A is a matrix of n rows and columns, as is G. The matrix K has, in the first r rows, non-vanishing elements only in the first r columns, these forming the matrix of r rows and columns, with numerical elements $K_{s,\rho}$, which has been made explicit above; in the r_1 rows of K which follow the first r rows, K has elements only

in the r_1 columns which follow the r-th column, these being formed
from ϵ_1 just as the elements in the first r rows and columns were
formed from ϵ; succeeding rows of K similarly break up into batches,
corresponding to the other branch places which occur for the value
of x considered, there being zero elements in the first $r + r_1$ columns,
in the first $r + r_1 + r_2$ columns, and so on. The matrix δ is the simple
diagonal matrix, whose non-vanishing elements, all in the diagonal,
are $1, t, \ldots, t^{r-1}, 1, t_1, \ldots, t_1^{r_1-1}$, etc., in order. Last, the matrix A is
one of which every element is a power series, these being series in t
in the first r rows, series in t_1 in the succeeding r_1 rows, and so on;
and, what is important, these series are such that when all of
$t, t_1, \ldots$ are made to vanish, the elements of A reduce to those of the
determinant A_0. It was proved that the determinant A_0 is not zero.
Further, in discussing the first r rows of the matrix K, we remarked
an equation which was written $K_{s,s} = M_{s-1}/M_s$; it follows that the
determinant of the matrix K is the product

$$M_1^{-1} . (M_1/M_2) . \ldots . (M_{r-1}/M_r) . (M_1^{(1)})^{-1} . (M_1^{(1)}/M_2^{(1)}) . \ldots$$
$$. (M_{r_1-1}^{(1)}/M_{r_1}^{(1)}) . \ldots,$$

where $M_1^{(1)}, M_2^{(1)}, \ldots$ are formed from ϵ_1 just as were $M_1, M_2, \ldots$,
from ϵ, and so for succeeding factors. Whence the determinant of
the matrix K is $[M_r M_{r_1}^{(1)} \ldots]^{-1}$.

Many consequences follow from the equation $KG = \delta A$, as will be
seen. One consequence, already remarked, relates to the determinant of the matrix G. When x is near to the value $x = c$ for which
the branch places are those discussed, it follows from $KG = \delta A$ that
the determinant of G is the product of a factor which does not
vanish when $x = c$ by the factor $t^{1+2+\cdots+(r-1)} . t_1^{1+2+\cdots+(r_1-1)} \ldots$, which
is $t^{\frac{1}{2}r(r-1)} t_1^{\frac{1}{2}r_1(r_1-1)} \ldots$. We have spoken of the determinant of the
matrix $\bar{G}G$; this is the square of the determinant of G. It is a
polynomial in x, having, we see, the factor $(x-c)^{(r-1)+(r_1-1)+\cdots}$. We
denote the polynomial in x, which is the determinant of $\bar{G}G$, by Δ;
for every finite value of x, we have separated Δ into a vanishing
factor, and a non-vanishing factor; we thus see that, save for a
constant, Δ is the product of all factors $(x-a)^{k-1}$, arising for values
$x = a$ at which there is a branch place of index k, there being repetitions of the factor $x - a$, with appropriate exponents, when several
branch places occur for the same value $x = a$. Now put $w = \Sigma(r-1)$,
the summation extending to all existing branch places (since we
have arranged that there shall be no branch places for $x = \infty$); then
Δ is a polynomial in x of order w. We have seen, however, (p. 150
above), that we may write G in the form $x^{(\sigma_1+1)+(\sigma_2+1)+\cdots}H$, where
H is a matrix, of n rows and columns, with i-th row consisting of
$1, h_1^{(i)}, \ldots, h_{n-1}^{(i)}$, in which $h_s = x^{-(\sigma_s+1)}g_s$, and h_s is a function
becoming infinite only for $x = 0$ (this being the definition of h_s

previously employed), and we can suppose $x = 0$ chosen so that no branch place arises for $x = 0$. Then, by what has been proved here for G, it follows that the determinant of the matrix H does not vanish for $x = \infty$. Wherefore, Δ is a polynomial in x which becomes infinite for $x = \infty$ to order $2 (\sigma_1 + 1 + \sigma_2 + 1 + \ldots)$, that is, to order $2p + 2n - 2$, if p, as above, be defined as $\sigma_1 + \ldots + \sigma_{n-1}$. In other words, the polynomial Δ is of this order. Thus we have the equation $w = 2n + 2p - 2$. It is not, in fact, necessary to choose $x = \infty$ so that no branch places occur for this value. By a similar argument it may be shewn that, if W denote the sum $(r - 1) + (r_1 - 1) + \ldots$, extended to all the branch places at $x = \infty$, of respective indices r, r_1, $\ldots$, then the order of the polynomial Δ is $2n + 2p - 2 - W$. Thus already it is shewn, if we recur to a result previously found (Chap. IV, p. 83, cf. Chap. V, p. 131), that p as defined here agrees with p as used before; but we shall continue to suppose p defined as here.

The importance of the matrix G, or of the polynomial Δ, may be brought out by brief reference to some of its properties. Let $\eta_1, \ldots, \eta_n$ be any n integral functions whatever, subject to the condition that there exists (in virtue of the fundamental equation $f(x, y) = 0$) no identity $U_1 \eta_1 + \ldots + U_n \eta_n = 0$, in which $U_1, \ldots, U_n$ are polynomials in x. Denoting by $\eta_k^{(1)}, \ldots, \eta_k^{(n)}$ the n conjugate values of η_k, for a general finite value of x, we may consider the square of the determinant, of n rows and columns, whose general row is $\eta_1^{(i)}, \eta_2^{(i)}, \ldots, \eta_n^{(i)}$. Each of the functions η_k, by the properties of the fundamental integral functions $g_1, \ldots, g_{n-1}$, is expressible linearly in terms of these latter, with coefficients which are polynomials in x (that is homogeneously in terms of $1, g_1, \ldots, g_{n-1}$). Hence it follows that the square of the determinant $| \eta_k^{(i)} |$ is of the form $\nabla^2 \Delta$, where Δ is as before, and ∇ is a polynomial in x. It is usual to call the square of the determinant $| \eta_k^{(i)} |$ the discriminant of the system $(\eta_1, \ldots, \eta_n)$; it appears then that Δ, which is a polynomial in x, is the highest common factor of the discriminants of all possible systems of n linearly independent integral functions. In general, though, when $u_1, \ldots, u_n$ are polynomials in x, any function $u_1 \eta_1 + \ldots + u_n \eta_n$ is an integral function, it is not conversely the case that every integral function can be so expressed; there may exist integral functions whose expression in terms of $\eta_1, \ldots, \eta_n$ is of the form $(u_1 \eta_1 + \ldots + u_n \eta_n)/D$, where D, like $u_1, \ldots, u_n$, is a polynomial in x. It can be shewn that the only polynomials D that can arise in this way, as denominators in the expression of integral functions by means of $\eta_1, \ldots, \eta_n$, are factors of the particular polynomial ∇, whose square arises in the equation just given for the discriminant of $\eta_1, \ldots, \eta_n$.

The functions reciprocal to the fundamental integral functions. We proceed now to build up, from the fundamental integral functions $g_1, \ldots, g_{n-1}$, a set of n important rational functions, which will be denoted by $\gamma_0, \gamma_1, \ldots, \gamma_{n-1}$. We recall the expression of the polynomial Δ as a determinant whose elements are the polynomials in x which we denoted by s_k, s_{ij} ($s_0 = n$); consider the minors of this determinant, which may be briefly represented by such notations as $\partial \Delta / \partial s_k$ and $\partial \Delta / \partial s_{ij}$. These are polynomials in x; the function $\partial \Delta / \partial s_k + g_1 \partial \Delta / \partial s_{k,1} + \ldots + g_{n-1} \partial \Delta / \partial s_{k,n-1}$ is thus an integral function; it is this function, divided by Δ, which is the function γ_k. By what has been proved in regard to Δ, this function, for finite values of x, is only infinite at branch places. To find the character of this infinity, and to make clear the simplicity of the matter, it will be best to approach the definition of γ_k in another way, then shewn to be equivalent to that just given. For this we define a matrix Γ, of n rows and columns, of which the general row is represented by $\gamma_0^{(i)}, \gamma_1^{(i)}, \ldots, \gamma_{n-1}^{(i)}$, and prove that the elements of a column of this, $\gamma_k^{(1)}, \gamma_k^{(2)}, \ldots, \gamma_k^{(n)}$, are the n conjugate values of a single function γ_k, for all the places corresponding to a general finite value of x. The definition in question is that Γ is the inverse of the matrix obtained from G by transposition of its rows and columns, or $\Gamma = (\bar{G})^{-1}$. This definition is equivalent to $\Gamma^{-1} = \bar{G}$, and, therefore, equivalent to either of the two equations $\Gamma \bar{G} = 1$, $\bar{G} \Gamma = 1$. From the former of these we have $\Gamma \bar{G} G = G$, so that $\Gamma = G(\bar{G} G)^{-1}$, and $\bar{G} G$ is the matrix whose general element is $s_{i,j}$, with $s_{0,0} = n$, $s_{0,j} = s_j$, $s_{i,j} = s_{j,i}$, $(i, j = 0, 1, \ldots, n-1)$, which we met with earlier (p. 153); from this, combining the i-th row of G with the $(k+1)$-th column of $(\bar{G} G)^{-1}$, we obtain the element $\gamma_k^{(i)}$ in the form

$$M_k + g_1^{(i)} M_{k,1} + \ldots + g_{n-1}^{(i)} M_{k,n-1},$$

where $M_{k,h}$ is $\Delta^{-1} \partial \Delta / \partial s_{k,h}$, $M_k = \Delta^{-1} \partial \Delta / \partial s_k$; these are rational in x only; for the n values of i, therefore, the elements $\gamma_k^{(i)}$ are the conjugate values of the function γ_k defined as at first. We have assumed the well-known result that the (i, j) element of the inverse of a matrix $(a_{i,j})$ is obtained by dividing the cofactor of $a_{j,i}$, in the determinant of $(a_{i,j})$, by this determinant itself.

Remark. The definition $\Gamma \bar{G} = 1$ is equivalent with the n^2 equations such as

$$\gamma_0^{(i)} + \gamma_1^{(i)} g_1^{(i)} + \ldots + \gamma_{n-1}^{(i)} g_{n-1}^{(i)} = 1,$$
$$\gamma_0^{(i)} + \gamma_1^{(i)} g_1^{(j)} + \ldots + \gamma_{n-1}^{(i)} g_{n-1}^{(j)} = 0,$$

for $i, j = 1, \ldots, n$, $i \neq j$. These we may abbreviate by

$$\gamma_0 + \gamma_1 g_1 + \ldots + \gamma_{n-1} g_{n-1} = 1, \quad \gamma_0 + \gamma_1 g_1' + \ldots + \gamma_{n-1} g'_{n-1} = 0,$$

for a value of x for which there is no branch place. And if y, y' be such that $f(x, y) = 0$, $f(x, y') = 0$, the same or different, and $f'(y)$

denote $\partial f/\partial y$, these abbreviated equations are included in the formal equation

$$\gamma_0 + \gamma_1 g_1' + \dots + \gamma_{n-1} g'_{n-1} = lim_{y'=y}\,[f(x, y') - f(x, y)]/(y' - y) f'(y).$$

Hence it may be seen that the function γ_k is of the form $\psi_k/f'(y)$, where ψ_k, or $\psi_k(y, x)$, is a polynomial in x and y. And, when the integral functions $g_1, \dots, g_{n-1}$, say $g_1(y, x), \dots, g_{n-1}(y, x)$, are given in explicit form, each as a quotient of a polynomial in x and y divided by a polynomial in x only, the polynomial $\psi_k(y, x)$, and hence γ_k, is determined by the equation

$$\psi_0(y, x) + \psi_1(y, x)\,g_1(y', x) + \dots + \psi_{n-1}(y, x)\,g_{n-1}(y', x)$$
$$= [f(x, y') - f(x, y)]/(y' - y),$$

regarded as an identity in y and y'. All the polynomials $\psi_k(y, x)$, and the numerator polynomials in $g_k(y, x)$, are understood to be reduced by $f(x, y) = 0$, so as not to involve any power of y higher than the $(n-1)$-th. For a particular instance, if $g_1, \dots, g_{n-1}$ were $y, y^2, \dots, y^{n-1}$, the polynomials $\psi_0, \dots, \psi_{n-1}$ would be the coefficients of the powers of y' in $[f(x, y') - f(x, y)]/(y' - y)$, that is, ψ_k would be the coefficient of $(y')^k$, and of order $n - 1 - k$ in y.

The preceding definition of the matrix Γ, if we use the equation, proved above, $KG = \delta A$, enables us to determine the maximum possible orders of infinity of the rational functions γ_k at a branch place; we have seen that (for finite values of x) these functions are only infinite at a branch place. For this equation gives $\bar{G}\bar{K} = \bar{A}\bar{\delta}$, while $\bar{\delta} = \delta$, and hence $(\bar{K})^{-1}(\bar{G})^{-1} = \delta^{-1}(\bar{A})^{-1}$, so that Γ, equal to $(\bar{G})^{-1}$, is equal to $\bar{K}\delta^{-1}(\bar{A})^{-1}$. Herein $\bar{K}$ is a matrix of constant numerical quantities, and $(\bar{A})^{-1}$, the inverse of a matrix, $\bar{A}$, of which the determinant does not vanish at a branch place, as we have seen, is likewise of non-vanishing determinant at the branch place. But δ^{-1} is the diagonal matrix whose elements are $1, t^{-1}, \dots, t^{-(r-1)}, 1, t_1^{-1}, \dots, t_1^{-(r_1-1)}$, etc., wherein $t, t_1, \dots$ are the local parameters for the places arising for the value of x under consideration, these being branch places of respective indices $r, r_1, \dots$. Thus it is clear that, at the first of these branch places, no one of the functions $\gamma_0, \gamma_1, \dots, \gamma_{n-1}$ becomes infinite to order higher than the $(r-1)$-th; and so, then, for any other branch place.

This statement is for a finite value of x; we can easily deduce the corresponding fact for $x = \infty$. For this, we consider the rational functions, denoted as before by $h_1, \dots, h_{n-1}$, in terms of which all rational functions, which are infinite only for a finite value of x, say $x = c$, can be linearly expressed. We can suppose c taken so that there are no branch places for $x = c$. From these functions $h_1, \dots, h_{n-1}$ we can deduce functions $\mathfrak{I}_0, \mathfrak{I}_1, \dots, \mathfrak{I}_{n-1}$, just as $\gamma_0, \gamma_1, \dots, \gamma_{n-1}$ were deduced from $g_1, \dots, g_{n-1}$; as will appear, these functions are very simply connected with $\gamma_0, \ldots, \gamma_{n-1}$. Putting z for $(x-c)^{-1}$, the behaviour of $\vartheta_0, \vartheta_1, \ldots, \vartheta_{n-1}$ for $z=0$ follows from what has been found for $\gamma_0, \gamma_1, \ldots, \gamma_{n-1}$ for any finite value of x; but $z=0$ when $x=\infty$; thus the behaviour of $\gamma_0, \ldots, \gamma_{n-1}$ for $x=\infty$ is easily found. In fact, we have seen above that $h_k=g_k(x-c)^{-(\sigma_k+1)}$; hence, if the matrix H be that defined from $h_1, \ldots, h_{n-1}$ just as was G from $g_1, \ldots, g_{n-1}$, namely to have for its i-th row the elements $1, h_1^{(i)}, h_2^{(i)}, \ldots, h_{n-1}^{(i)}$, we have $H=GC$, where C is the diagonal matrix whose elements in turn are $1, (x-c)^{-(\sigma_1+1)}, (x-c)^{-(\sigma_2+1)}, \ldots, (x-c)^{-(\sigma_{n-1}+1)}$. If then the matrix Θ be expressed in terms of the functions $\vartheta_0, \vartheta_1, \ldots, \vartheta_{n-1}$, to be defined, just as Γ was expressed in terms of the functions $\gamma_0, \gamma_1, \ldots, \gamma_{n-1}$, so that the i-th row of Θ is $\vartheta_0^{(i)}, \vartheta_1^{(i)}, \ldots, \vartheta_{n-1}^{(i)}$, and $\Theta=(\bar{H})^{-1}$, we have $\Theta=(\bar{G})^{-1}C^{-1}=\Gamma C^{-1}$, and C^{-1} is the diagonal matrix whose elements are $1, (x-c)^{\sigma_1+1}, \ldots, (x-c)^{\sigma_{n-1}+1}$. Hence we infer that $\vartheta_0=\gamma_0$ and $\vartheta_k=(x-c)^{\sigma_k+1}\gamma_k$, for $k=1, \ldots, n-1$. The equation $\gamma_k=z^{\sigma_k+1}\vartheta_k$ shews that if, near a place for which $z=0$, and $z=t^r$, the function ϑ_k be infinite like $t^{-(r-1)}$, then γ_k vanishes to order $r\sigma_k+1$, which is >0 even if $\sigma_k=0$; for greater ease of statement we have supposed that for $x=\infty$ (or $z=0$), there are no branch places, and then the conclusion is that γ_k vanishes, for $x=\infty$, to order σ_k+1 at least. For γ_0 (which is ϑ_0), the statement is that, at a branch place at $x=\infty$, of index r, the function γ_0 is not infinite to order greater than $r-1$; and is finite if there be no such branch place.

General modular form of the everywhere finite integrals.
The reason for introducing the functions γ_k is that they enable us to express the everywhere finite integrals in a simple explicit form. Suppose that $R(x, y)$, or R, is such a rational function of x and y, subject to the fundamental equation $f(x, y)=0$, that the product $R\, dx/dt$, for the neighbourhood of every place (including places at $x=\infty$), when expressed by powers of the parameter t appropriate to that place, involves no negative powers of t. If R be infinite at a place, for which x is finite, which is a branch place of index r, this involves that R cannot be there infinite to order greater than $r-1$. Let g_k be any one of the integral functions $g_1, \ldots, g_{n-1}$, and, as before, let $g_k^{(1)}, \ldots, g_k^{(n)}$ be the conjugate values of g_k for any general finite value of x; and similarly $R^{(1)}, \ldots, R^{(n)}$ be the conjugate values, thereat, of the rational function $R(x, y)$. Then the sum $R^{(1)}g_k^{(1)}+\ldots+R^{(n)}g_k^{(n)}$ is expressible as a rational function of x only; and this sum can only be infinite for a finite value of x if $R(x, y)$ be infinite for this value; and, moreover, by the property of R, if such infinity arise for a branch place of index r, the sum can be infinite there at most to order $r-1$. Thus, if Π denote the product of the factors $x-c$, extended to all different finite values of c for which branch places exist, the product $[R^{(1)}g_k^{(1)}+\ldots+R^{(n)}g_k^{(n)}]\Pi$

is a rational function of x only, not becoming infinite for any finite value of x, and indeed vanishing for every value c of x for which branch places exist. Thus it is a polynomial in x which divides by Π; we denote it by $U_k\Pi$, so that U_k is a polynomial in x. Hence, with the usual notation, by which the n linear functions

$$a_{i1}x_1 + \ldots + a_{in}x_n$$

(for $i=1, \ldots, n$) are represented by $a(x_1, \ldots, x_n)$, where a is the matrix (a_{ij}), we have, if we recall the definition of the matrix G, the equation $\bar{G}(R^{(1)}, \ldots, R^{(n)}) = (U_0, \ldots, U_{n-1})$; and hence, from $\Gamma\bar{G}=1$, we have $(R^{(1)}, \ldots, R^{(n)}) = \Gamma(U_0, \ldots, U_{n-1})$. This is the same as $R(x, y) = U_0\gamma_0 + U_1\gamma_1 + \ldots + U_{n-1}\gamma_{n-1}$, where $U_0, \ldots, U_{n-1}$ are polynomials in x only. Conversely, if we take any rational function of this form, since no one of $\gamma_0, \gamma_1, \ldots, \gamma_{n-1}$ is infinite at a branch place of index r, occurring for a finite value of x, to higher than the $(r-1)$-th order, this form ensures that $R\,dx/dt$ is finite for the neighbourhood of every finite value of x. To examine, however, the implication of the condition that $R\,dx/dt$ is finite also for $x=\infty$, we proceed as before. We put $(x-c)^{-1}=z$, where c may be chosen so that no branch places occur for $x=c$, and discuss $R\,dx/dt$, as a function of z, for $z=0$. Let $m_0, m_1, \ldots, m_{n-1}$ be the respective orders of the polynomials in x, $U_0, \ldots, U_{n-1}$; put $V_i = z^{m_i}U_i$, or $(x-c)^{-m_i}U_i$, so that V_i is a polynomial in z of order m_i. Also, as above, let $\gamma_k = z^{\sigma_k+1}\mathfrak{S}_k$, $\gamma_0 = \mathfrak{S}_0$. Then the preceding form for R, obtained by considering finite values of x, leads, if we introduce $\sigma_0, = -1$, to

$$R\frac{dx}{dt} = -\left(\sum_{i=0}^{n-1} z^{\sigma_i-1-m_i} V_i \mathfrak{S}_i\right)\frac{dz}{dt};$$

the condition for $R\,dx/dt$, at $x=\infty$, requires then, as a necessary and sufficient condition, in order that the right side should, for $z=0$, have the form, as a function of z, above found for $R\,dx/dt$ for finite values of x, that $z^{\sigma_i-1-m_i}V_i$ should be a polynomial in z. This is satisfied if, and only if, V_0 be identically zero, and $m_i \leqslant \sigma_i-1$. Thus we have proved that there exist everywhere finite integrals, namely $\int[(x,1)^{\tau_1-1}\gamma_1 + \ldots + (x,1)^{\tau_{n-1}-1}\gamma_{n-1}]\,dx$, where $(x,1)^k$ is a polynomial in x of order k, and $\tau_i \leqslant \sigma_i$; and no others. When $\sigma_i=0$, the corresponding term is absent. The number of homogeneously entering arbitrary coefficients in the most general everywhere finite integral is therefore $\sigma_1 + \ldots + \sigma_{n-1}$, the number we denoted by p. Such integrals do not exist for $p=0$. This is a second verification that p, as defined here, is the same as p of the earlier chapters. The intimate connexion thus established between the everywhere finite integrals, and the fundamental integral functions $g_1, \ldots, g_{n-1}$, is most striking.

Use of the reciprocal functions for algebraic integrals in general. A particular combination. The functions γ_0, γ_1, ..., γ_{n-1}, with the integral functions g_1, ..., g_{n-1}, can also be used to give detailed expressions for rational functions of assigned poles, and for the algebraic integrals which have infinities. To this end, denoting by (x) and (ξ) any two places of $f(x, y)=0$, we study the character of the expression following, which we constantly denote by (x, ξ), namely

$$(x,\ \xi)=[\gamma_0(x)+\gamma_1(x)g_1(\xi)+ \ldots +\gamma_{n-1}(x)g_{n-1}(\xi)]\,(x-\xi)^{-1}\,dx/dt,$$

where $\gamma_k(x)$ denotes the value of the function γ_k at the place (x), and $g_k(\xi)$ the value of the integral function g_k at the place (ξ). This expression is explicitly a rational function of (ξ). Its value at a place (x) is contemplated as derived by a limiting process; if (x_0) be this place, regarded as the origin of a particular branch, or element, which is given in terms of the parameter t for this branch by forms $x-x_0=t^r$, $y-y_0=$ power series in t, or forms derived from these by putting x^{-1} in place of $x-x_0$, or y^{-1} in place of $y-y_0$, or both, then these forms are to be substituted for x and y in (x, ξ); and the limit found as t vanishes. The notation has already been used. In the first place, we study (x, ξ) regarded as depending on a variable position of (x), for a definite position of (ξ); afterwards we regard (x) as definite, and study (x, ξ) as a rational function of (ξ). We recall that, if (x) be one of the places corresponding to a finite value of x, this being such that the n corresponding places are distinct, there being no branch place, and, if $(x^{(1)})$ be another of these n places, then (p. 158 above)

$$\gamma_0(x)+\gamma_1(x)g_1(x)+ \ldots +\gamma_{n-1}(x)g_{n-1}(x)=1,$$
$$\gamma_0(x)+\gamma_1(x)g_1(x^{(1)})+ \ldots +\gamma_{n-1}(x)g_{n-1}(x^{(1)})=0.$$

Also, if (x) be in the neighbourhood of a *finite* place (ξ), so that none of $g_1(\xi)$, ..., $g_{n-1}(\xi)$ is infinite, and this be a branch place of index r, there being expressions $x-\xi=t^r$, $y-\eta=$ power series in t, then, as before, we write $g_k(x)=a_k+a_k't+\tfrac12 a_k''t^2+ \ldots$. We can associate with (x) the $(r-1)$ other places, $(x^{(2)})$, ..., $(x^{(r)})$, which, when $t=0$, all coincide with (x) at (ξ), and correspondingly denote (x) by $(x^{(1)})$, so that we have $g_k(x^{(i)})=a_k+a_k't_i+\tfrac12 a_k''t_i^2+ \ldots$, in which $t_i=\epsilon^{i-1}t$, with $\epsilon=\exp(2\pi i/r)$; it will be sufficient also to use $\gamma_k(x)g_k(x')$ as an abbreviation for $\gamma_0(x)+\gamma_1(x)g_1(x')+ \ldots +\gamma_{n-1}(x)g_{n-1}(x')$. Then, by what is recalled above, we have, assuming for simplicity that the other $n-r$ places conjugate to (ξ) are simple, the r equations

$$\gamma_k(x)g_k(x^{(1)})=1,\quad \gamma_k(x)g_k(x^{(2)})=0,\ \ldots,\quad \gamma_k(x)g_k(x^{(r)})=0;$$

hence, if λ be any positive integer,

$$\frac{1}{r}\gamma_k(x)[g_k(x^{(1)})+\epsilon^{-\lambda}g_k(x^{(2)})+\epsilon^{-2\lambda}g_k(x^{(3)})+ \ldots +\epsilon^{-(r-1)\lambda}g_k(x^{(r)})]=\frac{1}{r};$$

substituting the expansions in powers of t, this is the same as

$$\gamma_k(x)\left[a_k p_0 + a_k{}' p_1 t + \frac{a_k{}''}{2} p_2 t^2 + \dots\right] = \frac{1}{r},$$

where, in virtue of $\sum\limits_{i=0}^{r-1} \epsilon^{im} = r$ or 0, according as the integer m divides by r, or does not, we easily see that $p_\lambda = 1$, $p_{\lambda+r} = 1$, ..., the other p's being zero. Hence we have

$$\gamma_k(x)\left[a_k + \frac{a_k{}^{(r)}}{r!} t^r + \frac{a_k{}^{(2r)}}{(2r)!} t^{2r} + \dots\right] = \frac{1}{r},$$

$$\gamma_k(x)\left[a_k{}' t + \frac{a_k{}^{(r+1)}}{(r+1)!} t^{r+1} + \dots\right] = \frac{1}{r},$$

and so on, the general equation being

$$\gamma_k(x)\left[\frac{a_k{}^{(i)}}{i!} t^i + \frac{a_k{}^{(r+i)}}{(r+i)!} t^{r+i} + \dots\right] = \frac{1}{r}.$$

Now let (x) approach to (ξ). We have shewn that none of the functions $\gamma_k(x)$ contains any higher negative power of t than $t^{-(r-1)}$. Hence we infer that the limits respectively of $\gamma_k(x) a_k$, $\gamma_k(x) a_k{}'$, $\gamma_k(x) a_k{}''$, ..., are $\frac{1}{r}, \frac{1}{r} t^{-1}, \frac{1}{r} 2 t^{-2}, \dots$. With a notation previously employed (p. 138), this is the same as saying that, if (ξ) be a branch place of index r, the limits, respectively, as (x) approaches to (ξ), of the functions

$$\gamma_0(x) + \gamma_1(x) g_1(\xi) + \dots + \gamma_{n-1}(x) g_{n-1}(\xi),$$

$$\gamma_1(x) D g_1(\xi) + \dots + \gamma_{n-1}(x) D g_{n-1}(\xi),$$

$$\gamma_1(x) D^2 g_1(\xi) + \dots + \gamma_{n-1}(x) D^2 g_{n-1}(\xi), \text{ etc.;}$$

are $\frac{1}{r}, \frac{1}{r} t^{-1}, \frac{2}{r} t^{-2}$, and so on, the second, third, ... of these becoming infinite; here $D g_k(\xi), D^2 g_k(\xi), \dots$ are the values of the differential coefficients, in regard to t, for $t = 0$. The second, third, ... results will be useful below; the first shews that, as (x) approaches to (ξ), the limit of (x, ξ), wherein we put $x - \xi = t^r$, $dx/dt = r t^{r-1}$, is t^{-1}. Namely (x, ξ) has a pole of the first order at (ξ), with multiplier (residue) equal to 1.

Next suppose that beside the branch place at (ξ), of index r, there is, among the $n - r$ conjugate places (that is, for the same value of ξ, if the place (ξ) have $x = \xi$, $y = \eta$), another branch place (ξ_1) of index r_1. We consider the limit of (x, ξ_1) as (x) approaches, not to (ξ_1), but to (ξ). Corresponding to the place (x), in the neighbourhood of (ξ), there will be, near the place (ξ_1), arising for the same value of x, places for which the values of x may be denoted by $x^{(1)}, x^{(2)}, \dots, x^{(r_1)}$, and the value of ξ by ξ_1, so that numerically $x^{(1)} = x$, $\xi_1 = \xi$, and, if we put $x^{(1)} - \xi_1 = t_1{}^{r_1}$, we have $t_1{}^{r_1} = t^r$. Using $(x^{(1)}), (x^{(2)}), \dots$ for places near to (ξ_1), we can put $g_k(x^{(1)}) = b_k + b_k' t_1 + \frac{1}{2} b_k'' t_1^2 + \ldots$, $g_k(x^{(2)}) = b_k + b_k' \epsilon_1 t_1 + \frac{1}{2} b_k'' \epsilon_1^2 t_1^2 + \ldots$, where $\epsilon_1 = \exp(2\pi i/r_1)$. As the place (x) does not coincide with any of the places in the neighbourhood of (ξ_1), here denoted by $(x^{(1)}), \ldots, (x^{(r_1)})$, we may suppose the equations denoted by $\gamma_k(x) g_k(x^{(s)}) = 0$, to hold, namely

$$\gamma_k(x)[b_k + b_k' \epsilon_1^{s-1} t_1 + \frac{1}{2} b_k'' \epsilon_1^{2(s-1)} t_1^2 + \ldots] = 0;$$

adding these r_1 equations we obtain

$$\gamma_k(x) \left[b_k + \frac{1}{(r_1)!} b_k^{(r_1)} t_1^{r_1} + \frac{1}{(2r_1)!} b_k^{(2r_1)} t_1^{2r_1} + \ldots \right] = 0,$$

and therein we can replace $t_1^{r_1}$ by t^r. Also, when (x) approaches to (ξ), the function $\gamma_k(x)$ can have no higher negative power of t than $t^{-(r-1)}$. Thus we infer that $\gamma_k(x) b_k = 0$ in the limit; and this is what is expressed by saying that $\gamma_k(x) g_k(\xi_1)$ vanishes as (x) approaches to (ξ). Hence also, $\gamma_k(x) g_k(\xi)$ vanishes as (x) approaches to (ξ_1); namely, expressed in powers of t_1, it must contain at least the first power of t_1 as factor. When this is so, however, $(x-\xi)^{-1} dx/dt$ is infinite only as $r_1 t_1^{-1}$. We conclude, therefore, that (x, ξ) remains finite as (x) approaches to (ξ_1), notwithstanding the vanishing denominator $x - \xi$.

We can now specify the character of the expression (x, ξ), given by

$$(x, \xi) = [\gamma_0(x) + \gamma_1(x) g_1(\xi) + \ldots + \gamma_{n-1}(x) g_{n-1}(\xi)](x-\xi)^{-1} dx/dt,$$

regarded as depending on (x), for all finite positions of (x), the place (ξ) being a given finite place. The expression can become infinite only on account of the vanishing of $x - \xi$, or on account of infinities of the functions $\gamma_0(x), \gamma_1(x), \ldots, \gamma_{n-1}(x)$. These latter are, at most, to order $r-1$, at a branch place of index r, and are therefore ineffective, on account of the factor dx/dt. The conclusion then is that only at the single place (ξ) can there be infinity, the function (x, ξ) being there infinite like t^{-1}, if t is the parameter for the neighbourhood of (ξ).

When (x) is at infinity, it follows from what we have proved in regard to the everywhere finite integrals, that each of the products $\gamma_1(x) dx/dt, \ldots, \gamma_{n-1}(x) dx/dt$ remains finite. It is also the case that, if u_2 denote a quadratic polynomial in x, the function $\gamma_0(x) u_2^{-1} dx/dt$ remains finite; for, by the substitution $x = c + z^{-1}$, previously employed, this takes the form $-z^2 \vartheta_0 v_2^{-1} . z^{-2} dz/dt$, where, if $u_2 = (x, 1)_2$, then $v_2 = (1, z)_2$, ϑ_0 being the function for $x = c$ previously used; and this is finite for $z = 0$, v_2 not vanishing.

We can thus state the important result: If (c) be any finite place, other than the finite place (ξ), the difference $(x, \xi) - (x, c)$ is not infinite for any position of (x), finite or infinite, other than the two places (ξ) and (c); and the integral, in regard to (x),

$$\int [(x, \xi) - (x, c)] dt,$$

where t is the parameter for (x), has no infinities except at (ξ) and (c), being infinite at these respectively like $\log t_1$ and $-\log t_0$, if t_1, t_0 be the respective parameters for the neighbourhoods of these places. An integral, in (x), of this behaviour, can be constructed when (ξ), or (c), is at infinity. It will be sufficient, however, to obtain this by first transforming the infinite places in question to finite places, the resulting changes of the functions $g_k(\xi)$ being simple, as has been explained (p. 160). The discussion of (x, ξ) when (ξ) is at infinity arises below.

It may be remarked, as an incidental consequence of the fact that the limit of $\gamma_0(x) + \gamma_1(x)g_1(\xi) + \dots + \gamma_{n-1}(x)g_{n-1}(\xi)$, when (x) approaches to the branch place (ξ), of index r, is $1/r$, that *not all of* $\gamma_0(x), \gamma_1(x), \dots, \gamma_{n-1}(x)$ *can vanish at the same place.* (Cf. p. 71.)

It may also be stated to follow, without great difficulty, from the character we have found for the integral

$$P_{\xi, c}^{x, a} \text{ or } \int_{(a)}^{(x)} [(x, \xi) - (x, c)] dt,$$

that the integral, in regard to (x),

$$E_{\xi}^{x, a}, = \int_{(a)}^{(x)} D_\xi(x, \xi) . dt$$

is infinite only at the (finite) place (ξ), and then has a pole of the first order, becoming infinite like $-t_\xi^{-1}$, where t_ξ is the local parameter for the neighbourhood of the place (ξ). In this, the notation D_ξ scarcely needs explanation: When (ξ) is not a branch place, and the function is not infinite at (ξ), D_ξ means $d/d\xi$; in general, when (ξ) is a branch place, we take (ξ') in its neighbourhood, and express (x, ξ') in powers of the parameter for the neighbourhood of (ξ); then D_ξ means that we take the coefficient of the first positive power of this parameter. The notation has already been employed several times. More generally, an algebraic integral, having only an algebraic infinity, at the branch place (ξ), of index r, being there infinite like $-[A_1 t_\xi^{-1} + \dots + A_{r-1} t_\xi^{-(r-1)} + A_r t_\xi^{-r}]$, is given by

$$rA_r \int_{(a)}^{(x)} \rho(x, \xi)(x - \xi)^{-2} dx + \int_{(a)}^{(x)} (x - \xi)^{-1} U\rho(x, \xi) dx,$$

where $U = A_1 D_\xi + \dots + \dfrac{1}{(r-2)!} A_{r-1} D_\xi^{r-1} + \dfrac{1}{(r-1)!} A_r D_\xi^r$,

and $\rho(x, \xi) = \gamma_0(x) + \dots + \gamma_{n-1}(x)g_{n-1}(\xi)$.

Another use of this combination to build a fundamental rational function. We now consider the character of the rational function of (ξ) given by (x, ξ), for finite positions of (x). The function being, save for $(x - \xi)^{-1}$, linear in the integral functions of (ξ), $1, g_1(\xi), \dots, g_{n-1}(\xi)$, can only become infinite, for finite positions of (ξ), when $\xi = x$. It follows from what has preceded that such infinity only arises when the place (ξ) approaches the place (x); and that, if t_x denote the parameter for the neighbourhood of the place (x), equal to $\xi - x$ when (x) is not a branch place, then, when (ξ) is near to (x), this infinity is that of $-t_x^{-1}$. Another proof of this is indicated below. More intricate is the behaviour of (x, ξ) when (ξ) passes to infinity: Introduce the definition

$$u_{i, m}^{(x)} = \int^{(x)} x^{m-1} \gamma_i(x) \, dx;$$

if, as before, $\sigma_i + 1$ is the dimension of the function $g_i(x)$, this is an everywhere finite integral for integer positive values of m such that $m \leqslant \sigma_i$, as was proved; consider then the function $G_k(\xi)$, of (ξ), for a definite (x), given by

$$G_k(\xi) = \xi^{-1} g_k(\xi) \, du_{k, 1}/dt + \xi^{-2} g_k(\xi) \, du_{k, 2}/dt + \dots$$
$$+ \xi^{-\sigma_k} g_k(\xi) \, du_{k, \sigma}/dt;$$

this is equal to

$$\gamma_k(x) g_k(\xi) (\xi^{-1} + x\xi^{-2} + \dots + x^{\sigma_k - 1} \xi^{-\sigma_k}) \, dx/dt,$$

namely

$$-\gamma_k(x) g_k(\xi) [1 - (x/\xi)^{\sigma_k}] \, (x - \xi)^{-1} \, dx/dt.$$

Hence we can write

$$(x, \xi) = [\gamma_0(x) + \sum_k \gamma_k(x) g_k(\xi) (x/\xi)^{\sigma_k}] (x - \xi)^{-1} \, dx/dt - \sum_{k=1}^{n-1} G_k(\xi);$$

here, since $\xi^{-(\sigma_k + 1)} g_k(\xi)$ is finite when (ξ) is at infinity, the first part on the right is finite when (ξ) is at infinity. Thus the poles of (x, ξ), as a rational function of (ξ), which arise at infinity, are given by the expression $-\sum_{k=1}^{n-1} G_k(\xi)$.

The alternative proof of the nature of the infinity of (x, ξ), as a function of (ξ), when $\xi = x$, which has been referred to, is, in outline, as follows: When (ξ) is approaching to the branch place (x), of index r, for the neighbourhood of which the parameter is denoted by t, the function $\gamma_0(x) + \gamma_1(x) g_1(\xi) + \dots + \gamma_{n-1}(x) g_{n-1}(\xi)$ can, by what has preceded, be expressed as U/D, where U, D are certain determinants; these we put down, supposing (x) to be not at first actually at the branch place, but to be given by a parameter t_0, appropriate to this branch place, using ϵ for $\exp(2\pi i/r)$; namely U is

$$U = \begin{vmatrix} 1, & g(t) & , & \dots \\ 1, & g(\epsilon t_0) & , & \dots \\ \multicolumn{4}{c}{\dotfill} \\ 1, & g(\epsilon^{r-1} t_0), & \dots \\ \multicolumn{4}{c}{\dotfill} \end{vmatrix},$$

wherein $g(t)$ denotes a power series of the form $a + a't + \frac{1}{2} a'' t^2 + \dots$, and $g(\epsilon^i t_0)$ denotes what this becomes when $\epsilon^i t_0$ is substituted for t;

and D denotes what this becomes when t_0 is put for t. As we have previously seen, the determinant D may be reduced by taking the determinant M_s, of s rows and columns, of which the general (λ, μ)-th element is $\epsilon^{(\lambda-1)(\mu-1)}$; denoting the quotient of the cofactor of $\epsilon^{(m-1)(s-1)}$, in this, by M_s itself, by $K_{s,m}$; then forming the array

$$\begin{pmatrix} 1 & & & \\ K_{2,1}, & K_{2,2} & & \\ \multicolumn{4}{c}{\dotfill} \\ K_{r,1}, & K_{r,2}, & \dots & K_{r,r} \end{pmatrix},$$

and operating on the first r rows of D as if we were multiplying D by the matrix indicated by this array. When this is done, a factor $t_0 t_0^2 \dots t_0^{r-1}$ can be removed, and we are able to infer that

$$D = M_r t_0^{\frac{1}{2}r(r-1)} A_1,$$

where A_1 is a determinant which, for $t_0 = 0$, would become the determinant A_0 considered above (p. 152), there proved not to vanish. A similar process is applicable to the 2nd, 3rd, ..., r-th rows of the determinant U. For this, we begin with the determinant of $s-1$ rows and columns which is the minor, save for sign, of the $(1, s)$-th element of M_s; denoting this determinant by N_s, we reach the equation $U = N_r t_0^{\frac{1}{2}(r-1)(r-2)} A_2$, where A_2 is the determinant

$$A_2 = \begin{vmatrix} 1, & g(t) & , & \dots \\ 1, & g(\epsilon t_0) & , & \dots \\ 0, & a' + \dots & , & \dots \\ \multicolumn{4}{c}{\dotfill} \\ 0, & \dfrac{a^{(r-2)}}{(r-2)!} + \dots, & \dots \end{vmatrix};$$

if then, in A_2, we put $t_0 = 0$, and then subtract the 2nd, 3rd, ..., r-th rows, respectively multiplied by $1, t, \dots, t^{r-2}$, all from the first row, we obtain (after placing the resulting first row into the r-th row), the result $A_2 = (-1)^{r-1} t^{r-1} A_1$.

We can, however, see at once, by adding all the rows of M_r to the first row, and then dividing the rows by powers of ϵ, that $M_r = r N_r^{\frac{1}{2}r(r-1)}$; and this identity is obtainable also by proving that the inverse of the matrix of which M_r is the determinant is r^{-1} times the matrix obtainable from M_r by changing ϵ into ϵ^{-1}, while $(-1)^{r-1} N_r / M_r$ is the $(r, 1)$-th element of this inverse matrix. Thus finally, the value of U/D is to be obtained from

$$U/D = \frac{(-1)^{r-1}}{r} (t^{r-1}/t_0^{r-1})(-1)^{r-1}, \text{ or } \frac{1}{r}(t/t_0)^{r-1};$$

and hence (x, ξ), or $(U/D)(x-\xi)^{-1} dx/dt$, as (ξ) approaches to (x), is obtainable from $\dfrac{1}{r}(t/t_0)^{r-1} \cdot r t_0^{r-1} \cdot (x-\xi)^{-1}$; or, now supposing (x) to be actually at the branch place, from $t^{r-1}(x-\xi)^{-1}$. Thus, finally, $(x, \xi) = -t^{-1}$, which is the result we obtained otherwise above. A similar investigation may be made for the approach of (ξ) to a place, other than (x), for which x has the same value as at (x).

We pass now to exhibit the use of the rational function of (ξ) denoted by (x, ξ). For this, we write (x) in place of (ξ), and, denoting by (z) an arbitrary finite place, we consider the rational function of (x), (z, x).

The differential coefficient of the general everywhere finite integral, which we have proved to have the form

$$\Sigma (x, 1)^{\sigma_k - 1} g_k(x) \, dx/dt,$$

contains p homogeneously entering coefficients. It is thus possible to choose p places, say $(c_1), \ldots, (c_p)$, of such generality that this differential coefficient does not vanish in all of these; and to choose one such differential coefficient uniquely by the conditions that it vanishes in all of $(c_1), \ldots, (c_p)$ except (c_i), but reduces to unity at (c_i). So chosen, let it be denoted by $\omega_i(x)$; and let the corresponding integral, $\int_{(a)}^{(x)} \omega_i(x) \, dt$, be denoted by $V_i^{x,a}$. Then consider the rational function of (x) given by $(z, x) - \lambda_1(c_1, x) - \ldots - \lambda_p(c_p, x)$, where $\lambda_1, \ldots, \lambda_p$ are independent of (x). By what has been proved in this chapter, the poles of this function, for finite positions of (x), occur only at $(z), (c_1), \ldots, (c_p)$, the infinite parts of the function at these places being, respectively, $-t_z^{-1}, \lambda_1 t_{c_1}^{-1}, \ldots, \lambda_p t_{c_p}^{-1}$, where $t_z, t_{c_1}, \ldots, t_{c_p}$ are the local parameters for the neighbourhoods of these places respectively. If, for brevity, we write, temporarily, $\mu_{k,m}(x) = x^{m-1} \gamma_k(x) \, dx/dt$, the infinite part of the function

$$(z, x) - \lambda_1(c_1, x) - \ldots - \lambda_p(c_p, x),$$

when (x) is at infinity, is, by what we have proved,

$$-T(x, z) + \sum_{k=1}^{p} T(x, c_k),$$

where $T(x, z)$ denotes

$$\sum_{i=1}^{n-1} [x^{-1} g_i(x) \mu_{i,1}(z) + \ldots + x^{-\sigma_i} g_i(x) \mu_{i, \sigma_i}(z)],$$

and $T(x, c_k)$ is obtained from this by putting (c_k) for (z); namely, this infinite part is

$$-\sum_{i=1}^{n-1} g_i(x) \sum_{m_i=1}^{\sigma_i} x^{-m_i} [\mu_{i, m_i}(z) - \lambda_1 \mu_{i, m_i}(c_1) - \ldots - \lambda_p \mu_{i, m_i}(c_p)].$$

We can choose $\lambda_1, \ldots, \lambda_p$, independent of (x), so that this infinite part vanishes; we recall that no rational functions of x only, $U_0(x), \ldots, U_{n-1}(x)$, exist, other than all zero, for which there is an identity $U_0(x) + U_1(x) g_1(x) + \ldots + U_{n-1}(x) g_{n-1}(x) = 0$; thus, the necessary, and obviously sufficient, conditions for $\lambda_1, \ldots, \lambda_p$ are that every one of the coefficients $\mu_{i, m_i}(z) - \lambda_1 \mu_{i, m_i}(c_1) - \ldots - \lambda_p \mu_{i, m_i}(c_p)$ should vanish, these conditions, since $\sigma_1 + \ldots + \sigma_{n-1} = p$, being independent. Now, in fact, the equations

$$\mu_{i, m_i}(z) - \omega_1(z) \mu_{i, m_i}(c_1) - \ldots - \omega_p(z) \mu_{i, m_i}(c_p) = 0$$

are true; for the left side, regarded as a function of (z), is the differential coefficient of an everywhere finite integral, but vanishes when (z) is at any one of the p independent places $(c_1), \ldots, (c_p)$; it is thus identically zero. Thus the values of $\lambda_1, \ldots, \lambda_p$ are $\omega_1(z), \ldots, \omega_p(z)$. We thus have the conclusion that the rational function of (x) given by $(z, x) - \omega_1(z)(c_1, x) - \ldots - \omega_p(z)(c_p, x)$, is not infinite when (x) is at infinity, but is infinite to the first order at each of the $p+1$ general places $(z), (c_1), \ldots, (c_p)$, with respective infinite parts, $-t_z^{-1}, \omega_1(z) t_{c_1}^{-1}, \ldots, \omega_p(z) t_{c_p}^{-1}$. This function we shall denote by $\psi(x, z; c_1, \ldots, c_p)$, or, more often simply by $\psi(x, z)$. It is a function of fundamental importance; and, in case the original curve $f(x, y) = 0$ have a simple form, it is easy to form the explicit expression of the function without recurring to the integral functions $g_1(x), \ldots, g_{n-1}(x)$ which we have employed; or conversely, from such an explicit expression, we can deduce the particular differentials $\omega_1(z), \ldots, \omega_p(z)$, by considering the poles $(c_1), \ldots, (c_p)$. In regard to the places $(c_1), \ldots, (c_p)$ we may add the remark: These places were chosen so that there is no everywhere finite integral whose differential coefficient vanishes at all of them; there can, also, be no rational function having poles of the first order at each of these p places, and no other infinity. For if $U(x)$ were such a function, with infinite parts $a_1 t_{c_1}^{-1}, \ldots, a_p t_{c_p}^{-1}$ at these places, and $V_i^{(x)}$, as before, be the integral $\int^{(x)} \omega_i(x) dt$, the application of the general theorem proved (Chap. III, p. 47), expressed by $[U(x) dV_i^{(x)}/dt]_{t^{-1}} = 0$, would give the equation $a_1 \omega_i(c_1) + \ldots + a_p \omega_i(c_p) = 0$, for all the p values of i; and hence $a_k = 0$ $(k = 1, \ldots, p)$. The function $\psi(x, z)$ is infinite at (z), as well as at $(c_1), \ldots, (c_p)$.

Application to the general theory of rational functions. The Riemann-Roch theorem. We proceed now, independently of what has been said previously, to consider the expression of any rational function, of (x), as a linear aggregate of functions $\psi(x, z)$, with different positions of (z), but always the same places $(c_1), \ldots, (c_p)$. In the first instance we suppose that all the poles of the function to be expressed are of the first order. Let this function have for poles the places $(z_1), \ldots, (z_m)$, supposed all different from $(c_1), \ldots, (c_p)$, its infinite parts at these places being respectively $k_1 t_{z_1}^{-1}, \ldots, k_m t_{z_m}^{-1}$. If $R(x)$ denote the function, the aggregate

$$R(x) + k_1 \psi(x, z_1) + \ldots + k_m \psi(x, z_m)$$

is a rational function of (x), with no poles at $(z_1), \ldots, (z_m)$, and poles only at $(c_1), \ldots, (c_p)$. By the remark just made, it is therefore an absolute constant, say k. By employing once more the theorem $[R(x) dV_i{}^{(x)}/dt]_{t^{-1}} = 0$, or by putting down the infinite parts of $\psi(x, z_1), \ldots, \psi(x, z_m)$ at the place (c_i), (from the explicit form of $\psi(x, z_r)$ given above), we see, as $R(x)$ is not infinite at (c_i), that the coefficients (residues) $k_1, \ldots, k_m$, at the poles of any existing rational function $R(x)$, must satisfy the p equations

$$k_1 \omega_i(z_1) + \ldots + k_m \omega_i(z_m) = 0,$$

for $i = 1, \ldots, p$; and, conversely that, if $k_1, \ldots, k_m$ satisfy these equations, then there exists a rational function with $(z_1), \ldots, (z_m)$ as its poles, of the first order, with the respective infinite parts $k_1 t_{z_1}{}^{-1}, \ldots, k_m t_{z_m}{}^{-1}$. Hence we can deduce a theorem considered before (p. 78) under the name of the Riemann-Roch theorem for functions. For the p linear conditions for $k_1, \ldots, k_m$ may not be linearly independent, there being equations of the form

$$c_1 \omega_1(z) + \ldots + c_p \omega_p(z) = 0$$

satisfied for every one of the positions $(z_1), \ldots, (z_m)$ of (z); suppose that this is so for $r' + 1$ sets of values of $c_1, \ldots, c_p$, where $r' + 1 \geqslant 0$. Then the p linear equations for $k_1, \ldots, k_m$ are equivalent to $p - (r' + 1)$, and can be satisfied with $m - p + r' + 1$ of $k_1, \ldots, k_m$ taken arbitrarily, the other $p - (r' + 1)$ of $k_1, \ldots, k_m$ being determinate in terms of these. In other words, as the rational function contains a single additive arbitrary constant, k, we can say: *If places $(z_1), \ldots, (z_m)$ be such that there are $(r' + 1)$ differential co-efficients of everywhere finite integrals which vanish in all these places, then a rational function can be constructed having these places as poles of the first order, and no other poles, and this function contains a number, $r + 1$, of homogeneously entering arbitrary constants, given by $r = m - p + r' + 1$ (provided this $r \geqslant 1$).* In particular, when $r' + 1 = 0$, and $m = p + 1$, we have $r = 1$. The separate functions which thus arise (r in number, beside a constant), each multiplied by an arbitrary constant, are not necessarily, or generally, every one of them infinite in all the m places $(z_1), \ldots, (z_m)$; it is possible, however, to choose from $(z_1), \ldots, (z_m)$, a set of $m - r$ places, say $(z_1), \ldots, (z_{m-r})$, such that, for every one of the remaining r places, say (z'), there exists a set of $(m - r)$ coefficients, $A_1, \ldots, A_{m-r}$, some of which may be zero, such that the p equations

$$A_1 \omega_i(z_1) + \ldots + A_{m-r} \omega_i(z_{m-r}) - \omega_i(z')$$

are satisfied $(i = 1, \ldots, p)$; there exists then a definite rational function, actually infinite at (z'), and infinite at such of $(z_1), \ldots, (z_{m-r})$ as really enter in these p equations. The general function obtained is the sum of the r definite functions so existing, each multiplied by an arbitrary constant, with the addition of a further arbitrary constant; the general function is actually infinite at every one of the r places (z'); but there may be places among $(z_1), \ldots, (z_{m-r})$ at which it is not infinite.

As a simple example of this condition, suppose that $m = p + 1$, and that $(z_1), \ldots, (z_m)$ consist of a place (z_{p+1}) of general position, together with p places $(z_1), \ldots, (z_p)$ connected by the p equations

$$A_1 \omega_i(z_1) + \ldots + A_{p-1} \omega_i(z_{p-1}) - \omega_i(z_p) = 0, \qquad (i = 1, \ldots, p),$$

in which no one of $A_1, \ldots, A_{p-1}$ is zero; there is then a function $\lambda_1 \omega_1(z) + \ldots + \lambda_p \omega_p(z)$ which vanishes in all the p places $(z_1), \ldots, (z_p)$, but not in (z_{p+1}). Then $r' + 1 = 0$, and the formula $r = m - p$ gives $r = 1$. The set of independent places $(z_1), \ldots, (z_{m-r})$, spoken of above, must then consist of $(z_1), \ldots, (z_{p-1})$ and (z_{p+1}), but the A_p of the general formulation is zero. There is then a rational function with $(z_1), \ldots, (z_{p-1}), (z_p)$ as actual poles, but no function with all the given $(p+1)$ places as poles (cf. Chap. IV, p. 86).

Ex. 1. By considering the rational function expressible by the fraction whose numerator is the differential coefficient of the general everywhere finite integral, and denominator is the differential coefficient of a particular one such integral, prove that such a differential coefficient has $2p - 2$ zeros (we have seen, p. 165 above, that there is no zero common to all the p differential coefficients).

Ex. 2. The theorem obtained before (Chap. VI, p. 144), and called the converse of Abel's theorem, may also be proved here, under a certain limiting hypothesis. With the notation used, of $V_i^{x,a}$ for an everywhere finite integral (p. 168), suppose that $(a_1), \ldots, (a_m)$ and $(z_1), \ldots, (z_m)$, are places such that the p equations $V_i^{z_1, a_1} + \ldots + V_i^{z_m, a_m} \equiv 0$ hold, the sign $\equiv$ indicating the possible presence of a linear aggregate of the $2p$ periods of the integral V_i, wherein the multipliers of these periods are the same for all the p equations; make the hypothesis that in the immediate neighbourhoods of $(z_1), \ldots, (z_m)$, respectively, there are other places $(z_1'), \ldots, (z_m')$ also satisfying these equations, $(a_1), \ldots, (a_m)$ remaining unaltered. Then, for proper values of the differentials of the parameters $t_1, \ldots, t_m$ appropriate respectively for the neighbourhoods of $(z_1), \ldots, (z_m)$, we have the p equations $\omega_i(z_1) . dt_1 + \ldots + \omega_i(z_m) . dt_m = 0$. Hence, if $\mu_1, \ldots, \mu_m$ be finite numbers which are in the ratios of $dt_1, \ldots, dt_m$, and μ be a constant, the rational function $\mu + \mu_1 \psi(x, z_1) + \ldots + \mu_m \psi(x, z_m)$ has $(z_1), \ldots, (z_m)$ for poles of the first order, and no other poles. The theorem receives another proof below.

Explicit form of relation for interchange of argument and parameter.

We add now some detailed investigations which furnish in particular; (i), a theorem, in explicit algebraic form, for the interchange of the limits and the infinities in an algebraic integral which has two logarithmic infinities. This was one of the achievements of the early theory in the elliptic case $(p = 1)$; and, for Riemann's transcendentally defined integral $\Pi_{z,c}^{x,a}$, is true in the form $\Pi_{z,c}^{x,a} = \Pi_{x,a}^{z,c}$ (cf. p. 144, Chap. VI above); (ii), also in explicit algebraic form, a function which may serve as a *prime function* for rational functions of x and y (subject to $f(x, y) = 0$), enabling us to express such a function in factors, each factor vanishing to the first order in one of the poles, or in one of the zeros, of the rational function. The former theorem goes back to Abel (for the hyperelliptic case, cf. *Œuvres*, 1881, I, p. 49) and to Weierstrass; the prime function to Kronecker (in an entirely different way) and Weierstrass (cf. Weierstrass, *Werke*, IV, p. 253).

Consider the rational function of (z) which is given by

$$R(z) = \psi(z, \xi)\frac{d}{dz}\psi(z, x),$$

where (x), (ξ) are two arbitrary places; and apply to this the fundamental theorem $[R(z)\,dz/dt]_t{}^{-1} = 0$, which we may express here by $[\psi(z, \xi)\,d\psi(z, x)/dt]_t{}^{-1} = 0$, the parameter t being that by which (z) is expressed in the neighbourhood of a place considered. Infinities arise only when (z) is at (ξ), or at (x), or at one of the places $(c_1), \ldots, (c_p)$ which were used in the definition of the function $\psi(z, \xi)$ or $\psi(z, x)$. To express the result of the theorem we use a notation, already several times employed, of which we repeat the definition here: If, for a rational function $R(z)$, with or without a pole at a place (c), we form the expansion, for the neighbourhood of (c), in terms of the parameter t appropriate to this neighbourhood, the coefficient of t in this expansion is denoted by $D_c R(c)$. Thence; (a), when (z) is near to (ξ), the expansions of $\psi(z, \xi)$, $\psi(z, x)$ give $\psi(z, \xi) = -t^{-1} + P(t)$, $d\psi(z, x)/dt = D_\xi\psi(\xi, x) + P_1(t)$, where $P(t), P_1(t)$ are power series; and the contribution to the summation involved in the theorem is $-D_\xi\psi(\xi, x)$; (b), when (z) is near to (x), the corresponding expansions give $\psi(z, \xi) = \psi(x, \xi) + tD_x\psi(x, \xi) + \ldots$, $d\psi(z, x)/dt = t^{-2} + P_1(t)$; and the resulting contribution is $D_x\psi(x, \xi)$; when (z) is near to (c_i), we have

$$\psi(z, \xi) = t^{-1}\omega_i(\xi) + M + tD_{c_i}\psi(c_i, \xi) + \ldots,$$
$$\psi(z, x) = t^{-1}\omega_i(x) + N + tD_{c_i}\psi(c_i, x) + \ldots,$$

where M, N are independent of t, and the corresponding contribution in the theorem is $\omega_i(\xi)D_{c_i}\psi(c_i, x) - \omega_i(x)D_{c_i}\psi(c_i, \xi)$; thus, altogether, the theorem gives

$$-D_\xi\psi(\xi, x) + D_x\psi(x, \xi) + \sum_{i=1}^{p}[\omega_i(\xi)D_{c_i}\psi(c_i, x)$$
$$-\omega_i(x)D_{c_i}\psi(c_i, \xi)] = 0.$$

We can express this result in terms of functions (z, a) defined above, p. 162. For we have $\psi(x, \xi) = (\xi, x) - \sum_k\omega_k(\xi)(c_k, x)$, and hence $D_x\psi(x, \xi) = D_x(\xi, x) - \sum_k\omega_k(\xi).D_x(c_k, x)$; in particular

$$D_{c_i}\psi(c_i, \xi) = D_{c_i}(\xi, c_i) - \sum_{k=1}^{p}\omega_k(\xi)D_{c_i}(c_k, c_i),$$

where, in $D_{c_i}(c_k, c_i)$, for $k = i$, $D_{c_i}(c_i, c_i)$ must be remembered to mean the value of $D_x(c_i, x)$ for $(x) = (c_i)$. Thus we obtain the formula, which we can express in four ways:

(I) The expression

$$D_\xi(x, \xi) + \sum_k \omega_k(x)[D_{c_k}(\xi, c_k) - D_\xi(c_k, \xi)] - \sum_k \sum_i \omega_i(x)\omega_k(\xi)D_{c_i}(c_k, c_i),$$

where $k = 1, \ldots, p$ and $i = 1, \ldots, p$, is unaltered by the interchange of (x) and (ξ).

(II) Putting

$$U_\xi^x = D_\xi(x, \xi) - D_x(\xi, x) + \sum_i \omega_i(x)[D_{c_i}(\xi, c_i) - D_\xi(c_i, \xi)]$$

we have

$$U_\xi^x = \sum_k \omega_k(\xi) U_{c_k}^x.$$

(III) If, as before, we put $E_z^{x,a} = \int_{(a)}^{(x)} D_z(x, z)\, dt_x$, and integrate the result in (I) in regard to (x), we have

$$E_\xi^{x,a} = \sum_k \omega_k(\xi) E_{c_k}^{x,a} + \psi(x, \xi) - \psi(a, \xi) - \sum_k V_k^{x,a} U_{c_k}^\xi.$$

Here the function $E_\xi^{x,a}$ does not depend on $(c_1), \ldots, (c_p)$. The right side, as a function of (x), becomes infinite when $\psi(x, \xi)$ becomes infinite, or when one of $E_{c_k}^{x,a}$ becomes infinite, and these do not depend on (ξ). Thus $E_\xi^{x,a}$ becomes infinite at (ξ) like $\psi(x, \xi)$, namely like $-t_\xi^{-1}$, and only then. This property of $E_\xi^{x,a}$ can be established directly, and hence the equation (III) deduced, as remarked below.

(IV) If, as before, we define $P_{z,k}^{x,a}$ as $\int_{(a)}^{(x)} [(x, z) - (x, k)]\, dt_x$, then, by integration, in regard to (ξ), of the result in (III), we obtain a result which expresses that the function

$$P_{z,k}^{x,a} + \sum_j V_j^{x,a}\{E_{c_j}^{z,k} - [(c_j, z) - (c_j, k)]\} - \sum_j \sum_i V_i^{x,a} V_j^{z,k} D_{c_i}(c_j, c_i)$$

is unaltered when we interchange the places (x), (a) respectively with the places (z), (k). Here the summation in regard to j and i is from 1 to p. This is then the generalisation of the formula in the theory of elliptic functions known as the interchange of the argument and parameter.

The direct investigation of the character of the function $E_\xi^{x,a}$, referred to under (III), and the consequent direct proof of (III) is as follows: The expression $D_\xi(x, \xi)\, dx/dt$ is $H\, dx/dt$, where

$$H = (x - \xi)^{-1} \sum_i \gamma_i(x) D_\xi g_i(\xi) + [\gamma_0(x) + \sum_i \gamma_i(x) g_i(\xi)] D_\xi(x - \xi)^{-1};$$

now, first, when (ξ) is not a branch place, we have $D_\xi g_i(\xi) =$ the differential coefficient $g_i'(\xi)$, and $\gamma_i(x) = \gamma_i(\xi) + (x - \xi)\gamma_i'(\xi)$, as (x) approaches to (ξ); while, from the fundamental relation $\gamma_0(x) + \sum \gamma_i(x)g_i(x) = 1$, we have $\gamma_0'(\xi) + \sum \gamma_i'(\xi)g_i(\xi) = -\sum \gamma_i(\xi)g_i'(\xi)$. Thus we see at once that $H\, dx/dt$, when (x) approaches to (ξ), is the same as $(x-\xi)^{-2}$, and the integral $\int D_\xi(x, \xi)\,dx$ becomes infinite like $-(x-\xi)^{-1}$; and has no other infinity. Next, when (ξ) is a branch place of index r, and (ξ_1) is within the neighbourhood of this, the expansion

$$(x-\xi_1)^{-1} = (x-\xi-t^r)^{-1} = (x-\xi)^{-1}[1+(x-\xi)^{-1}t^r + \ldots]$$

contains no term in t, and $D_\xi(x-\xi)^{-1}=0$; also $\Sigma\gamma_i(x_1)D_\xi g_i(\xi)$, which in a previously used notation (p. 152) is $\Sigma\gamma_i(x)a_i'$, was proved to be given by $\frac{1}{r}t^{-1}$. Thus, for $D_\xi(x, \xi)\,dx/dt$ the infinite part is $\frac{1}{r}t^{-1}.t^{-r}.rt^{r-1}$ or t^{-2}; and hence the integral $E_\xi^{x,a}$ is infinite like $-t^{-1}$; and, as before, has no other infinity.

Now, from the character of the rational function of (x) given by $\psi(x, \xi)$, which was defined as $(\xi, x)-\underset{k}{\Sigma}\omega_k(\xi)(c_k, x)$, we see that the difference $E_\xi^{x,a}-\psi(x, \xi)$, as a function of (x), has poles only at $(c_1), \ldots, (c_p)$; at these also it is infinite, respectively, like $-\omega_1(\xi)t_{c_1}^{-1}, \ldots, -\omega_p(\xi)t_{c_p}^{-1}$. Whence, the function of (x) defined by

$$K(x) = E_\xi^{x,a} - \Sigma\,\omega_k(\xi)E_{c_k}^{x,a} - [\psi(x, \xi)-\psi(a, \xi)]$$

has no infinities. But dK/dx is a rational function of (x). Thus $K(x)$ is an everywhere finite integral, and we may hence write

$$E_\xi^{x,a} = \underset{k}{\Sigma}\,\omega_k(\xi)E_{c_k}^{x,a} + \psi(x, \xi)-\psi(a, \xi)+\underset{k}{\Sigma}\lambda_k V_k^{x,a},$$

where λ_k does not depend on (x). Differentiating this equation in regard to (x), and then putting (x) at (c_i), we find the result given in (III).

Weierstrass's prime function for algebraic functions.
We now proceed to another result. The function $P_{z,c}^{x,a}$ is the integral in regard to (x), of a function which is rational in (z). We consider now the integral as a function of (z). If (a), (k) denote arbitrary places, and we put

$$U_{z,k}^{x,a} = \int_{(a)}^{(x)} [(x-z)^{-1}-(x-k)^{-1}]\,dx,$$

and recall that $\gamma_0(z)+\Sigma\gamma_i(z)g_i(z)=1$, and $\gamma_0(k)+\Sigma\gamma_i(k)g_i(k)=1$, we can put $P_{z,k}^{x,a} = U_{z,k}^{x,a} + V$, where

$$V = \int_{(a)}^{(x)} [F(x, z)-F(x, k)]\,dx,$$

and $F(x, z)$, $F(x, k)$ denote functions given by

$$(x-z)\,F(x, z) = \gamma_0(x)-\gamma_0(z)+\Sigma\,[\gamma_i(x)-\gamma_i(z)]g_i(z),$$
$$(x-k)\,F(x, k) = \gamma_0(x)-\gamma_0(k)+\Sigma\,[\gamma_i(x)-\gamma_i(k)]g_i(k).$$

The subject of integration in V remains finite when (x), in the course of integration, passes through the place (z); when x approaches z, but (x) does not approach (z), we know that $P_{z,k}^{x,a}$ remains finite. We infer, then, that $P_{z,k}^{x,a}$, regarded as a function of (z), is single valued except when (z) approaches the path of integration from (a) to (x).

This path we may regard as a barrier for the variation of (z); the function of (z) given by $P^{x,\,a}_{z,\,k}$ will have different values at opposite points on the two sides of this barrier; in fact, as $U^{x,\,a}_{z,\,k}$ is equal to

$$\log\left[(x-z)(x-k)^{-1}/(a-z)(a-k)^{-1}\right],$$

these values differ by $2\pi i$ (cf. Weierstrass, *Werke*, IV, p. 379). We now interchange the places (x), (a) respectively with (z), (k), and infer that the integral

$$\int_{(k)}^{(z)}\left[\psi(x,z)-\psi(a,z)\right]dz,$$

which is $\qquad P^{z,\,k}_{x,\,a}-\sum_i V_i^{z,\,k}\left[(c_i,x)-(c_i,a)\right],$

regarded as a function of (x), is single valued, save for additive integer multiples of $2\pi i$; also, by its definition, or by the equation (IV) obtained above, this function is infinite when (x) is at (z) like $\log t_z$, and when (x) is at (k) like $-\log t_k$, and also has a pole of the first order at each of the places $(c_1), \ldots, (c_p)$, but is not otherwise infinite. Hence it follows that the function of (x) defined by

$$E\,(x,a;\,z,k)=\exp\left\{P^{z,\,k}_{x,\,a}-\sum_i V_i^{z,\,k}\left[(c_i,x)-(c_i,a)\right]\right\}$$

has the following properties:

(i), It is a single-valued function of (x), it being understood that the path of integration for (z), from (k) to (z), does not pass through the place (x); (ii), The function vanishes to the first order when (x) is at (z), and has a pole of the first order when (x) is at (k), being capable of expression by integral powers of the local parameter thereat; (iii), At each of the places $(c_1), \ldots, (c_p)$ the function has an essential singularity.

Now let $(z_1), \ldots, (z_m)$ be the zeros, and $(t_1), \ldots, (t_m)$ be the poles of a rational function $R(x)$, all these being of the first order. Then the quotient

$$Q,=E(x,a;\,z_1,k)\ldots E(x,a;\,z_m,k)/E(x,a;\,t_1,k)\ldots E(x,a;\,t_m,k)$$

is equal to $\exp(W)$, where

$$W=\sum_{r=1}^{m}P^{z_r,\,t_r}_{x,\,a}-\sum_{i=1}^{p}\sum_{r=1}^{m}V_i^{z_r,\,t_r}\left[(c_i,x)-(c_i,a)\right];$$

but, by Abel's theorem, if the paths of integration in the integrals $V_i^{z_r,\,t_r}$ be properly chosen, we can suppose that $\sum_{r=1}^{m}V_i^{z_r,\,t_r}$ vanishes, for each of the p values of i; while also (see Chap. III, p. 54) we have

$$\sum_{r=1}^{m}P^{z_r,\,t_r}_{x,\,a}=\log\left[R(x)/R(a)\right].$$

Thus it appears that $R(x)/R(a)$ is expressed by the quotient Q above, in which there is a factor corresponding to every zero and pole. (Cf. Weierstrass, *Werke*, IV, p. 384. The function $\psi(x, x')$ is essentially that which Weierstrass denotes by $H(x, y; x', y')$; *v. loc. cit.* p. 73. For the factors that may arise when the paths of integration involved are arbitrary, *v. loc. cit.* pp. 367, 390). Conversely, suppose that $(z_1), \ldots, (z_m)$ and $(t_1), \ldots, (t_m)$ are such places that, with proper paths of integration, we have the p equations $\sum\limits_{r} V_i^{z_r, t_r} = 0$, for $i = 1, \ldots, p$. Then the quotient which, for brevity, we have denoted by Q, has $(z_1), \ldots, (z_m)$ for zeros of the first order, has $(t_1), \ldots, (t_m)$ for poles of the first order, and is analytic in (x), being expressible in the neighbourhood of any place by a series of integral powers of the appropriate parameter t. By the definition

$$E(x, a; z, k) = \exp\left\{P_{x,a}^{z,k} - \sum_i V_i^{z,k}[(c_i, x) - (c_i, a)]\right\},$$

the quotient has, when (x) is near to (c_i), an essential singularity depending on $\exp\left\{t_{c_i}^{-1} \sum\limits_{r} V_i^{z_r, t_r}\right\}$, which, however, in virtue of the p equations assumed, is ineffective. The quotient is thus a rational function of (x). This furnishes another proof of the converse of Abel's theorem (cf. Weierstrass, *Werke*, IV, p. 418). The proof originally given here, by use of Riemann's elementary normal integral of the third kind (Chap. VI, p. 144), was based on a knowledge of the periods of this function. By what we have proved here in regard to the function $P_{x,a}^{z,k}$, regarded as a function of (x), the periods of the integral $P_{z,k}^{x,a}$, as a function of (x), save for additive multiples of $2\pi i$ when (x) makes a circuit about (z), or (k), can be found from equation (IV) above.

On the whole then, it appears sufficiently that the theory of this chapter can be made the complete basis of a theory of the linear series upon the curve $f(x, y) = 0$, and of the algebraic integrals associated therewith; but, in order that all the algebraic formulae should be explicitly realisable, it is necessary that the fundamental integral functions $g_1, \ldots, g_{n-1}$ should be explicitly known. To determine these is a problem entirely representative of the problem which presents itself in the earlier geometrical theory, of determining explicitly the conditions of adjointness of a polynomial at a singular point of whatever intricacy. For a summary of a procedure for determining $g_1, \ldots, g_{n-1}$, derived from Hensel, reference may be made to the writer's *Abel's Theorem*, Cambridge, 1897, pp. 105–112. A systematic introduction to the theory is given in a volume, *Theorie der algebraischen Functionen*, u.s.w., by Hensel u. Landsberg, Leipzig, 1902, pp. 1–702; and for other points reference may be made to the *Enzyk. d. Math. Wiss.* Bd. II, 3. 5. That the theory may be developed parallel with Weierstrass's earlier theory (but with greater explicitness when $g_1, \ldots, g_{n-1}$ have been computed) is sufficiently indicated here. For an alternative method of dealing with the fundamental difficulty of the multiple points, the papers of J. C. Fields, of which the first is *Theory of the algebraic functions of a complex variable*, Berlin, Mayer u. Müller, 1906, pp. 1–186, may be consulted.

Some indications of the many applications of the theory of this chapter may be added:

Ex. 1. For the case when the fundamental equation is $y^2 - (x, 1)_{2p+2} = 0$, it is easy to prove that any rational function which becomes infinite only for $x = \infty$ can be expressed as $u + vy$, where u, v are polynomials in x. Thus the integral function g_1 is y, for which $\sigma_1 = p$. The equations $\gamma_0 + \gamma_1 y = 1$, $\gamma_0 - \gamma_1 y = 0$ (p. 158) determine $\gamma_0 = \frac{1}{2}$, $\gamma_1 = 1/2y$. The everywhere finite integrals are thus $\int (x, 1)_{p-1} dx/2y$. Also, the function (x, ξ) (of p. 162) is given by $(y + \eta)[2y(x - \xi)]^{-1} dx/dt$, if (ξ, η) be the place (ξ). This case is characterised by the existence of a rational function of order 2.

Ex. 2. If there exist a rational function of order 3, say x, and the fundamental algebraic equation $f(x, y) = 0$ contain y^3, as highest power of y, the fundamental integral functions g_1, g_2 will be such that the integral functions $g_1 g_2$, g_1^2, g_2^2 are expressible in the respective forms $w + vg_1 + ug_2$, $w_1 + v_1 g_1 + u_1 g_2$, $w_2 + u_2 g_1 + v_2 g_2$, where w_i, u_i, v_i are polynomials in x; herein the dimensions of v and u will not be greater respectively than those of g_2 and g_1; thus, using $g_1 - u$, $g_2 - v$ in place of g_1, g_2, respectively, we can suppose $g_1 g_2 = w$. Forming, from these expressions, the forms for $g_1^2 g_2$ and $g_1 g_2^2$ (since an identity $U + Vg_1 + Wg_2 = 0$ is impossible, wherein U, V, W are polynomials in x), we can infer $w = u_1 u_2$, $w_1 = -u_1 v_2$, $w_2 = -u_2 v_1$. Thence, also forming g_1^3, we deduce $g_1^3 - v_1 g_1^2 + u_1 v_2 g_1 - u_1^2 u_2 = 0$; or, if we use y for g_1/u_1, the fundamental equation may be taken to be $u_1 y^3 - v_1 y^2 + v_2 y - u_2 = 0$; the two fundamental integral functions are then $g_1 = u_1 y$ and $g_2 = u_2 y^{-1}$. Also, in the most general case, from $g_1^2 = w_1 + v_1 g_1 + u_1 g_2$, etc., we infer that the dimensions of u_1, u_2, v_1, v_2, in terms of σ_1, σ_2, are respectively, $2\sigma_1 - \sigma_2 + 1$, $2\sigma_2 - \sigma_1 + 1$, $\sigma_1 + 1$, $\sigma_2 + 1$. The maximum possible number of arbitrary coefficients (moduli), in the fundamental equation, if we use one constant as a multiplier for y, and three constants in a linear transformation of x, is, therefore, found to be $2(\sigma_1 + \sigma_2) + 1$, or $2p + 1$. For the hyperelliptic case the number was $2p - 1$; for the case of the fundamental equation most general for its genus it was $3p - 3$ (p. 94 above); in the general case, a rational function can be found (p. 92 above) of order $1 + \frac{1}{2}p$, or $1 + \frac{1}{2}(p + 1)$; and the least value of p for which this is > 3 is $p = 5$; the existence of a rational function of order 3 involves $3p - 3 - (2p + 1)$ or $p - 4$ conditions for the general moduli. Passing now to the computation of the functions γ_0, γ_1, γ_2, conjugate to the integral functions 1, g_1, g_2, we find from $\gamma_0 + \gamma_1 g_1 + \gamma_2 g_2 = 1$, $\gamma_0 + \gamma_1 g_1' + \gamma_2 g_2' = 0$, $\gamma_0 + \gamma_1 g_1'' + \gamma_2 g_2'' = 0$ (see p. 158 above) that $\gamma_0 = (u_1 y^2 - v_1 y)/f'(y)$, $\gamma_1 = y/f'(y)$, $\gamma_2 = 1/f'(y)$, where $f(x, y) = u_1 y^3 - v_1 y^2 + v_2 y - u_2$; and hence that the expression (x, ξ) is capable of the form $(u_2 - \omega_2)[(x - \xi)(y - \eta)f'(y)]^{-1} dx/dt$, where (ξ, η) is the place (ξ), and ω_2 is the polynomial u_2 when x is ξ. We return to this case again in Ex. 5, to introduce a general theorem.

Ex. 3. Prove that a general plane quartic curve can be transformed so as to be represented by an equation $y^3 + my^2 + yxu + x^2 v = 0$, where m is a constant, and u, v are respectively quadratic and cubic polynomials. Prove that, for this equation, $g_1 = y$ and $g_2 = y(y + m)/x$, with $\sigma_1 = 1$, $\sigma_2 = 2$; and that the functions γ_0, γ_1, γ_2 are respectively given by

$$(y^2 + my + xu)/f'(y), \quad y/f'(y), \quad x/f'(y).$$

Ex. 4. For the equation of a general quartic curve with a double point at $(0, 0)$, namely $(y, x)_4 + (y, x)_3 + (y, x)_2 = 0$, obtain the integral functions g_1, g_2, g_3, the values of $\sigma_1, \sigma_2, \sigma_3$, and the functions $\gamma_0, \gamma_1, \gamma_2, \gamma_3$.

Ex. 5. Suggested by the everywhere finite integral, for the equation in Ex. 2, which is $\int[(x, 1)^{\sigma_1-1}y + (x, 1)^{\sigma_2-1}] dx/f'(y)$, consider the curve of order $2p - 2$, in space $[p-1]$, in which the coordinates are

$$(\xi_0, \dots, \xi_{\sigma_1-1}, \eta_0, \dots, \eta_{\sigma_2-1}),$$

which is given by

$$\xi_0/yx^{\sigma_1-1} = \xi_1/yx^{\sigma_1-2} = \dots = \xi_{\sigma_1-1}/y = \eta_0/x^{\sigma_2-1} = \eta_1/x^{\sigma_2-2} = \dots = \eta_{\sigma_2-1}/1.$$

This is the canonical curve considered in an earlier chapter (p. 81 above). Evidently it lies on the rational ruled surface obtained by joining the general point $(\theta^{\sigma_1-1}, \theta^{\sigma_1-2}, \dots, 1)$ of the rational normal curve of order $\sigma_1 - 1$ in space $(\xi_0, \xi_1, \dots, \xi_{\sigma_1-1})$, to the corresponding point $(\theta^{\sigma_2-1}, \dots, \theta, 1)$ of the rational normal curve of order $\sigma_2 - 1$ in space $(\eta_0, \dots, \eta_{\sigma_2-1})$; this surface is of order $p - 2$, and every generator contains three points of the curve (corresponding to the three values of y in the equations determining the curve). The canonical curve lies on the $\frac{1}{2}(p-2)(p-3)$ quadrics expressed by $\xi_i/\xi_{i+1} = \dots = \eta_j/\eta_{j+1} = \dots$ $(i = 0, \dots, \sigma_1 - 2; j = 0, \dots, \sigma_2 - 2)$, which, however, do not determine this curve, since they all contain the ruled surface; the curve may in fact be determined on the surface as the residual intersection of this with a cubic primal described through $p - 4$ generators. (Cf. p. 96 above.) [It will in fact be proved later that an algebraic curve on a ruled surface of order n_1, whose prime section is of genus p_1, which meets each generator k times, and is of order ν, has a genus π such that $k(2p_1 - 2) = 2\pi - 2 - (2\nu - n_1 k)(k-1)$; if herein we put $n_1 = p - 2$, $p_1 = 0$, $k = 3$, $\nu = 2p - 2$, we find $\pi = p$.]

Ex. 6. A series g_1^n is found in the general case where the everywhere finite integral is of the form $\int[(x, 1)^{\sigma_1-1}\gamma_1 + \dots + (x, 1)^{\sigma_{n-1}-1}\gamma_{n-1}] dx$; for simplicity of statement, we assume every one of $\sigma_1 - 1, \dots, \sigma_{n-1} - 1$ is $\geqslant 0$, and that $n \leqslant p$. The canonical curve lies then in a space $[p-1]$ wherein the coordinates are $\xi_0, \dots, \xi_{\sigma_1-1}; \eta_0, \dots, \eta_{\sigma_2-1}; \dots; \zeta_0, \dots, \zeta_{\sigma_{n-1}-1}$, and is given by

$$\xi_0/\gamma_1 x^{\sigma_1-1} = \dots = \xi_{\sigma_1-1}/\gamma_1 = \eta_0/\gamma_2 x^{\sigma_2-1} = \dots$$

$$= \eta_{\sigma_2-1}/\gamma_2 = \dots = \zeta_0/\gamma_{n-1} x^{\sigma_{n-1}-1} = \dots = \zeta_{\sigma_{n-1}-1}/\gamma_{n-1}.$$

Evidently the curve lies on the manifold which is the locus of the space $[n-2]$ which joins the corresponding points

$$[\theta^{\sigma_1-1}, \dots, \theta, 1], \ [\theta^{\sigma_2-1}, \dots, \theta, 1], \dots, [\theta^{\sigma_{n-1}-1}, \dots, \theta, 1],$$

of $n - 1$ rational normal curves lying respectively in the spaces

$$(\xi_0, \dots, \xi_{\sigma_1-1}), \ (\eta_0, \dots, \eta_{\sigma_2-1}), \dots, (\zeta_0, \dots, \zeta_{\sigma_{n-1}-1});$$

and the canonical curve is met by each of the spaces $[n-2]$ in n points, corresponding to the possible values of y in the functions $\gamma_1, \gamma_2, \dots$ which enter in the definitions of $\xi_0, \xi_1, \dots$. The order of the manifold is $\Sigma(\sigma_i - 1)$, or $p - n + 1$. We may prove this *ab initio* by remarking that, as the manifold is of dimension $n - 1$, the order is the number of the spaces $[n-2]$, each determined by one value of θ, which meet $(n-1)$ given primes of the space $[p-1]$; the values of θ are those which satisfy $n-1$ equations, each of the form $\lambda_1(\theta,1)_{\sigma_1-1}+\ldots+\lambda_{n-1}(\theta,1)_{\sigma_{n-1}-1}=0$, in which $\lambda_1,\ldots,\lambda_{n-1}$ are the same in all, and the coefficients in the polynomials $(\theta,1)_{\sigma_i-1}$ are given, and different. By elimination of $\lambda_1,\ldots,\lambda_{n-1}$ this leads to an equation of order $p-n+1$ for θ. As follows from the equations $\xi_i/\xi_{i+1}=\eta_j/\eta_{j+1}=\ldots=\zeta_k/\zeta_{k+1}$, where there are $p-n+1$ fractions, the manifold (or scroll) of order $n-1$, lies on $\frac{1}{2}(p-n+1)(p-n)$ quadrics; as is known, the canonical curve lies on $\frac{1}{2}(p-2)(p-3)$ quadrics; thus, beside those containing the scroll, there is a number $\frac{1}{2}(n-3)(2p-2-n)$ remaining; this vanishes when $n=3$, as in Ex. 5.

[It may be remarked that if in space $[r]$ we have a scroll of ∞^1 spaces $[s]$, of genus p_1, of order n_1, forming thus a manifold $M_{s+1}^{n_1}$; and thereon a curve of order ν and genus π, meeting each generating $[s]$ in k points; then the number of generating spaces, $2y$, which touch the curve is given by $2y=2\pi-2-k(2p_1-2)$, and the number of generating spaces, d, in which $s+1$ of the k points of the curve lying therein lie on a space $[s-1]$ is given by $d=\nu(k-1,s)-n_1(k,s+1)-y(k-2,s-1)$, when (h,l) means $h!/l!(h-l)!$; and we have

$$k(2p_1-2)=2\pi-2-\frac{2(k-1)}{s(s+1)}\{(s+1)\nu-n_1k\}+\frac{2d}{(k-2,s-1)}.$$

This agrees with our case if $n_1=p-n+1$, $s=n-2$, $k=n$, $\nu=2p-2$, $\pi=p$, $p_1=0$, $y=p-1+n$, $d=0$. For this formula, and references, see Segre, *Enzykl. Math. Wiss.* III, C. 7, p. 954.]

Ex. 7. It is assumed in Ex. 6 that every one of $\sigma_1-1,\ldots,\sigma_{n-1}-1\geqslant 0$. For an indication of other possibilities we may take the general plane curve of order n without multiple points (which, for $n>4$, is not general of its genus). For this case, a fundamental set of integral functions is given by $y,y^2,\ldots,y^{n-1}$, and we have $\sigma_1=0$, $\sigma_2=1$, $\ldots$, $\sigma_{n-1}=n-2$. The scroll in question is then that generated by spaces $[n-3]$, through a point, which contain corresponding points of a line, a conic, $\ldots$, a rational normal curve of order $n-3$. Another case is that of the general curve of genus 5, for which again $\sigma_1=1$. The scroll is that of planes, through a point in space $[4]$, which contain corresponding points of two lines, pointwise related.

Ex. 8. The matter may be dealt with from the theory of special series considered above (p. 96). We limit ourselves to the consideration of complete series (in the first case of Ex. 7, the series g_1^n obtained is not complete), and to the general case, when the curve is general of its genus. As a simple case consider, for $p=5$, on the canonical curve of order 8 in space $[4]$, which is the complete intersection of three quadrics, the series g_1^4, in which the sets lie on quadrisecant planes of the curve. The complementary series is then also a g_1^4. There are through any set A_1 of the series two linearly independent primes of the space $[4]$, determining, suppose, the complementary sets A_1', A_2' on the canonical curve; through A_1' there are two linearly independent primes, one of which, we may suppose, has A_1 for its residual intersection with the curve, the other having a second set, A_2, of the original series g_1^4. If the primes (A_1,A_1'), (A_2,A_1') be $\phi_1=0$, $\phi_2=0$, and the primes (A_1,A_2'), (A_2,A_2') be $\psi_1=0$, $\psi_2=0$, the series g_1^4 is determined as well by the primes $\lambda_1\phi_1+\lambda_2\phi_2=0$, as by the primes $\mu_1\psi_1+\mu_2\psi_2=0$, in which $\lambda_1,\lambda_2,\mu_1,\mu_2$ are parameters, and we can suppose ψ_1,ψ_2 multiplied by such constants that $\mu_1=\lambda_1$, $\mu_2=\lambda_2$. Both the sets A_1, A_2, as well as all other coresidual sets of the series $g_1^4=0$, thus lie on the quadric $\phi_1\psi_2 - \phi_2\psi_1 = 0$, which contains the plane $\phi_1 = 0$, $\psi_1 = 0$ containing the set A_1, and the plane $\phi_2 = 0$, $\psi_2 = 0$ containing the set A_2, and similarly the planes of all the sets of the linear series, as well as the planes of all the sets of the complementary series. This quadric is a cone, whose vertex, O, is the point of intersection $\phi_1 = \phi_2 = \psi_1 = \psi_2$; this point O is determined by the particular series g_1^4 considered. There are ∞^1 such series; and a particular set of a particular series is determinable by taking two points P, Q of the canonical curve, and determining a plane through PQ having two other intersections with the curve; then Q may be regarded as varying from set to set of the same series, and P as determining the particular series. There is thus a correspondence between the canonical curve, regarded as the locus of P, and the curve which is the locus of the points O determined by all the series g_1^4. When O is given there are 4 points P, since O lies on a quadrisecant plane of the canonical curve; and when P is given there are 5 positions of O, since 5 quadrisecant planes can be drawn through a given chord; if we assume there are no coincidences in either case, this establishes that the locus of O is a curve of genus 6 (see Vol. vi, Chap. i). That it is of order 10 follows since O is the locus of the vertices of quadric cones containing the complete intersection of three quadrics, say $U = 0$, $V = 0$, $W = 0$; and is therefore the locus for which all determinants of order 3 in the matrix of 3 rows and 5 columns, whose general column is $\partial U/\partial x_i$, $\partial V/\partial x_i$, $\partial W/\partial x_i$, simultaneously vanish. (Cf. p. 110, Ex. 2, above.)

Ex. 9. Now consider more generally a complete special series g_r^n, on the canonical curve in space $[p-1]$, the complementary series being a $g_{r'}^{n'}$, when $n' = 2p - 2 - n$ and r' given by $r = n - p + r' + 1$. The sets A_1, A_2, ... of the series g_r^n are the residual sets of intersections, of the canonical curve, with primes passing through the space $[n' - r' - 1]$ which contains any set A_j' of the series of sets A_1', A_2', ... of the complementary series $g_{r'}^{n'}$. We may denote the prime which contains the set A_i of g_r^n, and the set A_j' of $g_{r'}^{n'}$, by $[i, j'] = 0$. The general prime through A_j' is then of the form $\lambda_0[1, j'] + \lambda_1[2, j'] + ... + \lambda_r[r+1, j'] = 0$ and we can, choosing a suitable constant multiplier for every prime $[i, j']$, suppose that the multipliers $\lambda_0, ..., \lambda_r$ are independent of j. By similar reasoning for the complementary series, it is sufficient to consider only the sets A_j' for which $j = 1, 2, ..., r' + 1$. Suppose $r' \geqslant r$; then eliminating $\lambda_0, ..., \lambda_r$ from every $r + 1$ of these equations, we see that all the spaces $[n - r - 1]$, containing the sets $A_1, ..., A_{r+1}$, and hence, also, those containing all the sets A_i of g_r^n, lie on the manifold which reduces to zero every determinant of $(r+1)$ rows and columns in the matrix, of $r+1$ rows and $r' + 1$ columns,

$$\left\| \begin{array}{cccc} [1, 1'], & [1, 2'], & ..., & [1, r'+1] \\ \hdotsfor{4} \\ [r+1, 1'], & [r+1, 2'], & ..., & [r+1, r'+1] \end{array} \right\|$$

By an algebraical theorem considered below (in Vol. vi) this manifold is of dimension $n - 1$, and of order $(r' + 1, r)$, where (p, q) is used for $p!/q!(p-q)!$. The manifold evidently contains the space $[\tau - 1]$, where $\tau = p - (r+1)(r'+1)$, through which every one of the primes $[i, j'] = 0$ passes, if $\tau \geqslant 1$; it was shewn (p. 91) that, in general, $\tau \geqslant 0$. The spaces $[n' - r' - 1]$, containing the sets A_j' of $g_{r'}^{n'}$, which lie in primes $\mu_0[i, 1] + ... + \mu_{r'}[i, r'+1] = 0$, also contain this space $[\tau - 1]$. As there are ∞^τ series g_r^n, the spaces $[\tau - 1]$ describe a manifold of dimension $2\tau - 1$. The general condition $\tau \geqslant 0$ is not necessary for the existence of the manifold; for instance, when the canonical curve is that for the general plane quintic curve, the planes which contain the sets of the series g_2^5 generate a manifold of order 3 and dimension 4, the locus of the planes of the conics lying on a Veronese surface in space [5] (which represents the plane of the quintic curve). The theory applies in particular to the series g_1^n where $n = 1 + \frac{1}{2}p$, or $n = 1 + \frac{1}{2}(p+1)$, giving respectively $\tau = 0$ and $\tau = 1$; in the latter case the manifold is a cone, whose vertex describes a curve.

Ex. 10. Interesting applications of the theory of this chapter arise when, for the curve $f(x, y) = 0$, we consider rational functions whose poles are all at one place. The equation may then be transformed to an equation $F(\xi, \eta) = 0$, where ξ is the function whose pole at this place has the lowest possible order, say a; and η is the function whose pole has the next lower possible order, r, which is prime to a. Cf. Weierstrass, *Werke*, II, p. 235 (and the writer's *Abel's Theorem*, pp. 34, 93, 99).

CHAPTER VIII

ENUMERATIVE PROPERTIES OF CURVES

Part I. General formulae. In ordinary space of three dimensions, an algebraic surface is the locus represented by a single polynomial equation, $F(x, y, z, t) = 0$, homogeneous in the co-ordinates x, y, z, t. It is generally intended that the polynomial F is incapable of being written as the product of other polynomials, and the surface is then said to be *irreducible*. The *order* of the surface is the number of its intersections with an arbitrary line, and is the order of the polynomial F in x, y, z, t. Unless the contrary is stated we shall suppose the plane $t = 0$ to have no special relation with the surface, and shall often replace t by 1, and represent the equation by $F(x, y, z) = 0$.

An *algebraic curve* is most naturally regarded as the intersection of algebraic surfaces; two such surfaces meet in a curve. But it is not true conversely that any given algebraic curve is the *complete* intersection of two surfaces; a familiar example to the contrary is the rational cubic curve, which is the part intersection of two quadric surfaces having, beside, a line in common. Nor indeed is it clear that *three* algebraic surfaces can be drawn through a given curve so as to have this as their only common part; they may have, beside, points in common; in fact it will be seen below that a curve may be such that no three surfaces can be drawn through this which do not have also points in common not lying on this curve. But *four* surfaces can be drawn through a curve to have no common curve or point, beside the given curve; and this in infinitely many ways; it is sufficient to take the cones projecting the curve from four points of general position*.

The indeterminateness of the definition of an algebraic curve, as the intersection of surfaces, suggests that it may be desirable to define such a curve in another way. We take the definition as follows: An algebraic curve is a locus of which the (non-homogeneous) coordinates of a point, x, y, z, are rational functions of two parameters ξ, η, which are not independent, but are connected by a

---

* It is similarly true that any algebraic construct, lying in space of r dimensions, may be regarded as the complete intersection of $(r + 1)$ primals (loci represented each by a single polynomial equation in the $r + 1$ homogeneous coordinates). This remark seems to have been first made by Kronecker, *Werke*, II, 1881, p. 280. A geometrical exposition is given by Segre, "Introduzione...", *Ann. d. Mat.* XXII, 1894, p. 47. An account of Kronecker's point of view is given in the writer's *Multiply-Periodic Functions* (Cambridge, 1907), p. 273; a much more exhaustive account by Molk, *Acta Math.* VI, 1885, p. 159.

rational polynomial irreducible equation $f(\xi, \eta) = 0$; these parameters being both expressible rationally in terms of the coordinates, x, y, z, of the corresponding point of the curve. That is, we have equations of the forms $x = U(\xi, \eta)$, $y = V(\xi, \eta)$, $z = W(\xi, \eta)$, $f(\xi, \eta) = 0$; $\xi = u(x, y, z)$, $\eta = v(x, y, z)$, where U, V, W, u, v denote rational functions, and $f(\xi, \eta)$ is a rational polynomial. By this definition an algebraic curve in space is in $(1, 1)$ birational correspondence with the irreducible plane curve expressed by $f(\xi, \eta) = 0$.

That a curve as so defined may be given by the intersection of surfaces is clear enough; a formal proof that a curve defined by the intersection of surfaces is capable of the definition we have given, is contained in the literature referred to in the footnote preceding.

From the definition, from what is known for a plane curve, it appears that all points (x, y, z) of the space curve, which are in the neighbourhood of any point (x_0, y_0, z_0) of the curve, belong to one of a finite number of *branches*; the points of a branch are given, in terms of a parameter t, by equations of the forms

$$x - x_0 = A_1 t + A_2 t^2 + \dots, \quad y - y_0 = B_1 t + B_2 t^2 + \dots,$$
$$z - z_0 = C_1 t + C_2 t^2 + \dots,$$

in which the parameter t is such that its value is unique for every point (x, y, z) of the branch in the neighbourhood of (x_0, y_0, z_0); and the series converge for all sufficiently small values of t. The statement is made supposing all of x_0, y_0, z_0 to be finite, but can easily be modified to cover all cases. Consider now one of these branches in the neighbourhood of (x_0, y_0, z_0), writing, for brevity, x, y, z in place of $x - x_0, y - y_0, z - z_0$. Suppose that not every one of A_1, B_1, C_1 is zero, and, in particular, that A_1 is not zero. Using then, instead of the coordinates y, z, the coordinates $y - B_1 A_1^{-1} x$ and $z - C_1 A_1^{-1} x$, the expressions take forms

$$x = A_1 t + A_2 t^2 + \dots, \quad y = Q_2 t^2 + Q_3 t^3 + \dots, \quad z = R_2 t^2 + R_3 t^3 + \dots;$$

suppose here that Q_2 is not zero; then, using, instead of z, the coordinate $z - R_2 Q_2^{-1} y$, the expressions are reduced to the forms

$$x = A_1 t + A_2 t^2 + \dots, \quad y = Q_2 t^2 + Q_3 t^3 + \dots, \quad z = H_3 t^3 + H_4 t^4 + \dots.$$

For a branch expressed by three such equations, we can define a certain line, called the *tangent line* of the branch, and a certain plane, called the *osculating plane* of the branch, at the point considered. First, it is obvious that any plane through the point, of equation $px + qy + rz = 0$, meets the branch in one point there; for the power series equation in t

$$p(A_1 t + A_2 t^2 + \dots) + q(Q_2 t^2 + Q_3 t^3 + \dots) + r(H_3 t^3 + H_4 t^4 + \dots) = 0$$

has $t = 0$ as a simple root. If, however, $p = 0$, the root $t = 0$ is double, and every plane $qy + rz = 0$ meets the branch in two points coinciding at the origin considered. Thus the line $y = 0$, $z = 0$ meets the branch twice at the origin. It is this line which is called the *tangent line* at the point. Again, if both p and q be zero, the root $t = 0$ is triple, and the plane $z = 0$ meets the branch in three points coinciding at the origin. It is this plane which is called the *osculating plane* of the branch. It is easy to shew further that, if the origin be O, and A, B be two other points of the branch, in the neighbourhood of O, the line OA tends to the tangent line as A tends to O, and the plane OAB tends to the osculating plane as A, B tend both to O; and that the plane containing the tangent line and the point A, tends to the osculating plane as A tends to O.

More generally, a branch of the curve may be representable, by proper choice of coordinates, by three equations of the forms

$$x = a_1 t^{l+1} + a_2 t^{l+2} + \ldots, \quad y = b_1 t^{m+2} + b_2 t^{m+3} + \ldots,$$
$$z = c_1 t^{n+3} + c_2 t^{n+4} + \ldots,$$

where the integers l, m, n, which are all positive or zero, are such that $l \leqslant m \leqslant n$ and none of a_1, b_1, c_1 is zero. Unless l, m, n are all zero, in which case these expressions reduce to those considered above, the names tangent line and osculating plane (for $y = 0 = z$, and $z = 0$) are not strictly applicable in the preceding sense *at the origin.* But they are applicable at a point (t) near the origin, and may be obtained, if we wish, by writing $t + \tau$ for t in these power series, and rearranging the series in powers of τ. It is more convenient to use differential coefficients. Suppose for example we seek the osculating plane at the point (t); let the equation of this plane be written $x\eta + y\xi + z + \omega = 0$; substitute for x, y, z the above power series in t, and determine ξ, η, ω from the resulting equation and the two equations $x'\eta + y'\xi + z' = 0$, $x''\eta + y''\xi + z'' = 0$, where $x' = dx/dt$, $x'' = d^2x/dt^2$, etc. If X, Y, Z denote the minors of x, y, z in the determinant Δ, given by

$$\Delta = \begin{vmatrix} x, & y, & z \\ x', & y', & z' \\ x'', & y'', & z'' \end{vmatrix},$$

the result is $\xi = Y/Z$, $\eta = X/Z$, $\omega = -\Delta/Z$.

Now introduce numbers l', m', n' given by $l' = n - m$, $m' = n - l$, $n' = n$, which lead to $l = n' - m'$, $m = n' - l'$, $n = n'$; then l', m', n' are positive or zero, and such that $l' \leqslant m' \leqslant n'$, while the expressions for ξ, η, ω in terms of t may be written $\xi = a_1' t^{l'+1} + \ldots$, $\eta = b_1' t^{m'+2} + \ldots$, $\omega = c_1' t^{n'+3} + \ldots$, where

$$a_1' = -\frac{c_1}{b_1} \frac{(n+3)(m'+2)}{(m+2)(m'-l'+1)}, \quad b_1' = \frac{c_1}{a_1} \frac{(n+3)(l'+1)}{(l+1)(m'-l'+1)},$$
$$c_1' = -c_1 \frac{(l'+1)(m'+2)}{(l+1)(m+2)};$$

and the reverse formulae are of precisely similar forms, namely

$$a_1 = -\frac{c_1'}{b_1'}\frac{(n'+3)(m+2)}{(m'+2)(m-l+1)}, \quad b_1 = \frac{c_1'}{a_1'}\frac{(n'+3)(l+1)}{(l'+1)(m-l+1)},$$

$$c_1 = -c_1'\frac{(l+1)(m+2)}{(l'+1)(m'+2)},$$

and, as none of a_1, b_1, c_1 is zero (or infinite), also none of a_1', b_1', c_1' is zero or infinite.

A branch of a curve for which the three indices l, m, n are all zero may be called an *ordinary branch*, the name being justified because, as will sufficiently appear, the number of branches of an *algebraic* curve which are not ordinary is necessarily finite; it is for such a branch that we have defined the tangent line and the osculating plane. A branch which is not ordinary at the origin (or centre) of its coordinates, above denoted by $x=0$, $y=0$, $z=0$, is ordinary at points in the immediate neighbourhood of its origin.

The expressions we have just found serve to emphasize the duality which exists for a curve, regarded as defined by its points, and as defined by its osculating planes. An algebraic curve is a locus of points defined by the continuous variation of a parameter; we may dually consider an aggregate of planes similarly defined by a parameter. Just then as a curve has *chords*, each defined by two points of the curve; and the limit of a chord, when one end is the origin of an ordinary branch of the curve, and the other end tends to this, is the definite line called the tangent line; so, the aggregate of planes has *axes*, each defined by the intersection of two of the planes; and there is a similar limit theorem. Similarly, as the osculating plane of the curve is obtainable by passing to a limit, with three points of which one is the centre of an·ordinary branch, so, a point is obtained from three planes of the aggregate of planes; and the locus of such points is exactly a curve as originally considered. When the algebraic aggregate of planes is regarded as defining a *developable surface* (of which the planes are the tangent planes), the curve obtained is that which is called the *cuspidal edge* (or the edge of regression) of this surface; any algebraic curve is thus the cuspidal edge of the developable surface formed by its own osculating planes; and the tangent line of the curve is the ultimate position of the line of intersection of two consecutive planes of the developable surface (the *generating line* of the developable).

Further, we notice that, in the formulae above, an ordinary branch, for which all the indices l, m, n are zero, gives expressions for ξ, η, ω for which all the indices l', m', n' are zero, and conversely. Points which are origins of branches which are not ordinary are thus exceptional as well from the dual as from the original point of view.

Of branches which are not ordinary we shall generally consider only three kinds:

(i) Those for which $l=0$, $m=0$, $n=1$. Here the expressions of the coordinates, in terms of the parameter, are of the forms $x=a_1t+\dots$, $y=b_1t^2+\dots$, $z=c_1t^4+\dots$, and the difference from the ordinary case is that the plane $z=0$ has *four*, not *three*, intersections with the branch at the origin. The origin may therefore be called a point at which there is a branch of *stationary osculating plane*. The origin itself may perhaps be called a *stall*. The total number of such branches for the whole curve (which, as we have said, is finite for an algebraic curve) will finally be denoted by κ'. In the dual formulae, for the osculating plane at a point of the branch in the neighbourhood of the origin, the indices l', m', n' are given by
$$l', =n-m, =1, \quad m', =n-l, =1, \quad n', =n, =1.$$

(ii) The second kind of branch which we commonly recognise, which is not ordinary, is that for which $l=1$, $m=1$, $n=1$. Here the expressions for the coordinates are of the forms $x=a_1t^2+\dots$, $y=b_1t^3+\dots$, $z=c_1t^4+\dots$, and the striking difference from the ordinary case lies in the fact that every plane through the origin meets the branch in *two* points at the origin, and every plane, $qy+rz=0$, through the line $y=0=z$, meets the branch in *three* points at the origin. While, in the ordinary case, y/x^2 has a definite limit at the origin, it is here y^2/x^3 which becomes definite. This is similar to the behaviour of a plane curve at an ordinary cusp; thus we shall also say here that the branch has a *cusp*, or *stationary point*, at the origin. This case is evidently dual to the case (i); the total number of branches of the curve which have a cusp will finally be denoted by κ.

(iii) Last, we may consider branches for which the indices are given by $l=0$, $m=1$, $n=1$; these give also $l'=0$, $m'=1$, $n'=1$, and the singularity is dual to itself. The expressions for the coordinates are of the forms $x=a_1t+\dots$, $y=b_1t^3+\dots$, $z=c_1t^4+\dots$; in this case, a general plane through the origin meets the branch in only one point at the origin, but any plane through the line $y=0$, $z=0$ meets the branch in three points coincident at the origin. This is analogous to a point of inflexion on a plane curve; we shall therefore say that the branch has an *inflexion*, or *stationary tangent line*; and we shall denote the whole number of such branches, arising for the curve, by i.

Ex. 1. If for a branch of a *plane* curve, with expressions in terms of a parameter t given by
$$x=a_1t^{l+1}+\dots, \quad y=b_1t^{m+2}+\dots, \qquad l, m \geqslant 0, \ l \leqslant m,$$
the tangent line at a point in the neighbourhood of the origin be written $px+y+q=0$, prove that
$$p=a_1't^{l'+1}+\dots, \quad q=b_1't^{m'+2}+\dots, \qquad \text{with } l'=m-l, \ m'=m,$$
where
$$a_1'=-\frac{b_1}{a_1}\frac{m+2}{l+1}, \quad a_1=-\frac{b_1'}{a_1'}\frac{m'+2}{l'+1}, \quad b_1'=b_1\frac{l'+1}{l+1},$$
and
$$l', m' \geqslant 0, \ l' \leqslant m'.$$

Thus, just as $l+1$ is the number of points of the branch which lie on an arbitrary line passing through the origin $x=0$, $y=0$, so $l'+1$ is the number of tangent lines of the branch which pass through an arbitrary point lying on the line $p=0$, $q=0$, that is, the line $y=0$. And, as $m+2$ is the number of points of the branch (at $t=0$) lying on the line $y=0$, so $m'+2$ (equal to $m+2$) is the number of tangents of the branch containing the point $q=0$, that is the point $x=0$, $y=0$.

Similar remarks may be made, from the formulae of the text, for the osculating plane of the curve in space. Cf. Halphen, *Bull. Soc. Math. d. France*, VI, 1878, p. 10.

Ex. 2. The line-coordinates of the tangent line, at the point (t), of the branch of the curve in space expressed by $x=at^{l+1}+\dots$, $y=bt^{m+2}+\dots$, $z=ct^{n+6}+\dots$, if the line-coordinates be computed from

$$\lambda:\mu:\nu:\lambda':\mu':\nu'=x':y':z':yz'-y'z:zx'-z'x:xy'-x'y,$$

where $x'=dx/dt$, $y'=dy/dt$, $z'=dz/dt$, are, in order, only the first terms being written, for λ, μ, ν, $(l+1)a$, $(m+2)bt^{m-l+1}$, $(n+3)ct^{n-l+2}$, and, for λ', μ', ν', $bc(n-m+1)t^{m+n-l+4}$, $-ca(n-l+2)t^{n+3}$, $ab(m-l+1)t^{m+2}$. When $t=0$ these approach to the values $(1, 0, 0, 0, 0, 0)$, which we may regard as the line-coordinates of the tangent line of the branch. The points of the branch at which the tangent line meets this particular tangent line are given by the vanishing of a power series in t of which the first term (that in λ') is $bc(n-m+1)t^{m+n-l+4}$. Thus we say that the number of tangent lines of the branch, "consecutive with" the tangent at the origin, which meet this tangent, is $m+n-l+4$, which is also $m'+n'-l'+4$. In particular for an ordinary branch this number is 4.

The number of such tangents which pass through the origin is to be found by considering the equations $\lambda'=0$, $\mu'=0$, $\nu'=0$, and is the highest power of t entering as factor in all these; this, as we see at once, is $m+2$. In particular, for a stationary point, for which $l=m=n=1$, this is 3. The number of such tangents which lie in the plane $z=0$, is similarly found from $\lambda'=0$, $\mu'=0$, $\nu=0$, and is $n-l+2$ or $m'+2$. In particular for a stall, where there is a stationary osculating plane, for which $l=0$, $m=0$, $n=1$, this is also 3.

It may be remarked that an algebraic curve is completely determinate by the power series which express x, y, z on any one branch, all other points and the branches thereat being deducible by "analytical continuation" from these power series. This is a result following from the corresponding theorem for a Riemann surface (above, p. 144).

We assume now that a general plane meets the algebraic curve in a definite number of points, which we call the *order* of the curve. This result may be rigorously deduced from the definition we have adopted for the curve; it is the same as saying that, for the curve $f(\xi, \eta)=0$, the rational function $pU(\xi, \eta)+qV(\xi, \eta)+rW(\xi, \eta)+s$ has a definite number of zeros, independent of precise values of the ratios of p, q, r, s; naturally the number of intersections of the plane with the curve at points which are not ordinary must be properly estimated. Dually, with a similar proper interpretation of the phrase, there is a definite number of osculating planes of the curve passing through an arbitrary general point. This we call the *class* of the curve. We shall denote the order of the curve by n, and its class by n'. Further, the curve has ∞^1 tangent lines, and for a tangent line to meet a given line is one condition; thus an arbitrary general line is met by a definite number of tangent lines of the curve; this we call the *rank* of the curve, and denote by r. The number of tangent lines of the curve which meet a line of special position may of course be different; in particular, from Ex. 2 above, the number of other tangent lines which meet the tangent line of the curve at an ordinary point is $r-4$ (if $r \geqslant 4$). Likewise, the curve has ∞^2 chords, and for a line to pass through a given point two conditions must be satisfied; thus, through an arbitrary general point of the space, there will pass a definite number of chords of the curve; the tangents of the curve generate a locus consisting of ∞^2 points, a surface; and the arbitrary general point will not lie upon a definite surface. Thus the chords of the curve from an arbitrary general point will be proper chords, having two points of intersection with the curve which are distinct; the number of such chords we denote by h; it is sometimes spoken of as the number of *apparent double points* of the curve. Dually, we denote by h' the number of lines of intersection of two distinct osculating planes of the curve, or, as before, the number of *axes* of the curve, which lie in an arbitrary general plane. Another characteristic of the curve arises from what we may call the *nodal curve* of the given curve. As we have remarked, a tangent of the curve is met by a certain number of other tangents; the points of intersection generate another curve, which is what we call the *nodal curve*. The characteristic in question is the order of this nodal curve; it is the number of points, in an arbitrary general plane, at which two distinct tangents of the original curve intersect one another. This number we denote by ν. Dually, we denote by ν' the number of planes, through an arbitrary general point, which contain two distinct intersecting tangent lines of the original curve; this is the number of planes through the point which touch the curve at two points, say, the number of bitangent planes of the curve through the point. The bitangent planes of a curve are, clearly enough, the tangent planes of a developable surface, upon which the original curve is a cuspidal curve, and the nodal curve, of the original curve, is a double curve; we may call this the *bitangent developable* of the original curve; it is to be distinguished from the developable surface of which the tangent planes are the osculating planes of the original curve, sometimes called the *osculating developable* of the original curve. The curve order, n, is different from the branch index, n.

For simplicity of reference we tabulate the notations thus introduced: n is the order, r is the rank, n' is the class;

κ is the number of branches of the curve with cusps, or stationary points, where the expressions in terms of a parameter have the forms $x = at^2 + \ldots$, $y = bt^3 + \ldots$, $z = ct^4 + \ldots$, the indices l, m, n of the general description being all equal to 1;

κ' is the number of branches with stationary osculating plane, the number of stalls, where the expressions have forms $x = at + \ldots$, $y = bt^2 + \ldots$, $z = ct^4 + \ldots$, the indices l, m, n having the respective values 0, 0, 1;

i is the number of inflexions, with expressions of the forms $x = at + \ldots$, $y = bt^3 + \ldots$, $z = ct^4 + \ldots$, for which l, m, n have the respective values 0, 1, 1. In all these it is supposed that none of the coefficients a, b, c is zero, and it is provisionally assumed that κ, κ', i are all finite;

h is the number of chords of the curve through an arbitrary general point, and, dually, h' is the number of axes of the curve lying in an arbitrary general plane;

ν is the order of the nodal curve, the locus of points of intersection of two distinct tangents of the original curve; and, dually, ν' is the number of planes through an arbitrary general point which contain two distinct tangents of the original curve.

Evidently, on a curve which is general, regarded as a locus of points—there being ∞^1 points of the curve, and the condition for an osculating plane to be stationary being one-fold—there will as a rule be points of stationary osculating plane, namely κ' will not vanish. Similarly κ will not as a rule be zero for a curve which is general when regarded as an envelope of osculating planes. But, in either case, it is exceptional for i not to be zero; for the condition for a third consecutive point of the curve to lie in the tangent line is two-fold.

We may also allow for the possibility of actual double points of the curve, where the curve crosses itself. Such a point will not generally exist (as requiring 3 conditions to be satisfied by 2 parameters); but we shall be concerned below with curves which arise as the intersection of surfaces, and where two such surfaces touch their curve of intersection has a double point. The number of such actual double points for the curve will be denoted by δ. Dually, δ' will denote the number of planes which are osculating planes of the curve at two distinct points. Further, the number of still less probable lines which are tangent to the curve at two distinct points will be denoted by τ.

Recall now, that, for a plane algebraic curve, of order N, class N', and genus P, whose singularities consist of D double points, K cusps, T double tangents, and I points of inflexion, we have

$$D = \tfrac{1}{2}N(N-7) + N' - 3(P-1), \quad K = 2N - N' + 2P - 2,$$
$$T = \tfrac{1}{2}N'(N'-7) + N - 3(P-1), \quad I = 2N' - N + 2P - 2.$$

These equations lead to

$$N' = N(N-1) - 2D - 3K, \qquad N = N'(N'-1) - 2T - 3I,$$
$$I - K = 3(N' - N),$$
$$P = \tfrac{1}{2}(N-1)(N-2) - D - K, \quad = \tfrac{1}{2}(N'-1)(N'-2) - T - I.$$

We use these equations for a plane curve, to make inferences for the curve in space, first projecting this from an arbitrary general point upon an arbitrary general plane, and then by considering, upon an arbitrary general plane, the locus of the point in which this plane is met by the tangents of the given curve.

When we project the given curve in space on to a plane, we obviously obtain a curve of order N, $=n$, and of class N', $=r$. It is also easy to see that we obtain cusps of the plane curve from the stationary points of the curve in space, or that $K = \kappa$. Again, when an osculating plane of the space curve, at an ordinary point of this, passes through the point of projection, we very clearly obtain a point of inflexion on the plane curve; the three coinciding intersections of the space curve with the osculating plane, project into three coinciding intersections of the plane curve with a tangent line. We also obtain a point of inflexion of the plane curve from any existing inflexion of the space curve. Thus we have $I = n' + i$. Likewise, we evidently have a double point on the plane curve for every chord of the space curve which passes through the centre of projection; so that, allowing for actual double points of the space curve, we have $D = h + \delta$. It is also clear that the number of double tangents T, of the plane curve, is $T = \nu' + \tau$.

Next, consider the plane curve, in an arbitrary general plane, which is the locus of the point in which the tangent line, of the curve in space, meets this plane. The order, N, of the plane curve, is evidently the rank, r, of the space curve. The tangent line of the plane curve, obtainable as the limit of a chord, arises from an osculating plane of the space curve, this containing two coincident tangent lines; thus the class N', of the plane curve, is given by $N' = n'$. At a point where the space curve meets the plane of the section, supposed to be an ordinary point of the space curve, two, ultimately coinciding, tangent lines of the space curve will give the same point of the plane curve; this point will thus be a stationary point (cusp) of the plane curve. There will likewise be a cusp on the plane curve arising from the stationary character of the tangent line at any of the existing i inflexions of the space curve. Wherefore $K = n + i$. It may be seen, further, that a point of the space curve which is the origin of a branch with a stationary osculating plane (a stall on the space curve) gives rise to an inflexion on the plane curve, so that $I = \kappa'$. Also, it is clear that there is a double point of the plane curve at every point where the nodal curve of the space curve meets the plane of section—and further, also at the intersection of this plane with any actual double tangent line of the space curve; hence $D = \nu + \tau$. Finally, for the double tangents of the plane curve, we have, easily, $T = h' + \delta'$.

Tabulating these statements, supposing that we have enumerated all the possibilities, we have, for the two plane curves considered, obtained by projection and section:

	Order (N)	Class (N')	Cusps (K)	Inflexions (I)	Double points (D)	Double tangents (T)
By projection	n	r	κ	$n'+i$	$h+\delta$	$\nu'+\tau$
By section	r	n'	$n+i$	κ'	$\nu+\tau$	$h'+\delta'$

Hence, from the Plücker equations recalled above, applied to these curves respectively, we have, for the space curve

$$r = n(n-1) - 2(h+\delta) - 3\kappa,$$
$$n = r(r-1) - 2(\nu'+\tau) - 3(n'+i),$$
$$n'+i-\kappa = 3(r-n),$$

and

$$n' = r(r-1) - 2(\nu+\tau) - 3(n+i),$$
$$r = n'(n'-1) - 2(h'+\delta') - 3\kappa',$$
$$n+i-\kappa' = 3(r-n');$$

also, if P, P' be the respective genera of the two plane curves, obtained by projection and section,

$$P = \tfrac{1}{2}(n-1)(n-2) - (h+\delta+\kappa), \quad = \tfrac{1}{2}(r-1)(r-2) - (n'+i+\nu'+\tau),$$
$$P' = \tfrac{1}{2}(r-1)(r-2) - (n+i+\nu+\tau), \quad = \tfrac{1}{2}(n'-1)(n'-2) - (h'+\delta'+\kappa').$$

The first six equations, however, lead to

$$\kappa - \kappa' = 2(n-n'), \quad \nu - \nu' = n' - n,$$
$$h + \delta - (h'+\delta') = \tfrac{1}{2}(n-n')(n+n'-7),$$

from which in particular it appears that $P = P'$. This number is thus independent of the centre from which the projection is made, and of the plane on to which it is made; and is independent of the plane on which the section is taken. We denote this number by p, and call it the genus of the curve in space.

It is at once seen that, if n, n', p and r be supposed known, these equations give

$$i = -(n+n') + 2r + 2p - 2,$$
$$\kappa = 2n - r + 2p - 2,$$
$$\kappa' = 2n' - r + 2p - 2,$$
$$\nu + \tau = \tfrac{1}{2}(r-1)(r-6) + n' - 3p,$$
$$\nu' + \tau = \tfrac{1}{2}(r-1)(r-6) + n - 3p,$$
$$h + \delta = \tfrac{1}{2}(n-1)(n-6) + r - 3p,$$
$$h' + \delta' = \tfrac{1}{2}(n'-1)(n'-6) + r - 3p.$$

If τ, δ and δ' be zero, these seven equations determine the other characters $i, \kappa, \kappa', \nu, \nu', h, h'$ in terms of n, n', r and p. In any case all the equations found, and in particular the four forms for p, can be deduced from these seven equations. From these equations also the five following equations can be deduced:

$$n' + 2\kappa + i = 3(n + 2p - 2), \quad n + 2\kappa' + i = 3(n' + 2p - 2),$$
$$3\kappa + 2i + \kappa' = 4(n + 3p - 3), \quad \kappa + 2i + 3\kappa' = 4(n' + 3p - 3),$$
$$\kappa + 2i + \kappa' = 2(r + 4p - 4).$$

Connexion of previous results with general principles. It is an interesting fact that these five equations, and the first three of the preceding seven equations (those which do not involve ν, ν', h or h'), are all illustrations of a single principle. This we propose to explain now, with the warning to the reader that the ideas involved are somewhat less elementary than those so far used in this chapter. We deal first with the second and third of the first seven equations (those giving κ and κ') and the first four of the last five equations; then, in a still less elementary way, with the two remaining equations, namely with the two $n + i + n' = 2(r + p - 1)$, $\kappa + 2i + \kappa' = 2(r + 4p - 4)$.

Referring to Chap. IV, preceding, for the ideas involved, suppose that on a plane curve, of genus p, given by an equation $f(\xi, \eta) = 0$, we have a linear series of sets of points, of freedom s; that is, a series of sets given by the intersections of the curve with a system of variable curves whose equation is

$$\lambda\phi(\xi, \eta) + \lambda_1\phi_1(\xi, \eta) + \ldots + \lambda_s\phi_s(\xi, \eta) = 0,$$

in which $\phi, \phi_1, \ldots, \phi_s$ are definite rational polynomials, linearly independent on $f(\xi, \eta) = 0$; in addition to possible fixed intersections, common to all of $\phi(\xi, \eta) = 0, \ldots, \phi_s(\xi, \eta) = 0$, $f(\xi, \eta) = 0$, these curves have a number, say m, of intersections with $f(\xi, \eta) = 0$, which vary when $\lambda, \lambda_1, \ldots, \lambda_s$ vary. There exists then a set of the series of which s points are arbitrarily chosen on $f(\xi, \eta) = 0$. In particular, there exists a set with s points all coincident at any arbitrary general point, P, of $f(\xi, \eta) = 0$; as P takes its possible ∞^1 positions on this curve, the remaining $m - s$ points of the set vary also; that one of these $m - s$ points should coincide with P is one condition. Thus there is a finite number of positions of P at which there fall $(s + 1)$ coincident points of a set of the linear series. The principle which we assume is that the number of such $(s + 1)$-fold points is $(s + 1)[m + s(p - 1)]$. It is understood that, in this total, special points of coincidence may need to be counted more than once; the difficulty in the application of the principle lies, in fact, most often, in this circumstance. The formula is applicable for a partial series, of freedom s_1, where $s_1 < s$, chosen from the given linear series, given, say, by a system $\lambda\phi + \lambda_1\phi_1 + \ldots + \lambda_{s_1}\phi_{s_1} = 0$, and gives the number of $(s_1 + 1)$-fold coincidences contained in this partial series; in particular the principle has already been employed for a pencil ($s_1 = 1$), in connexion with the definition of the genus p (Chap. IV, p. 84). The proof of the formula is a simple application of a formula of coincidence for correspondence of points on a curve. For clearness we refer to a following chapter (Vol. VI, Chap. I) for this proof; other proofs may be given (cf. Segre, *Ann. d. Mat.* XXII, 1894, p. 85).

Precisely the same principle can be employed for an algebraic curve in space of any number, say r, of dimensions. Such a curve is, by definition, as in the case of a curve in space of three dimensions, in (1, 1) birational correspondence with a plane algebraic curve. On such a curve, in space in which the homogeneous coordinates are $x_0, x_1, \ldots, x_r$, a linear series of sets of points is given by the intersections of the curve with a linear system of loci expressed by a single equation $\lambda\phi + \lambda_1\phi_1 + \ldots + \lambda_s\phi_s = 0$, wherein $\phi, \ldots, \phi_s$ are definite homogeneous polynomials of the same order in the coordinates $x_0, \ldots, x_r$, and $\lambda, \lambda_1, \ldots, \lambda_s$ are variable parameters. Such a system evidently leads to a linear series on the representative plane curve, and conversely. And, if we define the genus of the space curve as that of the representative plane curve, the same formula remains valid as in the case of the plane curve; more properly, as, in the (1, 1) correspondence, coincidences correspond to coincidences, the formula itself defines the genus.

Of this principle, the formula $r + \kappa = 2(n + p - 1)$, for the curve of order n and genus p, in the space of three dimensions, is an obvious application. We have only to consider the linear series, of sets of n points, of freedom 1, cut upon the curve by variable planes drawn through an arbitrary general line of the space. Whenever a tangent line of the curve meets this arbitrary axis, there is a coincidence of two of the intersections of the curve with a plane through the axis; and there is a coincidence also for a plane, through the axis, which contains a cusp of the curve. It is assumed that these two cases exhaust all the coincidences which are possible.

The formula $r + \kappa' = 2(n' + p - 1)$ is the dual of the other, or may be similarly obtained. More generally, representing branches of the curve as before, at points which are not ordinary, by series of the forms $x = at^{l+1} + \ldots, y = bt^{m+2} + \ldots, z = ct^{k+3} + \ldots$, the same principle leads to the result $r + \lambda = 2(n + p - 1)$, where λ is the sum of all not-zero values of the index l which arise (the genus p being appropriately defined).

Likewise we can obtain the formula $n' + 2\kappa + i = 3(n + 2p - 2)$; for this we consider the series, of sets of n points, of freedom 2, on the curve, which is obtained by variable planes drawn through an arbitrary general point, O. It is only necessary to shew that the left side of the equation is the number of planes through O meeting the curve in sets of which three points coincide. Of such planes, there are clearly n' osculating planes; and there are i planes, each joining O to a tangent line at a point of inflexion of the curve. To complete the formula it must be shewn that the plane joining O to the tangent line at a cusp of the curve counts doubly, as a plane with three coincident intersections; this is most clearly seen by considering the number of three-fold intersections with the curve, of planes from O, which arise for a branch $x = at^{l+1} + \dots, y = bt^{m+2} + \dots, z = ct^{k+3} + \dots$. This number is found to be $l + m$. Thus, if λ, μ denote respectively the sum of all values of l and m which are not zero, arising at all points of the curve, the complete formula is

$$n' + \lambda + \mu = 3(n + 2p - 2);$$

this includes the original form. For the proof see below, Ex. 5. The dual formula $n + 2\kappa' + i = 3(n' + 2p - 2)$ may be treated in a similar way.

Now take the formula $3\kappa + 2i + \kappa' = 4(n + 3p - 3)$. Consider the linear series, of freedom 3, of sets of n points, on the curve, determined by a general plane of the space. The right side of the equation gives, by the principle enunciated, the number of such planes meeting the curve in four coincident points. The stationary osculating planes of the curve, in number κ', are evidently such planes; but to complete the formula it is necessary to shew, (a), that through an inflexion of the curve there is one such plane which counts twice, and, (b), that through a cusp there is such a plane counting three times. More generally, in terms of the indices l, m, n for branches of the curve which are not ordinary, the formula to be proved is $\Sigma (l + m + n) = 4(N + 3p - 3)$, where N is the order of the curve, and the summation extends to every branch of the curve for which l, m, n are not all zero. For the proof, in this general form, which includes the simpler case, see Ex. 5 below. The dual formula for $\kappa + 2i + 3\kappa'$ needs no independent treatment.

We pass now to the other two formulae referred to, namely $n + i + n' = 2(r + p - 1), \kappa + 2i + \kappa' = 2(r + 4p - 4)$. These we interpret as applications of the fact that a line in space of three dimensions can be represented by a point in space of five dimensions, lying on a quadric, Ω, of that space (*Principles of Geometry*, Vol. IV, pp. 40 ff.). The tangents of the original curve will thus be represented by the points of an algebraic curve lying on Ω. The order of this curve, which is equal to the number of its intersections with a general prime (or flat four-fold space) of the five-fold space [5], and therefore equal to the number of its intersections with a general tangent prime of Ω, is equal to the number of tangent lines of the original curve which meet a general line of the three-fold space [3]; thus the curve on Ω will be of order r. We have defined the genus of a curve in any space as that of a plane curve with which it is in (1, 1) birational correspondence, incidentally assuming that two plane curves in such correspondence have the same genus, as is known. Whence, the curve on Ω, being in (1, 1) correspondence with the tangents of the original curve, and hence with the curve itself, will be of genus p. For the representation upon the quadric Ω, a point of the original space is regarded as determined by the ∞^2 lines of that space which pass through it; and a (line) coordinate of one of these lines is a linear function of the corresponding line coordinates of three of these lines, of general position; that is, there are six equations of the form $\lambda = al_1 + bl_2 + cl_3$, with a, b, c the same in all. Thus, as the coordinates of a point of the quadric Ω are the line coordinates of the line, in the original space, which it represents, it follows that a point of the original space is represented, in the space [5], by a plane, of which every point lies on the quadric Ω. In particular, the lines through a point of the original curve contain a flat pencil of lines, lying in the osculating plane of the curve at this point, of which pencil the tangent line of the curve is one line. Also, a pencil of lines of the original space is represented, in the space [5], by the points of a line lying on Ω; and, two consecutive tangents of the original curve ultimately belong to the pencil in the osculating plane which they define. Thus, the tangent line of the curve on Ω, at any point of this, lies entirely on the quadric Ω, in the plane on Ω which represents a point of the original curve. There are, in fact, two systems of planes lying entirely on Ω, one system, which we denote as α-planes, representing the points of the original space; but, by similar reasoning, the planes of the original space are also represented by planes on Ω, say β-planes; and, through the tangent line of the curve on Ω, passes, not only the α-plane representing a point of the original curve, but also the β-plane representing the osculating plane of this curve at this point. Indeed, through any line lying entirely on Ω (which represents a pencil of lines of the original space), there passes a single α-plane and a single β-plane; and these two planes together are the complete intersection of Ω with a flat three-fold space, or *solid*, of the space [5]— this solid being the intersection of the polar primes, in regard to Ω, of any two points of the line. Consider now the meaning, for the curve on Ω, of the order, n, of the original curve: An arbitrary plane, of the space [3] of this curve, which meets this curve in n points, is represented by a β-plane on Ω; the condition for a point of the original space to lie on a given plane, is that the α-plane on Ω which represents the point should meet *in a line* the β-plane which represents the plane; in general, the α-planes through the tangent lines of the curve on Ω will not meet an arbitrary β-plane; but there are just n tangent lines for which the corresponding α-plane meets an arbitrary β-plane (and therefore meets this plane in a line, as we easily see. In general an α-plane of Ω does not meet a β-plane; when they do meet, they meet in a line). Thus n is the number of tangent lines of the curve on Ω which meet an arbitrary β-plane. Similarly the class, n', of the original curve, is the number of tangent lines of the curve on Ω which meet an arbitrary α-plane. If then we take an α-plane and a β-plane which have a line in common, $n+n'$ will be the number of intersections, of tangent lines of the curve on Ω, with the solid which is defined by these two planes. Assume now that this is the same as the number of intersections of tangent lines, of the curve on Ω, with any general solid of the space [5]; through a solid, in space [5], there passes a pencil (an ∞^1 aggregate) of primes (spaces of four dimensions); each of these meets the curve on Ω in r points; and, in the series of ∞^1 sets of r points on the curve, there are, by the principle we have already used, $2(r+p-1)$ sets for each of which two of the r points coincide. Such a coincidence arises when a tangent line, of the curve on Ω, meets the solid which is the base of the pencil of primes; and there are $n+n'$ such cases, as we have seen. Such a coincidence, however, can arise in another way: namely, an inflexion of the original curve, where there is a stationary tangent, gives rise to a stationary point of the representative curve on Ω; and the prime joining this point to the base of the pencil of primes meets the curve on Ω in a set of which two points coincide at this stationary point. Thus, with the hypothesis that we have enumerated all cases of such coincidence, we have

$$n+n'+i=2(r+p-1),$$

which is the first of the two equations we desired to interpret.

We proceed now to the other equation referred to. We have represented the tangent lines of a curve in ordinary space by the points of a curve on the quadric Ω in space [5]. More generally, any algebraic system of ∞^1 lines in ordinary space, in (1, 1) correspondence with the points of an algebraic curve, say the generators of a ruled surface, may be represented by the points of a curve on Ω. This will be more general than the curve we have so far considered, its tangent lines not lying on Ω; the condition that this should be so is that every consecutive two of the system of lines in ordinary space, with which we begin, should intersect one another, or that the ruled surface should be a developable surface. In the general case there will be a certain finite number of the tangent lines of the curve on Ω which lie wholly thereon. In the special case we have taken above, of a curve on Ω of which every tangent line lies on Ω, the osculating planes of the curve will not generally lie on Ω; but there will be a certain finite number of these osculating planes which do. The curve being of order r and genus p, this number in fact is $2(r+4p-4)-2K$, where K denotes the number of cusps of the curve. The curve being derived as before from a curve in space [8], we have already remarked that $K = i$, where i is the number of inflexions of the original curve. For the proof of this number cf. Ex. 3, below. Consider now what character of the original curve gives rise to the property that the osculating plane of the representative curve lies entirely on Ω. This requires that all points of the plane, determined by three points of the curve which are tending to coincidence, should lie on Ω; there must then be three consecutive tangents of the original curve ultimately passing through the same point of this curve, or else three such tangents ultimately lying in the same osculating plane of this curve. The former arises for a cusp of the original curve, the latter for a stationary osculating plane. Thus we reach the interpretation of the equation

$$\kappa + \kappa' = 2(r + 4p - 4) - 2i$$

which we desired.

Ex. 1. For a so-called rational curve, whose points have coordinates $x:y:z:1 = u:v:w:\omega$, wherein u, v, w, ω are general polynomials of order n in a parameter θ (which is thus conversely expressible as a rational function of x, y, z), prove, (i), that the line coordinates of the tangent line involve θ to order $2(n-1)$; (ii), that the equation of the osculating plane involves θ to order $3(n-2)$; (iii), that there is no point of the curve such that every plane through this meets the curve doubly thereat. Whence, for this curve, $n = n$, $r = 2(n-1)$, $n' = 3(n-2)$, $\kappa = 0$, $\delta = 0$, and, from $r + \kappa = 2(n + p - 1)$, also $p = 0$. Hence infer

$$i = 0, \quad \kappa' = 4(n-3), \quad \nu + \tau = 2(n-1)(n-3), \quad \nu' + \tau = 2(n-2)(n-3),$$
$$h = \tfrac{1}{2}(n-1)(n-2), \quad h' + \delta' = \tfrac{1}{2}(9n^2 - 53n + 80).$$

In particular, for $n = 3$, since $\tau = \delta' = 0$, as is obvious geometrically, we have $r = 4$, $n' = 3$, $\kappa = \delta = i = \nu = \nu' = 0$, $h = 1 = h'$; and for $n = 4$, assuming $\tau = 0$, the equation $\delta' = 0$ being again obvious, $r = 6$, $n' = 6$, $\kappa = \delta = i = 0$, $\nu = 6$, $\nu' = 4$, $h = 3$, $h' = 6$.

Ex. 2. We have deduced from Plücker's formulae for a plane curve the formulae for a space curve, by projection from a general point and section by a general plane. The characters of the plane curve are modified if either the point of projection, or the plane of section, be specially chosen. Prove the following results (cf. Cayley, *Papers*, VIII, p. 81) for the curve obtained by projection when the centre of projection has the respective positions stated, and the dual results for the curve obtained by section by the planes respectively stated:

For the curve obtained by projection

	Order (N)	Class (N')	Double points (D)
With the vertex of projection			
General	n	r	$h + \delta$
On a tangent of the curve ...	n	$r - 1$	$h - 1 + \delta$
On the curve	$n - 1$	$r - 2$	$h - n + 2 + \delta$
At a double point of the curve	$n - 2$	$r - 4$	$h - 2n + 5 + \delta$
At a cusp	$n - 2$	$r - 3$	$h - 2n + 6 + \delta$
On an inflexional tangent ...	n	$r - 2$	$h - 2 + \delta$
At a point of inflexion ...	$n - 1$	$r - 3$	$h - n + 1 + \delta$

For the curve obtained by projection (cont.)

With the vertex of projection	Cusps (K)	Double tangents (T)	Inflexions (I)
General	κ	$\nu' + \tau$	$n' + i$
On a tangent of the curve ...	$\kappa + 1$	$\nu' - \tau + 4 + \tau$	$n' - 2 + i$
On the curve	κ	$\nu' - 2\tau + 8 + \tau$	$n' - 3 + i$
At a double point of the curve	κ	$\nu' - 4\tau + 20 + \tau$	$n' - 6 + i$
At a cusp	$\kappa - 1$	$\nu' - 3\tau + 13 + \tau$	$n' - 4 + i$
On an inflexional tangent ...	$\kappa + 2$	$\nu' - 2\tau + 9 + \tau$	$n' - 4 + i$
At a point of inflexion ...	$\kappa + 1$	$\nu' - 3\tau + 14 + \tau$	$n' - 5 + i$

For the curve obtained by section

With plane of section	Order (N)	Class (N')	Double points (D)
General	τ	n'	$\nu + \tau$
Containing a tangent	$\tau - 1$	n'	$\nu - \tau + 4 + \tau$
An osculating plane	$\tau - 2$	$n' - 1$	$\nu - 2\tau + 8 + \tau$
Touching curve at two points ...	$\tau - 4$	$n' - 2$	$\nu - 4\tau + 20 + \tau$
A stationary osculating plane ...	$\tau - 3$	$n' - 2$	$\nu - 3\tau + 13 + \tau$
Containing an inflexional tangent	$\tau - 2$	n'	$\nu - 2\tau + 9 + \tau$
Osculating curve at an inflexion	$\tau - 3$	$n' - 1$	$\nu - 3\tau + 14 + \tau$

With plane of section	Cusps (K)	Double tangents (T)	Inflexions (I)
General	$n + i$	$h' + \delta'$	κ'
Containing a tangent	$n - 2 + i$	$h' - 1 + \delta'$	$\kappa' + 1$
An osculating plane	$n - 3 + i$	$h' - n' + 2 + \delta'$	κ'
Touching curve at two points ...	$n - 6 + i$	$h' - 2n' + 5 + \delta'$	κ'
A stationary osculating plane ...	$n - 4 + i$	$h' - 2n' + 6 + \delta'$	$\kappa' - 1$
Containing an inflexional tangent	$n - 4 + i$	$h' - 2 + \delta'$	$\kappa' + 2$
Osculating curve at an inflexion	$n - 5 + i$	$h' - n' + 1 + \delta'$	$\kappa' + 1$

Ex. 3. Prove that if, for a curve of order n and genus p, with κ cusps, lying on a quadric locus Ω in space of any number of dimensions, it be the case that the tangent lines of the curve, the osculating planes, the osculating solids, and so on, and finally the osculating $(h-1)$-folds, all lie entirely on Ω, then the number of the osculating h-folds of the curve which lie on Ω, is $2[n + 2h(p-1)] - 2\kappa$. The formula holds for $h = 1$, in which only the curve and not all its tangent lines, lie on Ω. A proof will be found in the *Proceedings of the Edinburgh Math. Soc.* Vol. I (2nd series), 1927, p. 19.

Ex. 4. In the text, beside ordinary points and double points, we have considered in detail only singular points which are cusps, or inflexions, or stalls. This makes easier the application of the ordinary Plücker formulae for a plane curve, to obtain the equations for a curve in space. But it is more fundamental to employ the general formulae for a branch of the curve, whether in a plane or in space. If we slightly change the notation, and write the expressions for a branch of the space curve in the forms $x = a_1 t^{l_1+1} + \ldots$, $y = a_2 t^{l_2+2} + \ldots$, $z = a_3 t^{l_3+3} + \ldots$, with l_1', l_2', l_3' in the dual formulae for the osculating plane, we shall have, instead of κ, i, κ', respectively, $\Sigma l_1, \Sigma(l_2 - l_1)$, or $\Sigma(l_2' - l_1')$, and $\Sigma l_1'$, or $\Sigma(l_3 - l_2)$, where the summations extend to all branches of the curve in space for which these are not zero. For a cusp, inflexion or stall these give respectively a contribution of unity. The formula for a space curve, $\tau + \Sigma l_1 = 2n + 2p - 2$, would then follow, as in the text, from the formula for a plane curve, $N' + \Sigma l_1 = 2n + 2p - 2$. The generalised form of the plane curve formula $2K + I = 3(n + 2p - 2)$ would be $\Sigma(l_1 + l_2) = 3(n + 2p - 2)$, where l_1, l_2 refer to this curve; this leads, as in the text, to the formula

$$n' + \Sigma(l_1 + l_2) = 3(n + 2p - 2)$$

for a space curve which projects into the plane curve.

Ex. 5. It is proper, however, to obtain the formulae directly for the curve in space; and, as this is almost as simple for the case of an algebraic curve in space of any number, say r, dimensions, as for a curve in ordinary space, for brevity's sake, we pass at once to this case. Let non-homogeneous coordinates in the space be denoted by $x_1, x_2, ..., x_r$, these being chosen so that the points of a particular branch of the curve are given by

$$x_1 = a_1 t^{l_1+1} + ..., \quad x_2 = a_2 t^{l_2+2} + ..., \quad ......, \quad x_r = a_r t^{l_r+r} + ...,$$

in terms of a parameter t, where $l_1 \leqslant l_2 \leqslant ... \leqslant l_r$. That is, by proper choice of the coordinates, about the origin, $(0, 0, ..., 0)$, of the branch, there is a line, p_1, through the origin, given by the equations $x_2 = 0 = ... = x_r$; then a plane, p_2, through the line p_1, given by the equations $x_3 = 0 = ... = x_r$; and, in general, a space p_s, of s dimensions, given by the equations $x_{s+1} = 0 = ... = x_r$, which contains the space p_{s-1}. Further, every prime (or space of $r - 1$ dimensions) passing through the origin, not containing the line p_1, has $l_1 + 1$ coincident intersections with the branch at the origin; then, every prime through the line p_1, but not containing the plane p_2, has $l_2 + 2$ coincident intersections with the branch at the origin; and so on. At an ordinary point of the curve, where every one of $l_1, ..., l_r$ is zero, the line p_1 is the tangent line, the plane p_2 the osculating plane, and so on; and, in general, we may speak of p_1 as the tangent line, p_2 as the osculating plane, and so on, with p_s as the osculating $[s]$ space.

We denote then by n the order of the curve, the total number of intersections of the curve with an arbitrary general prime, by p the genus of the curve. The ∞^1 lines p_1 will generate a surface, V_2; the order of this surface we denote by n_1. Thus n_1 is the number of lines p_1 which meet an arbitrary general space $[r-2]$, of $r - 2$ dimensions; and n_1 is also the number of primes, through an arbitrary general space $[r-2]$, for which the n intersections with the curve have a coincidence of two, lying on a tangent line p_1; but when the point of contact of this line p_1 is not ordinary, and this line p_1 has $l_2 + 2$ coincident intersections with the curve, we must count the corresponding branch as furnishing $l_2 + 1$ such two-fold coincidences. In this general case also, as any line through the origin of the branch has $l_1 + 1$ coincident intersections with the branch at the origin, the line p_1 is multiple on the surface V_2, with multiplicity $l_2 - l_1 + 1$. We speak of the number n_1 as the *first rank* of the curve. If we consider the series of ∞^1 sets of n points on the curve, obtained by the primes through a fixed arbitrary space $[r-2]$, there is, by a formula which has been quoted, a number, given by $2n + 2p - 2$, of coincidences of two of the points of a set. These coincidences include those, already explained, in number n_1, whose position on the curve depends on the particular space $[r-2]$ chosen for base of the pencil of primes. The number $2n + 2p - 2$ will, however, also include a contribution l_1 for every branch of the curve for which $l_1 > 0$; for an arbitrary prime through the origin of such a branch has $l_1 + 1$ intersections coinciding at this origin. Thus we have the formula $n_1 + \Sigma l_1 = 2n + 2p - 2$. More generally, the osculating spaces p_s at all branches of the curve generate a locus V_{s+1}, of dimension $s + 1$; the order of this locus we denote by n_s. Thus n_s is the number of spaces p_s which meet an arbitrary space $[r-s-1]$, of $r-s-1$ dimensions; and n_s is also the number of primes, drawn through a general space $[r-s-1]$, of which the n intersections with the curve have a coincidence, on a branch of the curve, of $(s+1)$ points lying on a space p_s. As the primes through a space $[r-s-1]$ are a linear system of freedom s, such a prime can be put through s coincident points of a p_s, which, being given by $x_{s+1} = \ldots = x_r = 0$, contains $l_{s+1}+s+1$ coincident points of the branch. Of the primes through the space $[r-s-1]$ there are, however, in general, others, having $(s+1)$ coincident intersections with the curve, on branches whose position on the curve does not depend on the particular $[r-s-1]$ chosen. For, consider a general branch of the curve; and the condition that it be possible, from among the primes passing through a given space $[r-s-1]$, to take a prime, through the origin of this branch and s other neighbouring points of this branch; if such space be determined by $r-s$ points with coordinates, relatively to the origin of the branch considered, given by $\xi_{1k}, \ldots, \xi_{rk}$ ($k=s+1, \ldots, r$), the condition in question is that the determinant, of r rows and columns,

$$\begin{vmatrix} x_1, & \ldots\ldots, & x_r \\ x_1', & \ldots\ldots, & x_r' \\ \ldots\ldots\ldots\ldots\ldots\ldots\ldots\ldots \\ x_1^{(s-1)}, & \ldots\ldots, & x_r^{(s-1)} \\ \ldots\ldots\ldots\ldots\ldots\ldots\ldots\ldots \\ \xi_{1k}, & \ldots\ldots, & \xi_{rk} \\ \ldots\ldots\ldots\ldots\ldots\ldots\ldots\ldots \end{vmatrix},$$

where $(x_1, \ldots, x_r)$ denotes a point of the branch in the neighbourhood of the origin, and (in the first rows) $x_h^{(\lambda)}$ denotes $d^\lambda x_h/dt^\lambda$, should, on substitution of the power series in t for $x_1, \ldots, x_r$, be divisible by t^{s+1}. It is easy to see that this determinant divides by t^μ with $\mu = l_1 + \ldots + l_s + s$; thus, when $l_1 + \ldots + l_s = 1$, there is just one such prime, with an $(s+1)$-fold intersection; and, when $l_1 + \ldots + l_s > 1$, we reckon the branch as furnishing $l_1 + \ldots + l_s$ such $(s+1)$-fold intersections. Thus, by the general formula of coincidences quoted in the text, we infer the result

$$n_s + \Sigma(l_1 + \ldots + l_s) = (s+1)[n + s(p-1)],$$

the summation extending to all branches of the curve. In particular, the $(r-1)$-th rank, most often called the *class* of the curve, is given by

$$n_{r-1} + \Sigma(l_1 + \ldots + l_{r-1}) = r[n + (r-1)(p-1)],$$

these equal numbers being the number of osculating primes of the curve which pass through an arbitrary general point of the space. The same argument proves that $\Sigma(l_1 + \ldots + l_r) = (r+1)[n + r(p-1)]$, these being the number of stalls of the curve, namely the number of branches where an osculating prime contains $(r+1)$ coincident points of the branch, or is *stationary* (for which $l_1 + \ldots + l_r = 1$), together with contributions for other branches at which $l_1 + \ldots + l_r > 1$.

The reader may consult Veronese, *Math. Annal.* XIX, 1882, pp. 161 ff., and Segre, *Ann. d. Mat.* XXII, 1894, pp. 86–8, where a proof of the formula for n_s, by *induction*, from the two cases of n_1 and n_2, will be found. The set of formulae can be built up, either from such as

$$n_k + \lambda_k - n_{k-1} = n - 2k + 2kp, \qquad k = 1, \ldots, r,$$

with $n_0 = n$, $n_r = 0$, where λ_k denotes the sum of the values of l_k for all branches of the curve; or from such as

$$n_k + n_{k-2} + \lambda_k - \lambda_{k-1} = 2n_{k-1} + 2p - 2, \qquad k = 1, \ldots, r;$$

and either of these may be obtained directly.

Ex. 6. In order to call attention to the relation of the results of the last example to the well-known theory of Weierstrassian points on a curve (see Chap. IV, p. 90; also the writer's *Abel's Theorem* (1897), pp. 34–46, and pp. 90 ff.), we may briefly consider the case of the canonical curve, of order $2p - 2$, in space $[p - 1]$, where the (non-homogeneous) coordinates are $x_1, ..., x_{p-1}$. By applying Clifford's theorem (above, p. 82) to the linear series, of sets of $2p - 2 - (l_s + s)$ points, which is determined by the ∞^{p-s-1} primes passing through the osculating space p_{s-1}, at a general branch of the curve, we see that $l_s < s$; in particular $l_1 = 0$, and the curve has no branch at which the origin is not simple. By the Riemann-Roch theorem, we find that the $l_s + s$ intersections of p_{s-1} with the branch determine a linear series of coresidual sets which has freedom l_s; there is thus a rational function, on the curve, which has a $(p_s + s)$-fold pole at the origin of the branch and no other pole; and in this function there enter l_s arbitrary constants (beside the additive constant). Further, in this case, with $r = p - 1$, the number $(r - 1)[n + r(p - 1)]$ is $(p - 1)p(p + 1)$, and this is the sum of the values of $l_1 + ... + l_r$ for all branches of the curve; hence, as $l_s < s$, the number of points of the curve which are not ordinary is certainly $> 2p + 2$. Also, the least value which $l_1 + ... + l_r$ can have, at a point which is not ordinary, arises when $l_1 = l_2 = ... = l_{r-1} = 0$ and l_r (i.e. l_{p-1}) $= 1$; there is then a stall, the osculating prime $x_{p-1} = 0$ having p coincident intersections with the branch at the origin. If all the not-ordinary points of the curve are stalls, their number is $(p - 1)p(p + 1)$, and the s-th rank of the curve is $n_s = (p - 1)(s + 1)(s + 2)$; in particular the class (i.e. the rank n_{p-2}) is $p(p - 1)^2$. The simplest case is that of the general plane quartic curve $(p = 3)$. The case $p = 4$ arises by the intersection of a general quadric surface and a general cubic surface in ordinary space; the sextic curve is then of rank 18, of class 36, and has 60 stalls, where the osculating plane meets the curve in four coincident points.

Part II. Curves which are the complete intersection of two surfaces. It appears necessary to deal in some detail with the case of a curve in ordinary space, which is the complete intersection of two algebraic surfaces. We denote the orders of these surfaces by m and M, so that the order of the curve is given by $n = mM$. Further, we suppose the number, κ, of stationary points, and the number, i, of stationary tangents, to be given. We have expressed all the characters of the curve in terms of four of these, for instance n, n', r, p; if we assume a knowledge of n, κ and i, it will be sufficient to find one other character, and the rest, of those we have recognised, can then be computed. We put down the formulae with the inclusion of the number, δ, of actual double points, and of, τ, actual double tangents, of the curve; it is well known that, where two surfaces touch, at ordinary points of the surfaces, their curve of intersection has a double point; we shall suppose the double points of the curve considered to arise in this way, and similarly the cusps to arise from such contacts, at which the two directions of the curve of intersection coincide; in the most general case δ, κ, i and τ are all zero.

We can give a direct deduction either of h, or r, or n'; of these the first involves perhaps the most elementary considerations. We give, however, also a direct deduction of r, to illustrate a method of subsequent importance; and also an indication of the direct deduction of n'.

We prove that $h = \frac{1}{2}mM(m-1)(M-1)$, assuming the following theorem of elementary algebra: Let

$$f = a_0 \lambda^m + (m, 1) a_1 \lambda^{m-1} + (m, 2) a_2 \lambda^{m-2} + \ldots + a_m,$$
$$F = A_0 \lambda^M + (M, 1) A_1 \lambda^{M-1} + (M, 2) A_2 \lambda^{M-2} + \ldots + A_M,$$

be two polynomials in the variable λ, of respective orders m and M, written with binomial coefficients $(m, 1)$, $(m, 2)$, etc. Writing, momentarily, also p_i for the coefficient $(m, i) a_i$, and P_j for $(M, j) A_j$, denote by Δ_k the determinant, of $m + M - 2k$ rows and columns,

$$\Delta_k = \begin{vmatrix} p_0, & p_1, & p_2, & \cdot & \cdot \\ 0, & p_0, & p_1, & \cdot & \cdot \\ 0, & 0, & p_0, & \cdot & \cdot \\ \cdot & \cdot & \cdot & & \\ P_0, & P_1, & P_2, & \cdot & \cdot \\ 0, & P_0, & P_1, & \cdot & \cdot \\ \cdot & \cdot & \cdot & & \end{vmatrix},$$

where there is first a set of $M - k$ rows, containing the elements $p_0, p_1, p_2, \ldots$, in order, beginning in turn in the first, second, ... columns, and then a set of $m - k$ rows, containing the elements $P_0, P_1, P_2, \ldots$, in order, beginning in turn in the first, second, ... columns, all other elements of the determinant being zero. Then, (i), a necessary and sufficient set of conditions for the polynomials f, F to have common a polynomial factor of order i in λ is the vanishing of the i determinants $\Delta_0, \Delta_1, \ldots, \Delta_{i-1}$; while, (ii), the determinant Δ_k is unaltered if for a_r and A_s we put, respectively, the values $\bar{a}_r$, $\bar{A}_s$ given by

$$\bar{a}_r = a_r + (r, 1) a_{r-1} \theta + (r, 2) a_{r-2} \theta^2 + \ldots + a_0 \theta^r,$$
$$\bar{A}_s = A_s + (s, 1) A_{s-1} \theta + (s, 2) A_{s-2} \theta^2 + \ldots + A_0 \theta^s,$$

where θ is an arbitrary parameter. This last fact is equivalent to saying that Δ_k is reduced to zero by the operator

$$a_0 \frac{\partial}{\partial a_1} + 2a_1 \frac{\partial}{\partial a_2} + \ldots + ma_{m-1} \frac{\partial}{\partial a_m} + A_0 \frac{\partial}{\partial A_1}$$
$$+ 2A_1 \frac{\partial}{\partial A_2} + \ldots + MA_{M-1} \frac{\partial}{\partial A_M}.$$

For (i), see Scheibner, *Leipziger Bericht.* XL, 1880, p. 1; Kronecker, *Werke*, II, 1881, p. 115; Netto, *Algebra*, I, 1896, p. 157. For (ii), see *Proc. Edinburgh Math. Soc.* II, 1980, p. 55.

Now suppose that $f_x{}^m = 0$, $F_x{}^M = 0$, are, in homogeneous variables x, y, z, t, the equations of two surfaces, of respective orders m and M. Taking a fixed point (ξ), and a point (x), substitute, in the equations of the surfaces, for x, y, z, t respectively, the values $\lambda\xi+x, \lambda\eta+y, \lambda\zeta+z, \lambda\tau+t$. We thus obtain, for the determination of the intersections of the line, joining (ξ) and (x), with the surfaces, the equations

$$\lambda^m f_\xi{}^m + (m, 1)\lambda^{m-1} f_\xi{}^{m-1} f_x + \ldots + f_x{}^m = 0,$$
$$\lambda^M F_\xi{}^M + (M, 1)\lambda^{M-1} F_\xi{}^{M-1} F_x + \ldots + F_x{}^M = 0;$$

these, as polynomials in λ, are of the forms considered above, with $a_i = f_\xi{}^{m-i} f_x{}^i$, $A_j = F_\xi{}^{M-j} F_x{}^j$, so that a_i, A_j are of respective orders i and j in (x, y, z, t).

The condition that the line joining (ξ) to (x) should meet the two surfaces in the same point, is that these two equations in λ should have a common root. This, however, is that (x) should satisfy the equation $\Delta_0 = 0$. Now, when (x) are current coordinates, the equation $\Delta_0 = 0$ represents a *cone*, with vertex at (ξ); for if, in $(\lambda\xi+x, \ldots, \lambda\tau+t)$, we put $(x+\theta\xi, \ldots, t+\theta\tau)$ for $(x, \ldots, t)$, this will be equivalent to considering the equations in λ which, in the notation used above, are

$$\bar{a}_0 \lambda^m + (m, 1)\bar{a}_1 \lambda^{m-1} + \ldots + \bar{a}_m = 0,$$
$$\bar{A}_0 \lambda^M + (M, 1)\bar{A}_1 \lambda^{M-1} + \ldots + \bar{A}_M = 0,$$

these being what are obtained from the original two by putting $\lambda + \theta$ in place of λ; the equation, $\bar{\Delta}_0 = 0$, obtained from these two equations in λ, is, however, we have said, the same as $\Delta_0 = 0$, whatever θ may be. Wherefore, $\Delta_0 = 0$, as an equation in (x), represents the cone joining (ξ) to the curve of intersection of the two surfaces. As this equation is, we easily see, isobaric in regard to the suffixes in a_i, A_j, and contains the term $a_0{}^M A_M{}^m$, it is of order mM in (x). By a similar argument, the equation $\Delta_1 = 0$ represents a cone, with vertex (ξ), of order $(m-1)(M-1)$. The lines common to $\Delta_0 = 0$, $\Delta_1 = 0$ are thence the lines from (ξ) which have two intersections with the curve of intersection of the surfaces; these lines are then the double lines of the cone $\Delta_0 = 0$, and their number is

$$\tfrac{1}{2}mM(m-1)(M-1).$$

If the two surfaces have a point of contact, the cone $\Delta_0 = 0$ will have a double generator through this point. But, unless the common tangent plane of the two surfaces, at this point, pass through the point (ξ), this generator, having a simple intersection with each surface at this point, will not meet the common curve of the surfaces in a further point, and will not lie on $\Delta_1 = 0$. Thus, no correction of the number $\tfrac{1}{2}mM(m-1)(M-1)$ arises for double points (or cusps) of the curve arising from contacts of the two surfaces; we have agreed not to consider other possible double points of the curve (as for instance when one surface passes through an actual conical point of the other surface). Hence we have

$$h = \tfrac{1}{2}mM(m-1)(M-1).$$

Ex. 1. For the quadric surfaces $f_x^2 = 0$, $F_x^2 = 0$, the equation $\Delta_1 = 0$ is

$$\begin{vmatrix} f_\xi^2, & f_\xi f_x \\ F_\xi^2, & F_\xi F_x \end{vmatrix} = 0,$$

and represents the plane through (ξ), drawn to the line of intersection of the polar planes of (ξ) in regard to the quadric surfaces. This plane meets the quadric surfaces in two conics, in regard to which the polar lines of (ξ) are the same line; two of the common chords of these conics intersect in (ξ), these being the two chords of the quartic curve, of intersection of the two quadric surfaces, which can be drawn from (ξ). These lines are also the generators through (ξ) of a particular quadric surface, $f_x^2 + kF_x^2 = 0$, which passes through (ξ).

Ex. 2. As an example of the algebraical theorem which has been used, the reader may prove that, if $f = a\lambda^2 + b\lambda + c$, $F = p\lambda^3 + q\lambda^2 + r\lambda + s$, while the constants $P, Q, ..., B$ are so chosen that

$$(P\lambda^2 + Q\lambda + R)f + (A\lambda + B)F = \Delta_0,$$

then, with Δ_1 as before,

$$\begin{vmatrix} A, & B, & 0 \\ 0, & A, & B \\ P, & Q, & R \end{vmatrix} = -\Delta_1^4,$$

this being an identity not requiring $\Delta_0 = 0$.

Ex. 3. As a further example, prove that the invariant Δ_k, of the two polynomials of orders m and M in λ,

$$(p_0\lambda^{m-1} + ... + p_{m-1})(\lambda - \theta), \quad (q_0\lambda^{M-1} + ... + q_{M-1})(\lambda - \theta),$$

when multiplied by a certain determinant whose elements depend only on θ, is equal to the invariant Δ_{k-1}, of the polynomials, of orders $m-1$ and $M-1$, $p_0\lambda^{m-1} + ... + p_{m-1}, q_0\lambda^{M-1} + ... + q_{M-1}$.

Note. That the chords from an arbitrary point to the complete curve intersection of two surfaces of orders m and M, meet this curve on a surface of order $(m-1)(M-1)$, is proved in Salmon, *Geom. of Three Dimensions* (1882, p. 331). That the chords lie on a *cone* of this order, was proved by Valentiner, *Acta Math.* II, 1883, p. 191; and by Noether, *Berlin. Abh.* 1882, p. 27. A converse theorem was given by Halphen, *J. d. l'École polyt.* LII, 1883, p. 106, namely: If, for a curve in space, of order mM, the chords from an arbitrary point lie on a cone of order $(m-1)(M-1)$, and the curve itself does not lie on a surface of order less than the lesser of m and M, then the curve is the complete intersection of two surfaces of orders m and M.

Having obtained the formula $h = \frac{1}{2}mM(m-1)(M-1)$, if we use the formulae, previously found for an algebraic curve of order n,

$$n' + i = 3(r - n) + \kappa,$$
$$r = n(n-1) - 2h - 2\delta - 3\kappa,$$
$$p = \frac{1}{2}(n-1)(n-2) - h - \delta - \kappa,$$

we can deduce, for the curve which is the complete intersection of two surfaces of orders m and M, that $n = mM$ and

$$n' + i = 3mM(m + M - 3) - 6\delta - 8\kappa,$$
$$r = mM(m + M - 2) - 2\delta - 3\kappa,$$
$$p - 1 = \frac{1}{2}mM(m + M - 4) - \delta - \kappa,$$

and thence, using the general formulae which express i, κ', $\nu + \tau$, $\nu' + \tau$, $h + \delta$, $h' + \delta'$ in terms of n, n', r, p, we infer

$$i = 0, \quad \kappa' = 2mM(3m + 3M - 10) - 3(4\delta + 5\kappa),$$
$$\nu + \tau = \tfrac{1}{2}\mu_2(\mu_2 - 2\rho - 4) + \tfrac{1}{2}\rho_2{}^2 + 4\delta + \tfrac{11}{2}\kappa,$$
$$\nu' + \tau = \tfrac{1}{2}\mu_2(\mu_2 - 2\rho - 10) + \tfrac{1}{2}\rho_2{}^2 + 10\delta + \tfrac{27}{2}\kappa + 4mM,$$
$$h' + \delta' = \tfrac{1}{2}\mu_3(9\mu_3 - 12\sigma - 22) + 2\sigma^2 + 22\delta + 28\kappa + \tfrac{5}{2}mM,$$

where, for brevity, $\mu_2 = mM(m + M - 2)$, $\mu_3 = mM(m + M - 3)$, $\rho = 2\delta + 3\kappa$, $\sigma = 3\delta + 4\kappa$.

We next develop a direct verification of the formula

$$r = mM(m + M - 2) - 2\delta - 3\kappa.$$

Denote the equations of the two surfaces now by $u = 0$, $U = 0$; let λ, μ be two arbitrary planes, say

$$a_1 x + b_1 y + c_1 z + d_1 t = 0, \quad a_2 x + b_2 y + c_2 z + d_2 t = 0.$$

Consider the locus of a point (x, y, z, t) whose polar planes, taken in regard to the two surfaces, meet in a line which intersects the line (λ, μ). The equation of this locus is $(u, U, \lambda, \mu) = 0$, where

$$(u, U, \lambda, \mu) = \begin{vmatrix} u_1, & u_2, & u_3, & u_4 \\ U_1, & U_2, & U_3, & U_4 \\ a_1, & b_1, & c_1, & d_1 \\ a_2, & b_2, & c_2, & d_2 \end{vmatrix},$$

u_1 denoting $\partial u / \partial x$, etc. This surface is of order $m + M - 2$. A point common to the surfaces $u = 0$, $U = 0$, at which both these have definite, different, tangent planes, will be on the surface $(u, U, \lambda, \mu) = 0$ if, and only if, the tangent line of the curve (u, U), at this point, meets the line (λ, μ). Thus, if the surfaces $u = 0$, $U = 0$ have ordinary points with different tangent planes at all points of the curve (u, U), then this curve is of rank $mM(m + M - 2)$. The determinant (u, U, λ, μ) vanishes, however, at any point where the surfaces $u = 0$, $U = 0$ touch, and the tangent lines (or line) of the curve (u, U), at such a contact, do not, as a rule, meet the chosen line (λ, μ). It can be proved that the tangent plane of the surface $(u, U, \lambda, \mu) = 0$, at such a contact, is definite in general, and meets both branches of the curve (u, U) at this point, when these are distinct, in a single point; but contains the tangent line of the curve (u, U) at this point, when this curve has a cusp. Such a contact thus diminishes the previous expression for the rank of the curve (u, U) by 2, or by 3, in these two cases. Thus we reach the result above given for the rank, r, of the curve, if we consider, as we have agreed to do, only double points or cusps of the curve

(u, U) which arise at simple contacts of the surfaces. It may be shewn, however, that double points of the curve (u, U) which arise by simple passage of, say, $u=0$, through a conical node of $U=0$, may be supposed to be included in the formula.

It is important to notice also, that, if the curve of intersection of the surfaces $u=0$, $U=0$ break up into two curves, one of order n_1, with δ_1 double points and κ_1 cusps, at ordinary contacts of $u=0$, $U=0$, the other similarly with n_2, δ_2, κ_2, these curves having t points of intersection at ordinary contacts of the surfaces, then the rank of one curve (n_1) is given by $r_1=n_1(m+M-2)-2\delta_1-3\kappa_1-t$; the proof is precisely as in the simpler case above.

We now indicate a direct verification of the equation

$$n'+i=3mM(m+M-3)-6\delta-8\kappa,$$

arising from a formula given by Hesse for the equation of the osculating plane at an ordinary point (x, y, z, t) of the curve (u, U). Using (x', y', z', t') for current coordinates, the equation of this osculating plane is

$$\omega(x'U_1+y'U_2+z'U_3+t'U_4)=\Omega(x'u_1+y'u_2+z'u_3+t'u_4),$$

where, as before, $u_1=\partial u/\partial x$, etc., and ω denotes the determinant

$$\omega=(m-1)^{-2}\begin{vmatrix} u_{11}, & u_{12}, & u_{13}, & u_{14}, & U_1 \\ u_{21}, & u_{22}, & u_{23}, & u_{24}, & U_2 \\ u_{31}, & u_{32}, & u_{33}, & u_{34}, & U_3 \\ u_{41}, & u_{42}, & u_{43}, & u_{44}, & U_4 \\ U_1, & U_2, & U_3, & U_4, & 0 \end{vmatrix},$$

wherein $u_{12}=\partial^2 u/\partial x\,\partial y$, etc., while Ω denotes what ω becomes when m, u, U are replaced by M, U, u. To verify that this is so, we may first verify that the given equation remains of the same form after any homogeneous linear (so-called projective) transformation of the coordinates; then we can suppose the coordinates to be three non-homogeneous coordinates x, y, z, and a fourth, t, equal to 1. By means of Euler's formulae for homogeneous polynomials, it then follows that, on the curve (u, U), the function ω may be replaced by

$$u_{11}(u_2U_3-u_3U_2)^2+\ldots+2u_{23}(u_3U_1-u_1U_3)(u_1U_2-u_2U_1)+\ldots;$$

if then we regard x, y, z on this curve as functions of a parameter, θ, it follows hence that ω may be replaced by

$$u_{11}\left(\frac{dx}{d\theta}\right)^2+\ldots+2u_{23}\frac{dy}{d\theta}\frac{dz}{d\theta}+\ldots,$$

and hence also by

$$u_1\frac{d^2x}{d\theta^2}+u_2\frac{d^2y}{d\theta^2}+u_3\frac{d^2z}{d\theta^2},$$

because $u_1 dx/d\theta + u_2 dy/d\theta + u_3 dz/d\theta$ vanishes on the curve. And a similar replacement is possible also for Ω. Wherefore, the plane given by the equation, which obviously contains the tangent line of the curve (u, U), contains also the point

$$\left(\frac{d^2x}{d\theta^2}, \frac{d^2y}{d\theta^2}, \frac{d^2z}{d\theta^2}, \ 0 \right),$$

and is therefore the osculating plane.

Now, in the original form, the function ω is a polynomial of order $3(m-2) + 2(M-1)$ in x, y, z, t; the whole equation is thus of order $3(m+M-3)$ in (x, y, z, t). The ordinary points of the curve (u, U) for which the osculating plane passes through an arbitrary point (x', y', z', t'), are thus upon a surface of this order; the number of such points is thus $3mM(m+M-3)$, save for a correction for points of the curve which are not ordinary. To complete the verification of the formula under discussion it must be shewn that this correction is one at an inflexion of the curve (u, U), six at a double point, and eight at a cusp.

To examine the case of a double point or cusp arising from a simple contact of the surfaces $u = 0$, $U = 0$, we may take non-homogeneous coordinates x, y, z, with origin at the point of contact, so that the equations of the surfaces are

$$u \equiv z + \tfrac{1}{2}(ax^2 + 2hxy + by^2) + gxz + fyz + \tfrac{1}{2}cz^2 + \ldots = 0,$$

$$U \equiv z + \tfrac{1}{2}(Ax^2 + 2Hxy + By^2) + Gxz + Fyz + \tfrac{1}{2}Cz^2 + \ldots = 0;$$

then, in the equation of the osculating plane (with (x', y', z') as current coordinates), namely

$$\omega[(x'-x)U_1 + (y'-y)U_2 + (z'-z)U_3]$$
$$= \Omega[(x'-x)u_1 + (y'-y)u_2 + (z'-z)u_3],$$

the terms of the lowest order in x, y, z are those multiplying z', and therein the lowest terms are

$$\Delta \equiv (a-A)l^2 + (b-B)m^2 + (c-C)n^2$$
$$+ 2(f-F)mn + 2(g-G)nl + 2(h-H)lm,$$

where $l = u_2 U_3 - u_3 U_2$, $m = u_3 U_1 - u_1 U_3$, $n = u_1 U_2 - u_2 U_1$.

For a contact of the surfaces in which the curve (u, U) has two branches, if we choose the axes so that the tangent line of one branch is $z = 0$, $y = 0$, we shall have $A = a$, and this branch will be given, in terms of a parameter θ, by expressions of the forms $x = \theta$, $y = q\theta^2 + \ldots$, $z = r\theta^2 + \ldots$. Along this branch, then, l, m, n are respectively proportional to $\theta, \theta^2, \theta^3$, and the terms of lowest order in Δ are the terms in ln, lm, which are proportional to θ^3. The surface arising from the equation of the osculating plane thus meets this branch in three coincident points at the origin; and meets the curve in all in six points.

When, however, the curve has a cusp, the branch of the curve is given by expressions of the forms $x = \theta^2$, $y = q\theta^3 + \ldots$, $z = r\theta^4 + \ldots$, and l, m, n are ultimately proportional, respectively, to $\theta^3, \theta^4, \theta^5$. In this case we have $h = H$ beside $a = A$. The lowest terms in Δ are the terms in m^2 and

nl, which are proportional to θ^8; thus there are now eight intersections with the curve at the origin.

A similar examination may be made for an inflexion of the curve; it is obvious however that the plane joining (x', y', z', t') to the inflexional tangent has three intersections with the curve at the inflexion; thus this point counts once. The formula $n' + i = 3mM(m + M - 3) - 6\delta - 8\kappa$ is thus verified, when only double points and cusps which arise from simple contacts of the two surfaces are taken into account.

Ex. Prove that the osculating plane at (x, y, z, t) on the curve of intersection of the quadric surfaces whose equations are

$$ax'^2 + by'^2 + cz'^2 + dt'^2 = 0, \quad Ax'^2 + By'^2 + Cz'^2 + Dt'^2 = 0,$$

is expressed by

$$l'mnx^3x' + m'nly^3y' + n'lmz^3z' + l'm'n't^3t' = 0,$$

where
$$l = bC - cB, \quad l' = dA - aD, \text{ etc.}$$

Part III. Curves which are the partial intersection of two surfaces.

Consider now the case when the two surfaces $u = 0$, $U = 0$, meet in a curve of order n_1, itself irreducible and simple on both surfaces, this being only part of the intersection, the residual part being another curve, of order n_2, also irreducible and simple on both surfaces.

We can put down equations by which, when the characters of the first curve (n_1) are known, and the surfaces do not touch at points of the second curve (n_2), the characters of this second curve can be found, as well as the number of intersections of the second curve with the first. Let the usual numbers for the two curves be $r_1, \delta_1, \kappa_1, \ldots$, and $r_2, \delta_2, \kappa_2, \ldots$, and the number of their intersections be t. We clearly have $n_1 + n_2 = mM$, and it has already been remarked (p. 206), considering the surface $(u, U, \lambda, \mu) = 0$, that

$$r_1 = n_1(m + M - 2) - 2\delta_1 - 3\kappa_1 - t, \quad r_2 = n_2(m + M - 2) - 2\delta_2 - 3\kappa_2 - t.$$

By a general property of curves, given in Part I, we have, however,

$$r_1 + \kappa_1 = 2n_1 + 2p_1 - 2, \qquad r_2 + \kappa_2 = 2n_2 + 2p_2 - 2,$$
$$h_1 + p_1 + \delta_1 + \kappa_1 = \tfrac{1}{2}(n_1 - 1)(n_1 - 2), \quad h_2 + p_2 + \delta_2 + \kappa_2 = \tfrac{1}{2}(n_2 - 1)(n_2 - 2);$$

thus we have

$$t = n_1(m + M - n_1 - 1) + 2h_1, \; = n_1(m + M - 4) - 2\delta_1 - 2\kappa_1 - (2p_1 - 2),$$

with precisely similar forms for t in terms of $n_2, h_2, \delta_2, \kappa_2, p_2$. Comparing the two forms of t we deduce

$$h_2 - h_1 = \tfrac{1}{2}(n_2 - n_1)(m - 1)(M - 1),$$
$$r_2 - r_1 = (n_2 - n_1)(m + M - 2) - (2\delta_2 + 3\kappa_2 - 2\delta_1 - 3\kappa_1),$$
$$p_2 + \delta_2 + \kappa_2 - (p_1 + \delta_1 + \kappa_1) = \tfrac{1}{2}(n_2 - n_1)(m + M - 4),$$

and, for the second curve to be entirely known when the first is known, the values of δ_2 and κ_2 are required.

The equations also give* the results

$$h_1 + h_2 + n_1 n_2 - t = \tfrac{1}{2} mM(m-1)(M-1),$$
$$p_1 + p_2 + t - 1 = 1 + \tfrac{1}{2} mM(m+M-4) - (\delta_1 + \kappa_1 + \delta_2 + \kappa_2).$$

Of these equations the former is obvious directly if we recall what was found above (p. 208) for the cones Δ_0 and Δ_1; the cone Δ_0 is formed by the lines from an arbitrary point (ξ, η, ζ, τ) to the points of the curve (or curves) of intersection of the surfaces $u=0$, $U=0$, and the cone Δ_1 by the lines from (ξ, η, ζ, τ) upon each of which two of the intersections with $u=0$ coincide with two of the intersections with $U=0$. The argument used does not assume that the curve (u, U) is irreducible, and is applicable to the case now under consideration. Upon the cone Δ_1 will be the h_1 proper chords from (ξ, η, ζ, τ) to the first curve (n_1), also the h_2 proper chords to the second curve (n_2); but there will also be upon the cone Δ_1 the lines from (ξ, η, ζ, τ) which meet both (n_1) and (n_2) in two different points, whose number is $n_1 n_2 - t$ (the cones joining (ξ) to the curves being of orders n_1 and n_2). Thus the former of the two equations is clear; it may if desired be replaced by the two equations

$$n_1 n_2 - t = n_1(m-1)(M-1) - 2h_1 = n_2(m-1)(M-1) - 2h_2.$$

The preceding formulae have been obtained, for simplicity, with the supposition that the curve (n_2) is irreducible. With suitable modifications this condition may be omitted.

Ex. 1. Suppose that, through the curve (n_1), a further surface, of order k is drawn; if the curve (n_2) is not degenerate, it will be met by this surface in points, not on (n_1), whose number is $kn_2 - t$. Various forms for this are

$$\begin{aligned} kn_2 - t &= mMk - n_1(m+M+k-4) + 2p_1 - 2 + 2\delta_1 + 2\kappa_1, \\ &= mMk - n_1(m+M+k-2) + r_1 + 2\delta_1 + 3\kappa_1, \\ &= mMk - n_1(m+M+k) + n_1(n_1+1) - 2h_1. \end{aligned}$$

---

* For any two curves, not assumed to form together the complete intersection of two surfaces, it may be proved, by elementary algebra, that the equations

$$p_1 = \tfrac{1}{2}(n_1-1)(n_1-2) - h_1 - [\delta_1 + \tfrac{1}{2}\Sigma s_1(s_1-1)],$$
$$p_2 = \tfrac{1}{2}(n_2-1)(n_2-2) - h_2 - [\delta_2 + \tfrac{1}{2}\Sigma s_2(s_2-1)],$$
$$p = \tfrac{1}{2}(n_1+n_2-1)(n_1+n_2-2) - (h_1+h_2+k) - [\delta_1 + \delta_2 + \tfrac{1}{2}\Sigma(s_1+s_2)(s_1+s_2-1)],$$
$$n_1 n_2 = t + k + \Sigma s_1 s_2,$$

lead to
$$p = p_1 + p_2 + t - 1,$$

as the expression for the genus of a curve which breaks up into two curves with t intersections. Notice that there is no term, in the definition of p, corresponding to these intersections. Cf. Picard et Simart, *Fonctions algéb.* II, p. 106. See Ex. 11, p. 215, below.

And it can be shewn that a rational curve of order r, in space of r dimensions, taken with p chords of this curve, is a possible form of the degeneration of a curve of order $r+p$, of genus p. Cf. Enriques e Chisini, *Teoria geometrica*, III, p. 396. For the connexion of this result with the theory of connected polygons of lines, see Severi, *Geom. Algeb.* 1921, p. 373 (Anhang G).

These various results, though not arising directly from the text, will be of subsequent interest.

Taking the last form we may say: If three surfaces, of orders m, M, k, have common a curve of order n, with h apparent double points, this curve absorbs a number $n(m+M+k-n-1)+2h$ of the mMk points generally common to three surfaces of these orders. This is often called the *point equivalence* of the curve for these surfaces; it depends, we see, only upon the sum of the orders of these surfaces; and it is assumed that these surfaces have no further common curve.

It may be proved that the formula for the point equivalence of the curve (n) remains true when this curve is itself an aggregate of two curves (ν_1), (ν_2), so that $n = \nu_1 + \nu_2$, provided h be replaced by the number of chords to this composite curve from an arbitrary point, that is by $\eta_1 + \eta_2 + \nu_1\nu_2 - \tau$, where η_1, η_2 are the numbers of apparent double points for the component curves, and τ the number of their intersections. Putting

$$\sigma = m+M+k, \quad \beta_1 = \nu_1{}^2 + \nu_1 - 2\eta_1, \quad \beta_2 = \nu_2{}^2 + \nu_2 - 2\eta_2, \quad \beta_{12} = \beta_1 + \beta_2 + 2\tau,$$

this is the same as saying that the equivalences of the curves (ν_1), (ν_2), ($\nu_1 + \nu_2$) are respectively $\nu_1\sigma - \beta_1$, $\nu_2\sigma - \beta_2$, $(\nu_1 + \nu_2)\sigma - \beta_{12}$.

Ex. 2. The formula $t = n(m+M-n-1)+2h$, for the number of contacts, on a given curve (n, h), of two surfaces $u=0$, $U=0$, drawn through this curve, can be obtained very simply in the case when the curve is the complete intersection of two surfaces $u_0 = 0$, $U_0 = 0$ and it is assumed that u, U are capable of the forms

$$u = Au_0 + BU_0, \quad U = Cu_0 + DU_0,$$

in which A, B, C, D are polynomials (Salmon, *Higher Algebra*, 1885, p. 297, where the so-called *rank* is what we have denoted by $r + 2n + 2\delta + 3\kappa$).

For then, the complete intersection of $u=0$, $U=0$ consists, beside the given curve (u_0, U_0), of a part lying on the surface

$$\begin{vmatrix} A, & B \\ C, & D \end{vmatrix} = 0;$$

if u_0, U_0 be of orders m_0, M_0, this surface is of order $m - m_0 + M - M_0$, and meets the given curve, which is of order n, $= m_0 M_0$, in $n(M+m) - n(m_0 + M_0)$ points; as $h = \frac{1}{2}m_0 M_0 (m_0 - 1)(M_0 - 1)$, this is the same as $n(M+m) - n(n+1) + 2h$.

Hence, also, a third surface, of order k, through the curve (u_0, U_0), intersects the residual curve (u, U) $-$ (u_0, U_0), which is of order $mM - m_0 M_0$, in a number of points, not on (u_0, U_0), given by

$$k(mM - m_0 M_0) - m_0 M_0 (m + M - m_0 - M_0),$$

which, as before, is $mMk - n(m+M+k) + n(n+1) - 2h$.

Ex. 3. If, in space of four dimensions, there be drawn, through a curve of order n, three primals of orders m, M, N, and these intersect in a further curve, this last intersects the former in $n(m+M+N) - \beta$ points, where β depends only on the original curve. This can be easily proved, in Salmon's manner, in the case where the original curve is the complete intersection of three primals in terms of which the primals (m), (M), (N) are linearly expressible. It will be proved below (in Vol. VI), that, when the original curve is without multiple points, the value of β is $3n + r$, where r, the rank of the curve, is the number of its tangent lines which meet an arbitrary plane. It follows at once that the point equivalence of the given curve for four primals, of orders m, M, N, k, passing through it, is $n\sigma - \beta$, where $\sigma = m + M + N + k$. It can be shewn, as in Ex. 1, that this formula for the point equivalence remains valid when the original curve is composite, if for β we put the sum of the values of β for the separate components increased by twice the total number of intersections of the components. It follows easily also that, if the curves (n, ν) and (n', ν') be, together, the complete intersection of the three primals (m), (M), (N), then $n(m + M + N) - 3n - r = n'(m + M + N) - 3n' - r'$.

Ex. 4. As a consequence of Ex. 3, prove that, if ten lines in space of four dimensions have in all fifteen intersections, the point equivalence of these lines, for four cubic primals drawn through them, is 60. Deduce from this that the number of quadric surfaces, in space of three dimensions, which pass through 5 given points and touch 4 given planes is 21 (Salmon, *loc. cit.*).

Ex. 5. It has been stated, at the beginning of this chapter, that there exist curves in space of three dimensions, such that no set of three surfaces exists, passing through the curve, without residual point intersections. We give two examples, one due to Vahlen (*Crelle*, cviii, 1891, p. 346), of a rational quintic curve, the other, given by Enriques-Chisini (*Teoria geometrica*, iii, p. 515), of a quintic curve of genus 1. A quintic curve is evidently not the complete intersection of two surfaces.

A rational quintic curve is in fact obtained by the residual intersection with a quadric surface of a quartic surface drawn through three skew generators of the quadric surface; but this curve has an infinite number of quadrisecants, or transversals meeting it in four points (generators of the quadric surface). There exists, however, a rational quintic curve with only one quadrisecant, and it is this curve which we proceed to consider. It may be defined as the residual intersection of two cubic surfaces which are drawn through a rational cubic curve and through a line which does not meet the cubic curve. For, a cubic surface contains such a rational cubic curve if it contain ten points of this curve, and contains a line if it contain four points of this line; and a cubic surface has twenty coefficients in its equation; we infer then that there are six linearly independent cubic surfaces through the cubic curve and the line. Take two of these cubic surfaces, and the quintic curve which is their remaining intersection. To see that this curve is rational, remark that an arbitrary plane through the line meets these two cubic surfaces, further, in conics, with four intersections, not generally on the line; and of these three are on the cubic curve; so that an arbitrary plane through the line gives one point on the quintic curve; and conversely. The curve is thus expressible by the parameter which determines the plane. Also, the quintic curve meets the line in four points, these, with the point of the curve on a plane through the line, being the five intersections with the plane. The quintic curve does not lie on a quadric surface, since, else, the quadrisecant of the curve would lie thereon, and, therefore, all other generators of the quadric surface of the same system would be quadrisecants of the curve (as we easily see); these generators, and hence the quadric surface, would then be upon every cubic surface containing the curve. Further, the curve will have no other quadrisecant, since this would also lie on all cubic surfaces containing the curve.

The quintic curve has a rank r given by $r = 2n + 2p - 2$, or 8; and the chords from an arbitrary point are in number given by

$$h = \tfrac{1}{2}(n - 1)(n - 2) = 6.$$

Thus the point equivalence of the curve for surfaces of order m, M, k, passing through it, is $5\sigma - \beta$, where $\sigma = m + M + k$, and $\beta = r + 2n$, $= n(n + 1) - 2h$, $= 18$. As the curve does not lie on a quadric surface, we can suppose $m = \lambda + 3$, $M = \mu + 3$, $k = \nu + 3$, where λ, μ, ν are not negative; the point equivalence is thus $5(\lambda + \mu + \nu) + 27$. If the three

surfaces have no further common curve, they intersect then in
$$(\lambda+3)(\mu+3)(\nu+3)-5(\lambda+\mu+\nu)-27$$
further points; this is easily seen to be
$$\lambda\mu\nu+3(\mu\nu+\nu\lambda+\lambda\mu)+4(\lambda+\mu+\nu),$$
and can only be zero if λ, μ, ν be all zero. While, in this case, the three cubic surfaces would have further points in common, since they would all contain the quadrisecant of the curve.

It is not possible then to put three surfaces through the rational quintic curve considered, which shall have no further points in common.

Next consider the quintic curve which is the residual intersection of two cubic surfaces drawn through a rational quartic curve. The formula $p_2-p_1=\frac{1}{2}(n_2-n_1)(m+M-4)$ shews that this quintic curve is of genus 1. Hence it can be proved that this curve does not lie on a quadric surface, on which existing quintic curves are either rational or of genus 2 (see below, Ex. 7). But three surfaces of orders $\lambda+3$, $\mu+3$, $\nu+3$ passing through the quintic curve, if without further common curve, intersect further, we easily see, in $(\lambda+3)(\mu+3)(\nu+3)-5(\lambda+\mu+\nu)-25$ points, which is $\lambda\mu\nu+3(\mu\nu+\nu\lambda+\lambda\mu)+4(\lambda+\mu+\nu)+2$, and is at least 2. It is not possible then to put three surfaces through this elliptic quintic curve to have no further common points.

Ex. 6. Prove that the rational quintic curve of the last Ex., the residual intersection of two cubic surfaces passing through a line and a rational cubic curve, beside meeting the line in 4 points, meets the cubic curve in 8 points. For these three curves, the numbers of chords through an arbitrary point are, respectively, $h_1=0$ (for the line), $h_2=1$ (for the cubic curve), and $h_3=\frac{1}{2}(5-1)(5-2)$ (for the quintic curve), or $h_3=6$. The orders are $n_1=1$, $n_2=3$, $n_3=5$. The numbers of intersections of pairs of the curves are $t_{23}=8$, $t_{31}=4$, $t_{12}=0$. Hence we have an example of the formula
$$h_1+h_2+h_3+n_2n_3+n_3n_1+n_1n_2-t_{23}-t_{31}-t_{12}=\tfrac{1}{2}mM(m+M-2),$$
(for $m=M=3$), which generalises the formula proved in the text for two curves (see below, Ex. 11).

Ex. 7. We may enumerate the simpler cases of existing curves:

A general cubic curve, in space of three dimensions, is the residual intersection of two quadric surfaces having a common generator, in infinitely many ways; and the curve is rational.

A quartic curve is either, (i), the intersection of two general quadric surfaces; it is then of genus 1, and meets every generator of any quadric surface, containing the curve, in two points; or, (ii), the intersection of two quadric surfaces which have a point of contact; then the curve has a double point, and is rational; or, (iii), the residual intersection with a quadric surface of a cubic surface containing two skew generators of the quadric surface; then the curve is rational. Either of the rational quartic curves is obtainable by projection from a rational quartic curve in space of four dimensions.

For quintic curves, two rational curves have been spoken of in Ex. 5; and also a curve of genus 1, the residual intersection of two cubic surfaces through a rational (non-singular) quartic curve. For this last case, prove that the quartic and quintic curves meet in 10 points; and that the quadric surface described through 9 of these points contains the quartic curve entirely, thus meeting the quintic curve further only in the tenth point. A quintic curve of genus 2 is obtainable as the residual intersection of a quadric surface with a cubic surface drawn through one of its generators. Prove that there are no other (non-singular) quintic curves, not lying in a plane.

It will be proved below (Part VI, p. 234) that, in space of three dimensions, no algebraic curve of order n exists with genus greater than $\frac{1}{4}(n-2)^2$ or $\frac{1}{4}(n-1)(n-3)$, according as n is even or odd. Further, Noether has proved that, upon a given surface, the greatest genus possible, for curves of given order, arises for curves which, if not complete intersections of this surface with other surfaces, are part of a complete intersection of which the other part (possibly composite) lies in a plane. When n is odd, there exists a curve upon a quadric surface meeting all generators of one system in $\frac{1}{2}(n-1)$ points, and all generators of the other system in $\frac{1}{2}(n+1)$ points, and this curve has the genus $\frac{1}{4}(n-1)(n-3)$; when n is even, there exists a curve of genus $\frac{1}{4}(n-2)^2$ upon a quadric surface, meeting every generator in $\frac{1}{2}n$ points (see the following example). Conversely, a curve of order n, in space of three dimensions, with the maximum possible genus, necessarily lies on a quadric surface. For curves of order n on a cubic surface, however, the maximum possible genus is $\frac{1}{6}(n-1)(n-2)$, or $\frac{1}{6}(n^2-3n+6)$, according to the form of n.

An enumeration of existing curves is given to a high order by Noether ("Zur Grundlegung u.s.w."), *Berl. Abh.* 1882 (also, *Crelle's Journ.* XCIII, 1882). To order 6 an enumeration is given in Pascal's *Repertorium*, II, 2 (Leipzig, 1922), p. 932. See also Halphen, *J. école polyt.* LII, 1882; and Valentiner, *Acta Math.* II, 1883.

Ex. 8. Consider algebraic curves lying on a quadric surface. Such a curve meets all generators of the surface of one system (a) in the same number of points, say in α points, and all generators of the other system (b) also in the same number, say β, of points, the order of the curve being $\alpha + \beta$. This is easily seen by considering the intersections of the curve with a general plane through a generator. Suppose $\alpha \geqslant \beta$; then the surfaces of lowest order, other than the quadric surface, which contain the curve, are of order α; these are surfaces containing $\alpha - \beta$ of the (a) generators of the surface. The general surface of order α through $\alpha - \beta$ such generators, gives a system of "coresidual" curves on the quadric surface, all of order α; each of these curves meets every (a) generator in α points, and every (b) generator in β points. A definite curve of the system is determined by the prescription of passing through $\alpha + \beta + \alpha\beta$ given points of the quadric surface; a surface of order α passes through the curve and $\alpha - \beta$ (a) generators, arbitrarily chosen. More generally, it may be shewn that, in the equation of a surface of order m, so far as its intersection with the quadric surface is concerned, there are effectively $(m+1)^2$ homogeneous coefficients; thus a surface of order α, through $\alpha - \beta$ (a) generators, gives a residual curve of intersection, with the quadric surface, which is determined by $(\alpha+1)^2 - 1 - (\alpha+1)(\alpha-\beta)$, or $\alpha + \beta + \alpha\beta$ points. A particular curve, of the system of coresidual curves of order $\alpha + \beta$ spoken of, is constituted by β generators of the (a) system, together with α generators of the (b) system; thus the curves of the system may be said to be $\equiv \beta u + \alpha v$, where u, v denote generators respectively of the (a) system and the (b) system; we may also speak of a curve of the system as an (α, β) curve. The genus, p, of the curves of the system is given by $p = (\alpha-1)(\beta-1)$, as may be proved, for example, by projecting the curve into a plane curve with an α-fold and a β-fold multiple point. With this system of curves is associated another system, called the *adjoint system* of this; this is the ($\alpha - 2$, $\beta - 2$) system; the curves of the adjoint system cut, upon a curve of the original system, a series of points, of freedom $p - 1$, of which every set consists of $2p - 2$ points. Any curve of the system (α, β) meets any curve of the system (α', β') in $\alpha\beta' + \alpha'\beta$ points, as we may see, for example, by taking curves consisting of sets of generators; in particular this is $2p-2$ when $\alpha'=\alpha-2$, $\beta'=\beta-2$; and a curve of the system $(\alpha-2, \beta-2)$ can be described through

$$(\alpha-2+1)(\beta-2+1)-1,$$

or $p-1$ arbitrary points. Thus, on any curve of the (α, β) system, the complete canonical series is determined by the curves of the adjoint system. Further, two curves of the (α, β) system cut in $2\alpha\beta$ points; and, when $\beta=2$, these curves are hyperelliptic. And, it may be noticed, in anticipation of a general result which we reach later, the freedom, $\alpha+\beta+\alpha\beta$, of the system of (α, β) curves, is $n-p+1$, where n, $=2\alpha\beta$, is the number of intersections of any two curves of the system.

For curves on quadric surfaces see Halphen, *Bull. Soc. Math. d. France*, I, 1872, p. 19.

Ex. 9. A theory almost as simple holds for curves upon a cubic surface. But the expression of a system of coresidual curves thereon is not by two elements (as the generators, u, v in the preceding example), but by seven elements, consisting of six skew lines of the surface, together with one rational cubic curve thereon which does not meet any of the six lines. The expression is then of the form $\lambda u - \lambda_1 a_1 - \ldots - \lambda_6 a_6$, where u denotes the cubic curve, $a_1, \ldots, a_6$ denote the lines, and $\lambda, \lambda_1, \ldots, \lambda_6$ are integers. Two curves of this system intersect in n points given by

$$n = \lambda^2 - \lambda_1^2 - \ldots - \lambda_6^2;$$

any curve of the system is of order m given by $m = 3\lambda - \lambda_1 - \ldots - \lambda_6$, and of genus p given by $p = 1 + \frac{1}{2}(n-m)$, while any particular curve of the system is determined by passing through r points, given by $r = \frac{1}{2}(n+m)$, which is $n-p+1$. More generally, two curves, respectively of the systems $(\lambda, \lambda_1, \ldots, \lambda_6)$ and $(\lambda', \lambda_1', \ldots, \lambda_6')$, intersect in a number of points given by $\lambda\lambda' - \lambda_1\lambda_1' - \ldots - \lambda_6\lambda_6'$. For the system $(\lambda-3, \lambda_1-1, \ldots, \lambda_6-1)$, the freedom r', given by $\frac{1}{2}(n'+m')$, is easily seen to be $\frac{1}{2}(n-m)$ or $p-1$, while a curve of this system meets a curve of the original system in $2p-2$ points. This system is the *adjoint system* of the original, determining the complete canonical series on any curve of the original system.

Proofs of these statements are easily obtained by considering the familiar representation of the cubic surface upon a plane, whereon cubic curves through six fundamental points represent the plane sections of the cubic surface, and the lines represent rational cubic curves on the cubic surface. Or, arguing directly on the cubic surface, a surface of order r, so far as its intersection with the cubic surface is concerned, may be proved to have, in its equation, a number of homogeneous coefficients given by $\frac{1}{2}(3r^2+3r+2)$; while, through any curve of order m lying on the cubic surface, there can be drawn a surface whose order is the greatest integer in $2m/3$. For this, and the preceding example, a note in the *Proc. Lond. Math. Soc.* XI, 1912, pp. 286, 290, may be consulted.

Ex. 10. Any curve in space of three dimensions may be defined as the partial intersection of a cone, of equation $f(x, y, t)=0$, and a surface with equation of the form $z\phi(x, y, t)=\psi(x, y, t)$, where f, ϕ, ψ are homogeneous in x, y, t. The cone $\phi(x, y, t)=0$ may be taken arbitrarily, save that it must contain the chords of the curve drawn from the (general) point $(0, 0, 1, 0)$. The common generators of the cones $f(x, y, t)=0$, $\phi(x, y, t)=0$, together with the curve, are the complete intersection of the cone $f(x, y, t)=0$ with the surface $z\phi(x, y, t)-\psi(x, y, t)=0$. In general, the curve not passing through $(0, 0, 1, 0)$, the common generators of $f=0$, $\phi=0$ will also lie on the cone $\psi=0$. The surface $z\phi-\psi=0$, having a point at $(0, 0, 1, 0)$ of multiplicity one less than its order, was called by Cayley a *Monoid* (*Papers*, V, 1862, p. 8).

Ex. 11. As has already been indicated in a simple example (Ex. 6, p. 212), the formula $h_1 + h_2 + n_1 n_2 - t = \frac{1}{2}mM(m + M - 2)$ can be extended to the case when the curve (u, U) breaks up into any number of curves, its general form being $\Sigma h_i + \Sigma n_i n_j - \Sigma t_{ij} = \frac{1}{2}mM(m + M - 2)$, where i, j refer to any two different component curves, these having t_{ij} intersections; the direct proof for two curves applies to the general case.

For two component curves, the formula (p. 209)

$$p_1 + p_2 + \delta_1 + \kappa_1 + \delta_2 + \kappa_2 + t - 1 = 1 + \frac{1}{2}mM(m + M - 4)$$

is at once deducible from the formula above for $h_1 + h_2$, using $p_1 = \frac{1}{2}(n_1 - 1)(n_1 - 2) - h_1 - \delta_1 - \kappa_1$, etc. This formula may be expressed by saying that, if, on a surface of order m, one surface of order M cuts an irreducible curve of genus p, with δ double points and κ cusps, and another surface, also of order M, cuts a couple of irreducible curves of genera p_1, p_2, with respectively δ_1, κ_1 and δ_2, κ_2 multiple points, and t intersections, then $p - 1 + \delta + \kappa = p_1 - 1 + \delta_1 + \kappa_1 + p_2 - 1 + \delta_2 + \kappa_2 + t$; and, as we see, if, instead of two irreducible curves, we have several, the second member is to be replaced by $\Sigma(p_i - 1 + \delta_i + \kappa_i) + \Sigma t_{ij}$.

In view of subsequent developments we consider this further, assuming also a familiarity with the notion of the completeness of a linear series upon a curve in space, which follows from its (1, 1) correspondence with a plane curve, for which the notion has been explained (p. 66, above). Suppose that the surfaces $u = 0$, $U = 0$, drawn through the irreducible curve (n_1), intersect again, not in a single irreducible curve of order (n_2), as on p. 208 above, but in an aggregate of curves, all irreducible, of orders n_{2i}, with multiple points $(\delta_{2i}, \kappa_{2i})$; we may put $n_2 = \Sigma(n_{2i})$. Let the component curve (n_{2i}) meet the curve (n_1) in t_i points, and meet the curve (n_{2j}) in t_{ij} points. Using r_{2i}, h_{2i}, p_{2i} for the curve (n_{2i}), as were r_2, h_2, p_2 for the curve (n_2), we prove, as in the simple case,

$$r_{2i} = n_{2i}(m + M - 2) - 2\delta_{2i} - 3\kappa_{2i} - t_i - \sum_j t_{ij},$$

$$r_{2i} + \kappa_{2i} = 2n_{2i} + 2p_{2i} - 2,$$

$$h_{2i} + p_{2i} + \delta_{2i} + \kappa_{2i} = \frac{1}{2}(n_{2i} - 1)(n_{2i} - 2),$$

and hence $t = \Sigma t_i = \Sigma[n_{2i}(m + M - 2) - 2\delta_{2i} - 3\kappa_{2i} - r_{2i}] - 2\sum_{i,j} t_{ij}$,

which is the same as

$$n_2(m + M - 4) - (2p_2 - 2 + 2\delta_2 + 2\kappa_2),$$

if we put $\quad p_2 - 1 + \delta_2 + \kappa_2 = \Sigma(p_{2i} - 1 + \delta_{2i} + \kappa_{2i}) + \Sigma t_{ij}$.

Also we have, from the preceding equations,

$$n_{2i}(m + M - 4) = 2\delta_{2i} + 2\kappa_{2i} + t_i + \sum_j t_{ij} + 2p_{2i} - 2.$$

Consider now the number of conditions in order that a surface of order $m + M - 4$ should contain all the component curves (n_{2i}). Such a surface will contain the multiple points of each separate component, and its intersections with the other components. We therefore consider first the surfaces of order $m + M - 4$ through the multiple points of all the components and through the mutual intersections. Upon the single curve (n_{2i}) such surfaces will have free intersections of number

$$n_{2i}(m + M - 4) - 2\delta_{2i} - 2\kappa_{2i} - \sum_j t_{ij},$$

which is the same as $t_i + 2p_{2i} - 2$. Thus, if we suppose $t_i > 0$, the series determined by these surfaces upon the curve (n_{2i}) is not special, though it may be incomplete; the freedom of this series is then of the form

$t_i + p_{2i} - 2 - \zeta_i$, with $\zeta_i \geqslant 0$; and the number of conditions for one of the surfaces of order $m + M - 4$, under consideration, to contain the curve (n_{2i}) is one more than this freedom. The number of conditions for a surface of order $m + M - 4$ to contain all the component curves (n_{2i}) is thus, if we allow for the possibility that the separate conditions are not in-dependent, $\Sigma t_{ij} + \Sigma(\delta_{2i} + \kappa_{2i}) - \epsilon + \Sigma(t_i + p_{2i} - 1) - \zeta$, with $\epsilon \geqslant 0$, $\zeta \geqslant 0$; and this, with the notation employed above, is $t + p_2 - 1 + \delta_2 + \kappa_2 - \sigma$, where $\sigma (= \epsilon + \zeta)$ is $\geqslant 0$. The number of surfaces of order $m + M - 4$ containing the aggregate of component curves (n_{2i}) is then obtained by subtracting this number from $(m + M - 1, 3)$.

Ex. 12. It may be proved that an algebraic curve of order ν, without multiple points, lying on a ruled surface of order n (or on a cone, but not through the vertex), and meeting every generator of the surface in k points, touches $(k-1)(2\nu - kn)$ generators, and has a genus π given by $2\pi - 2 = k(2p - 2) + (k-1)(2\nu - nk)$, where p is the genus of the plane section of the ruled surface. This will be proved below (Vol. VI), as a simple application of the theory of correspondence; but a proof of the expression for π can be given which is an application of the ideas we have illustrated in Exx. 8, 9, in dealing with the curves lying on a quadric or a cubic surface. For it can be shewn that such a curve can be regarded as coresidual with an aggregate representable by $kP + (\nu - nk)G$, where P denotes a plane section of the surface, and G a generator. Assuming this fact, it follows that two curves (k, ν), (k', ν') intersect in a number of points given by

$$kk'(P, P) + [k'(\nu - nk) + k(\nu' - nk')](P, G) + (\nu - nk)(\nu' - nk')(G, G),$$

where (P, P), the number of intersections of two plane sections of the surface, is n, and (P, G), the number of intersections of a plane section with a generator, is 1, while, similarly, $(G, G) = 0$. The number in question is thus $k\nu' + k'\nu - nkk'$; for example, for a quadric surface, putting $k = \alpha$, $\nu = \alpha + \beta$, in the notation used in Ex. 8, this gives the number, $\alpha\beta' + \alpha'\beta$, there found.

Applying now the formula, for the genus of a composite curve, given above in Ex. 11, namely $1 + \Sigma(p_i - 1) + \Sigma t_{ij}$, the genus of the composite component kP is $1 + k(p-1) + \frac{1}{2}nk(k-1)$, the genus of the composite component $(\nu - nk)G$ is $1 + (\nu - nk)(-1)$, and there is a contribution to the genus of the whole curve $kP + (\nu - nk)G$ arising from the intersections of kP and $(\nu - nk)G$, which is $k(\nu - nk)$; the genus of a curve $kP + (\nu - nk)G$, by the application of the same formula, is thus

$$\pi = 1 + k(p-1) + \frac{1}{2}nk(k-1) - (\nu - nk) + k(\nu - nk),$$

so that we have $2\pi - 2 = k(2p-2) + (k-1)(2\nu - nk)$, as was stated above. In connexion with this Ex. the reader may consult Severi, *Mem. Torino*, LIV, 1903, p. 23.

Two remarks may be added. (1), The formula for the number of intersections of the curves (k, ν), (k', ν'), on the ruled surface, would give, if there existed a curve with $k' = -2$, $\nu' = 2p - 2 - n$, the value $2\pi - 2 - N$, where N, equal to $2k\nu - nk^2$, is the number of intersections of two curves of the system (k, ν). Let this number N be called the *grade* of the system (k, ν); for the plane sections, $(1, n)$, its value is n. Further let $2\pi - 2 - N$ be called the *canonical number* of the curves of the system (k, ν). Then the result is that the canonical number of any curve of a system is equal to the number of intersections of the curve with a certain fictitious curve for which $k' = -2$; the order of this fictitious curve (as of any curve on the surface) is equal to the number of its intersections with a plane section, equal then to the canonical number, $2p - 2 - n$, of the plane sections. The value of k' is the canonical number of a generator. Also, the canonical number of a curve $(k_1 + k_2,\ \nu_1 + \nu_2)$ is the sum of the canonical numbers of the curves $(k_1,\ \nu_1)$ and $(k_2,\ \nu_2)$. The second remark, (2), is that the above formula for π, in terms of ν and k, holds equally when the ruled surface is in space of any number of dimensions, and also holds when the curve (ν, k) has double points, in number d, if we replace $2\pi - 2$ by $2\pi - 2 + 2d$ (in case the surface be in space of four dimensions, however, such double points of the curve as arise at the so-called accidental double points of the surface (see below, Chap. IV in Vol. VI) are not to be included in d).

Part IV. Linear series upon a curve in space. It is clear, from the definition we have adopted for an algebraic curve, that there is a correspondence between the sets of a linear series, on the curve, and the sets of a linear series on a corresponding plane curve. In this statement it is to be understood that the linear series on the curve is determined by a linear system of surfaces, given by an equation of the form $\lambda \Phi + \lambda_1 \Phi_1 + \ldots + \lambda_r \Phi_r = 0$, in which $\lambda, \ldots, \lambda_r$ are variable parameters; the surfaces $\Phi = 0, \ldots, \Phi_r = 0$ are supposed to be linearly independent on the curve, and may have intersections with the curve which are common to all. The linear series on the plane curve is similarly determined by a system of curves with an equation of the form $\lambda \phi + \lambda_1 \phi_1 + \ldots + \lambda_r \phi_r = 0$. The substitution, in either of these equations, for the coordinates entering therein, of the values of these coordinates in terms of the coordinates appropriate to the other curve, leads from this equation to the other. The *freedom*, r, of the two corresponding series, is thus the same; and, from the (1, 1) correspondence between the points of the curves, the number of points which vary with the parameters $\lambda, \ldots, \lambda_r$, in the sets of one series, is the same as for the other; this is the number called the *grade* of the series.

There is therefore (from the theory developed for a plane curve in Chap. IV), upon an algebraic curve in space whose genus is p, a *canonical series*, of sets of $2p - 2$ points, with freedom $p - 1$; and the Riemann-Roch theorem, for complete series, which gives the freedom of the series in terms of the grade and the degree of specialness, holds also for non-plane curves. But there remain questions needing discussion, in regard to the completeness of the series determined on the space curve by surfaces whose construction is prescribed.

A simple example may illustrate this: On the rational quartic curve which is the intersection of two quadric surfaces having a point of contact, a series of sets, of each four points, of freedom 3, is obtained by the intersections of the curve with variable planes. A series of sets, of each four points, is also obtained by the intersections of the curve with variable quadric surfaces prescribed to pass through the double point, O, of the curve, and also to pass through two arbitrary fixed points, A, B, of the curve; of such quadric surfaces there are seven which are linearly in-

dependent, of which two may be taken to be surfaces wholly containing
the curve. Thus, these quadric surfaces determine a linear series of sets,
of each four variable points, on the curve; and the freedom of this series
is 4. This series contains the previous series; namely of the five linearly
independent quadric surfaces through O, A, B, there are four which
break up into the plane OAB and a general plane. Thus, on the quartic
curve, the series determined by arbitrary plane sections is *not complete*,
being contained in another linear series of the same grade whose freedom
is greater. For the rational quartic curve which is the residual intersection of a quadric surface with a cubic surface, we similarly have a
linear series, of freedom 3, of sets of 4 points, obtained by plane sections;
but this is contained in a series of freedom 4, also of grade 4, obtainable
by the intersections with cubic surfaces drawn through the two generators.

It seems on the whole to be briefest and clearest to enunciate and
prove immediately the following general theorem*, which covers
many cases: let C be a given algebraic curve; through this curve let
two surfaces $F=0$, $\Phi=0$, of orders m and M, be put, so that the
curve is simple on both these surfaces. These surfaces may intersect
in further curves; to make an inclusive statement, suppose that
such a further curve, C_j, of intersection of $F=0$ and $\Phi=0$, is
j_1-fold on $F=0$ and j_2-fold on $\Phi=0$; and also that a k_1-fold conical
point, P_k, of $F=0$, is a k_2-fold conical point on $\Phi=0$, though this
will be of importance only when this point is on the curve C. Consider now surfaces, of given order, passing through all such common
curves C_j of $F=0$, $\Phi=0$ other than C, these surfaces being such that
every such curve C_j is (j_1+j_2-1)-fold thereon, and also such that
they meet every branch of the curve C which passes through a
common multiple point P_k of $F=0$, $\Phi=0$ in (k_1+k_2-2) points,
beside containing every point where the surfaces $F=0$, $\Phi=0$ touch
one another. The theorem is that, for every order for which such
surfaces are possible, these surfaces determine a complete series on
the curve C. In particular, there are surfaces with the specified
behaviour which are of order $m+M-4$; and these cut on the curve
C (of genus p) the complete canonical series, of freedom $p-1$, of
sets of $2p-2$ points.

As an illustration of the theorem, suppose C is the complete intersection of two non-singular surfaces, being simple on both. Then
the surfaces of given order through the points of contact of these
two surfaces (the double points and cusps of the given curve) cut a
complete series on the curve; in particular there are such surfaces,
of order $m+M-4$, determining the canonical series. That the
surfaces of order $m+M-4$ give a series of sets of $2p-2$ points, was
already clear from the formula found above

$$2p-2 = mM(m+M-4) - 2\delta - 2\kappa;$$

but, that these surfaces give a series of freedom $p-1$ is a new fact.

---

* A particular consequence is proved independently below (p. 227).

To prove this independently we should require to evaluate the number of surfaces of order $m + M - 4$ which contain the curve, and to consider whether the multiple points of the curve present independent conditions for surfaces of this order not containing the curve. Or (see below, p. 227) to invoke the Riemann-Roch theorem. The general theorem, however, gives also the striking result that a complete series is obtained by surfaces of any order, however low, which can be put through the multiple points of the curve. For a simple instance, on the quartic curve of intersection of two quadric surfaces which touch in one point, a complete series is obtained by the planes through this point; or, more generally, on the complete intersection of two non-singular surfaces which have no point of contact, supposed a simple curve on both surfaces, the series, of freedom 3, determined by arbitrary planes, is complete.

This result leads to the conclusion that the curve is *normal* in the space of three dimensions, that is, cannot be obtained by projection of a curve, of the same order, existing in higher space. It will be found indeed, speaking in general terms, that any manifold, in space of any dimensions, which is a complete intersection, and without multiple parts of dimension one less than its own, is likewise normal (Severi, "Su alcune questioni di postulazione", *Rend. Palermo*, XVII, 1903, § 2).

The general theorem we have stated is given by Noether, *Math. Annal.* VIII, 1875 (*Zur Theorie des eind. Entspr. alg. Geb.* § 7). We give Noether's proof:

Preliminary theorem. Any surface $\Psi = 0$, which passes through the complete intersection of two surfaces $F = 0$, $\Phi = 0$, in such a way that every component curve of this intersection which is j_1-fold on $F = 0$ and j_2-fold on $\Phi = 0$ is $(j_1 + j_2 - 1)$-fold on $\Psi = 0$, is expressible by taking $\Psi = AF + B\Phi$, where A, B are suitable polynomials in the coordinates. The condition for $\Psi = 0$ is that, on a general plane, it shall cut a curve having a $(j_1 + j_2 - 1)$-ple point at every common point of the sections of $F = 0$, $\Phi = 0$ by this plane, where these curves have respectively a j_1-ple and a j_2-ple point. No condition for the behaviour of $\Psi = 0$ at isolated multiple points of the curve (F, Φ) is postulated. The reader may prefer to pass over the proof of this preliminary theorem (pp. 220–24), to its application for the proof of Noether's general theorem (pp. 224–26).

We first prove* that if $f = 0$, $\phi = 0$ be two curves in a plane,

---

* The theorem which we prove for plane curves is sufficient for the purpose to which we apply it, the section of a figure in space by an arbitrary plane. For the case when the common multiple points (f, ϕ) are of more intricate kind, reference may be made to Severi, *Algebr. Geom.* (Leipzig, 1921), p. 116; and for the general theorem in higher space, to Bertini, *Geom. d. iperspazi*, 1907, pp. 262 ff. A further list of authorities is given below (p. 224).

respectively of orders m, n, and with respectively an h-fold and a k-fold point at an intersection, so that hk intersections of the curves are absorbed at this point, and $\Sigma hk = mn$, then any curve $\psi = 0$ in the plane, which has a $(h+k-1)$ point at such intersection, is given by taking $\psi = uf + v\phi$, where u, v are suitable polynomials. And this we first prove under the hypothesis that the order N, of the curve $\psi = 0$, is so high, that the prescribed conditions at the common points of $f = 0$, $\phi = 0$ are all linearly independent for the equation of $\psi = 0$; the number of these conditions is then $\Sigma(h+k, 2)$, that is $\frac{1}{2}\Sigma(h+k)(h+k-1)$. In this case the most general curve $\psi = 0$ satisfying these conditions contains $(N+2, 2) - \Sigma(h+k, 2)$ homogeneously entering independent coefficients. To shew then that the form $uf + v\phi$ is general enough for the representation of any curve ψ, it is sufficient to shew that, with general curves $u = 0$, $v = 0$, of respective orders $N - m$, $N - n$, chosen to have respectively a $(k-1)$-ple and a $(h-1)$-ple multiple point at the general intersection (f, ϕ) spoken of, for all such intersections, the polynomial $uf + v\phi$ contains as many as $(N+2, 2) - \Sigma(h+k, 2)$ homogeneously entering arbitrary coefficients*. Now, the form u, after the prescribed conditions are satisfied, will contain $(N-m+2, 2) - \Sigma(k, 2)$ arbitrary homogeneous coefficients; or more if the conditions at the intersections (f, ϕ) are not independent. Similarly v will contain $(N-n+2, 2) - \Sigma(h, 2)$, at least. The aggregate $uf + v\phi$ will thence contain a number of homogeneous arbitrary coefficients equal to the sum of the two preceding numbers, diminished by the number of terms which can enter simultaneously in uf and $v\phi$. But, assuming that f, ϕ have no common factor, an identity $u_1 f = v_1 \phi$, where u_1, v_1 denote respectively aggregates of terms in u and v, requires that v_1 divides by f, and hence $N - n \geqslant m$. Conversely, if $N \geqslant m + n$, by taking $v_1 = wf$, and hence v of the form $v_0 + wf$, where the terms $v_0 \phi$ are independent of the terms uf, and w is a general polynomial, of order $N - m - n$, the possible reduction does arise; in this case the form $uf + v\phi$ contains, at least, a number of arbitrary homo-

---

* That is, we assume that two linear systems of linearly independent curves, $\lambda_0 U_0 + \ldots + \lambda_r U_r = 0$, $\mu_0 V_0 + \ldots + \mu_s V_s = 0$, are identical if they are of the same order, have the same base points (simple or multiple), and the same freedoms (i.e., if $r = s$). In the application made the parameters $\mu_0, \ldots, \mu_s$ on the right are the coefficients entering in the polynomials u and v. We could also consider the theorem by analysing both sides of the equation into the individual monomials $x^i y^j z^k$ which enter; the coefficients on the right would then be *linear functions of* the coefficients in the polynomials u and v (the coefficients in these linear functions being taken from the given coefficients in f and ϕ). When N is so small ($N < m + n$) that no identity $u_1 f + v_1 \phi = 0$ is possible, these linear functions are independent; and so the identity $\psi = uf + v\phi$ would follow, for proper values of u and v, from shewing that the prescribed conditions for u, v and ψ, at the intersections of $f = 0$, $\phi = 0$, involved that the two sides of the equation contained the same monomials. Similarly in general.

geneous coefficients given by

$$(N-m+2, 2)+(N-n+2, 2)-(N-m-n+2, 2)-\Sigma(h, 2)-\Sigma(k,2).$$

Remarking then the identities

$$(N+2, 2)-(N-m+2, 2)-(N-n+2, 2)+(N-m-n+2, 2)=mn,$$
$$(h+k, 2)-(h, 2)-(k, 2)=hk,$$

and recalling the hypothesis $\Sigma hk=mn$, the conclusion is that the form $uf+v\phi$ contains enough arbitrary coefficients to represent the form ψ, however general the latter may be, subject to the conditions it is to satisfy. Such general ψ will be a linear aggregate of particular forms ψ satisfying these conditions, and these particular forms will occur in $uf+v\phi$, each multiplied by an arbitrary coefficient. Thus any particular form ψ, for which the curve $\psi=0$ has the prescribed multiple points, is expressible in the form $uf+v\phi$, with u, v as particular curves of the prescribed behaviour.

This is on the hypothesis that ψ is of sufficiently high order. We can, however, deduce from this that the expression in question equally holds whatever be the order of ψ, provided this has the prescribed behaviour at the multiple points (f, ϕ). For assume that it has been proved that all curves of order M which have a $(h+k-1)$-ple point at a point which is h-ple for $f=0$ and k-ple for $\psi=0$ are expressible in the form $u\phi+v\psi=0$. Let $\psi=0$ be any curve of order $M-1$ with the like behaviour at every common point (f, ϕ), and $\zeta=0$ be an arbitrary line. Then the composite curve $\zeta\psi=0$ is of order M, and has the like behaviour. By linear change of coordinates, let all the curves be expressed homogeneously in terms of x, y, ζ; and denote ζ by z; in the changed coordinates denote f, ϕ, ψ by $f_1(x, y, z)$, $\phi_1(x, y, z)$, $\psi_1(x, y, z)$. Then, by the assumption made,

$$z\psi_1(x, y, z)=u(x, y, z) f_1(x, y, z)+v(x, y, z) \phi_1(x, y, z),$$

with proper polynomials u and v. Hence we have the identity

$$u(x, y, 0)f_1(x, y, 0)+v(x, y, 0) \phi_1(x, y, 0)=0,$$

and we can suppose that the general line $z=0$ has been chosen so that $f_1(x, y, 0), \phi_1(x, y, 0)$ have no common factor. Thus we can infer that $u(x, y, 0)=\theta(x, y) \phi_1(x, y, 0)$, where $\theta(x, y)$ is a homogeneous polynomial in x and y; and hence

$$u(x, y, z)=\theta(x, y) \phi_1(x, y, z)+zu_1(x, y, z),$$

where u_1 is a polynomial homogeneous in x, y, z. With this, the equation $z\psi_1=uf_1+v\phi_1$ takes the form $z\psi_1=(\theta\phi_1+zu_1)f_1+v\phi_1$, which is $z(\psi_1-u_1f_1)=(\theta f_1+v) \phi_1$; and, from this identity it follows that $\psi_1-u_1f_1$ divides by ϕ_1, or we have an equation $\psi_1=u_1f_1+v_1\phi_1$, where v_1 is a homogeneous polynomial. We can now return to the original variables x, y, z, and hence have an expression for ψ, in terms of f and ϕ, of the form specified, though ψ is only of order $M-1$. The reduction can be continued in the same way, to any form ψ, whatever its order, which satisfies the condition at the intersections (f, ϕ).

There are remarks, not of importance for our immediate purpose, which may usefully be made here. It follows from what is proved that, for any value of N, the number of independent conditions which are involved, for a (ternary) form ψ of order N, in order that the curve $\psi=0$ should have a multiple point of order $h+k-1$ at every (h, k) common point of the curves $f=0$, $\phi=0$ (supposed to be multiple points of the most general kind), is given by

$$(N+2, 2) - (N-m+2, 2) - (N-n+2, 2) + (N-m-n+2, 2) \\ + \Sigma(h, 2) + \Sigma(k, 2),$$

where m, n are the orders of $f=0$, $\phi=0$; for this is equal to the number of terms in a general polynomial ψ, less the number of independent terms in the form $uf+v\phi$. In this statement, however, the term $(N-m-n+2, 2)$ is to be omitted if $N < m+n$, as we see by recalling the enumeration above made of the terms in $uf+v\phi$; this term vanishes identically if $N = m+n-1$, or $N = m+n-2$. Thus, recalling an identity remarked above, the number of conditions in question is $(h+k, 2)$ when $N > m+n-3$; and, for values of N of the form $N = m+n-3-q$, the number is $(h+k, 2) - (q+2, 2)$. We may easily see that in all cases the number is $\sigma + \gamma$, where σ denotes $\Sigma(h, 2) + \Sigma(k, 2)$, and γ denotes the coefficient of t^N in the expansion, in ascending powers of t, of $(1-t^m)(1-t^n)(1-t)^{-3}$. We may call this number the *postulation* of the (h, k) base points for curves of order N. In particular, if mn points in a plane be the complete intersection of two curves of orders m and n, all curves of order $m+n-3-q$, through $mn - (q+2, 2)$ of these intersections, pass through the remaining $(q+2, 2)$ intersections, in general (q being not greater than the less of $m-3$ and $n-3$). A more exact statement has already been given (Chap. IV, Note III, On the Cayley-Bacharach theorem). A particular consequence is, that curves of order $m+n-4$ passing through all but 3 of the mn intersections of two general curves of orders m and n, pass, as a rule, through these 3 also (but, when these 3 are collinear, it is necessary also to prescribe that the curve of order $m+n-4$ passes through one of them).

Ex. We have (pp. 218, 219) deduced from Noether's theorem two results, (*a*), that on the complete intersection of two non-singular surfaces of orders m and n, which do not touch (the curve of intersection being simple on both surfaces), general surfaces of order $m+n-4$ give the canonical linear series of freedom $p-1$; (*b*) that the series cut on the curve by arbitrary planes is complete. It may be interesting to see how the result (*b*) follows from (*a*), in virtue of the result just obtained, for plane curves of order $m+n-4$ through the complete intersection of two curves of orders m and n. In fact, by the Riemann-Roch theorem for the curve of order mn, the complete series on the space curve, which contains the sets of mn intersections by general planes of the space, has a freedom $mn-p+i$, where i is the number of canonical sets on the space curve which contain the intersections by an arbitrary plane. By the result (*a*), this number (i) is the number of surfaces of order $m+n-4$ (not containing the space curve) which pass through the mn plane intersections. The conditions for such a surface to contain these mn points are the conditions for the curve of order $m+n-4$, which is the plane section of the surface, to contain these mn points, whose number, by what is proved above, is $mn - 3$. Now all possible surfaces of order $m + n - 4$ may be regarded as made up of two linear aggregates, those which wholly contain the space curve, and those which do not; the number in the latter aggregate is one more than the freedom of the series cut on the space curve by surfaces of order $m + n - 4$, namely is p. Whence, the number of surfaces of order $m + n - 4$, not containing the space curve, which contain the mn intersections with a plane, is $p - (mn - 3)$. If we put this for i in the formula $mn - p + i$, we obtain 3 as the freedom of the complete series, on the space curve, defined by the plane sections; which is the freedom of the system of planes.

From the theorem now proved for curves in a plane, we can pass to the theorem for surfaces which we have called the preliminary theorem (p. 219). The proof depends on the obvious remark that, of surfaces (Ψ), through the intersection of the given surfaces (F), (Φ), having the specified relation to these on every arbitrary general plane section, *there is a least possible order*. For instance, as an extreme case, there is no *plane*, containing the common curve of (F) and (Φ) with the appropriate multiplicity, even if this common curve is wholly plane; for this curve, as coinciding with the plane section of the surface (F), which is of order m, must be M-fold on (F), in order that it may be of the order mM; and must similarly be of multiplicity m on (Φ); and it is not then $(m + M - 1)$-fold on the plane.

Let $t = 0$ be an arbitrary general plane; change the coordinates, by linear transformation, so that the surfaces are represented in terms of the homogeneous coordinates x, y, z, t. We can then suppose that $F = f + t f_1$, $\Phi = \phi + t \phi_1$, $\Psi = \psi + t \psi_1$, where f, ϕ, ψ are homogeneous *in x, y, z only*. In virtue of the supposed relation of (Ψ) to (F) and (Φ), we can then suppose, from the theorem proved above for plane curves, that $\psi = u f + v \phi$, where u, v are polynomials in x, y, z only. Whence $\Psi = u f + v \phi + t \psi_1 = u F + v \Phi + t(\psi_1 - u f_1 - v \phi_1)$, which we may write in the form $\Psi - (u F + v \Phi) = t \Psi_1$, where Ψ_1 is a homogeneous polynomial in x, y, z, t.

Assuming that the curves (u), (v), on the perfectly general plane $t = 0$, have the multiplicities $(j_2 - 1)$, $(j_1 - 1)$, respectively, at every typical common point (j_1, j_2) of the curves (f), (ϕ), it follows, from the last equation, that, at every common point of (F), (Φ) on $t = 0$, the section of (Ψ_1) by this plane has the behaviour specified for (Ψ) in the theorem under proof.

Whence, applying to (Ψ_1) the argument used for (Ψ), we infer an equation $\Psi_1 - (u_1 F + v_1 \Phi) = t \Psi_2$, where ($\Psi_2$) is a surface whose section by $t = 0$ has the specified relation to the sections of (F) and (Φ). This, however, leads to $\Psi = (u + t u_1) F + (v + t v_1) \Phi + t^2 \Psi_2$, and a similar process can be continued. But the surfaces (Ψ_1), (Ψ_2), ... are of diminishing order, and all in the specified relation to (F) and (Φ). Therefore, by the remark made at starting, there is some stage at which the corresponding surface (Ψ_s) is absent, so that we finally have an equation of the form $\Psi = AF + B\Phi$, where A, B are polynomials; and this is the preliminary theorem required.

It is important to remark that the argument does not require the assumption that the surface (Ψ) is irreducible.

The proposition we have obtained belongs to a theory which has a wide literature. We may make reference to *Gergonne's Annales*, XVII, 1827, p. 214; Jacobi, *Ges. Werke*, III, p. 292; Noether, *Math. Ann.* VI, 1872, p. 358; *ibid.* VIII, 1875, § 8; *ibid.* XL, 1891, p. 140; C. A. Scott, *Math. Ann.* LII, 1899, p. 593; Severi, *Rend. Lincei*, XI, 1902; *Rend. Palermo*, XVII, 1903; *Atti . . . Torino*, XLI, 1906; Bertini, *Geom. d. iperspazi*, 1907, p. 268; Picard-Simart, *Fonct. algéb.* II (Paris, 1906), p. 17; Enriques-Chisini, *Teoria geometrica*, III, p. 530; Castelnuovo-Enriques, *Enzyk. d. math. Wiss.* III, C., p. 645.

Having considered the preliminary theorem (p. 219), we pass now to the proof of the general theorem enunciated on p. 218. Through a given curve, C, are put two surfaces F_1, F_2, of orders N_1, N_2, upon each of which C is simple. We consider a surface, S, passing through the residual curve intersection of F_1 and F_2, so as to have any part of this, which is a curve C_j, of multiplicity j_1 on F_1 and j_2 on F_2, as a (j_1+j_2-1)-ple curve; and, also, so as to meet any branch of C containing a common multiple point of F_1 and F_2, k_1-ple on F_1 and k_2-ple on F_2, in (k_1+k_2-1) points coincident thereat; the surface 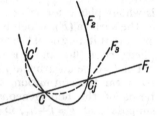 S, moreover, containing every point of contact of F_1 and F_2. We are to shew, first, that the series, cut on C by the general surface S of this behaviour, is independent of the surfaces F_1, F_2 which are used.

Describe, passing simply through C, another surface, F_3, of order N_3, to contain, beside C, all the residual curves of intersection, C_j, of F_1 and F_2, this curve C_j being j_3-fold on F_3. The surface F_3 may have further undesigned curves of intersection with the surface F_2; denote these by C'. It is supposed that C' is simple on F_2 and F_3.

Consider the surface SF_3, composed of the surface S spoken of together with F_3, in relation to the two surfaces F_1 and F_2; passing simply through C, and, with multiplicity at least j_1+j_2-1, through every other curve C_j common to F_1 and F_2, this surface is capable of expression in the form $SF_3 = TF_1 + AF_2$, where T, A are polynomials. This equation shews, if σ be the order of S, that T is of order $\sigma + N_3 - N_1$, so that if $\sigma = N_1 + N_2 - 4$, T is of order $N_2 + N_3 - 4$; and shews that $T = 0$ passes through the general curve C_j with multiplicity $j_1+j_2-1+j_3-j_1$, or j_2+j_3-1; and, further, that $T = 0$ passes through the curve (or curves) C' which are common to F_2 and F_3 but are not on F_1; also $T = 0$ passes through the points of contact of F_2 and F_3. Also, at any multiple k_2-fold point of F_2 which is k_3-fold on F_3 (and k_1-fold on F_1), any branch of C, which passes through it, meets SF_3 in $(k_1 + k_2 - 2 + k_3)$ points, and F_1 in k_1 points, and so meets $T = 0$ in $(k_2 + k_3 - 2)$ points.

On the whole, then, $T = 0$ is defined for the curve C, with the help of the surfaces F_2 and F_3 through the curve, just as S was defined with the help of the surfaces F_1 and F_2.

We can, however, prove that the surfaces S and T have the same intersections with the curve C. For this curve lies on the surface F_2, whereon, by the preceding identity, we have the identity $SF_3 = TF_1$. This shews, as the surface F_3 wholly contains C, that the intersections of the surface T with the curve C lie on the surface S; and, also, as F_1 contains C, that the intersections of the surface S with the curve C lie on the surface T.

The definition of the series on the curve C by the surfaces S, with the use of the surfaces F_1 and F_2, can thus be modified by the substitution, for F_1, of any other surface F_3, (which we have chosen, however, in the argument, so general as to have with F_2, outside the curves (F_1, F_2), only simple intersections); and it has been shewn that if S is of order $N_1 + N_2 - 4$, then T is of order $N_2 + N_3 - 4$. Thus, for F_1 and F_2, any two general surfaces through C can be chosen. In particular, we can (as already remarked, Ex. 10, p. 214) suppose these surfaces to be K and L, where K is a cone projecting C from a suitably general point, and L is a surface, of order m, say, having a $(m-1)$-fold point at this point of projection; the equations of these can be supposed to be

$$K \equiv f_n(x_1, x_2, x_3) = 0, \quad L \equiv x_0 f_{m-1}(x_1, x_2, x_3) + f_m(x_1, x_2, x_3) = 0.$$

Beside C, these surfaces intersect in the lines common to the cones $f_n = 0, f_{m-1} = 0$, and, if $(1, 0, 0, 0)$ be a general point, not lying on the curve C, such a line must equally lie on the cone $f_m = 0$, being r-fold for this if r-fold for $f_{m-1} = 0$. The surfaces S will then be surfaces ϕ, cones of vertex $(1, 0, 0, 0)$, passing $(q + r - 1)$ times through every q-fold line of f_n which is r-fold for f_{m-1}; so that, from what was proved above, ϕ is of the form $Bf_n + Af_{m-1}$, where B, A are homogeneous in x_1, x_2, x_3. In particular when, in the notation used above, the surface S was of order $N_1 + N_2 - 4$, the cone ϕ will be of order $m + n - 4$; and its intersections with the cone f_n, other than those on f_{m-1}, will be on A, which is then of order $m + n - 4 - (m - 1)$, or $n - 3$; a q-fold generator of f_n which is r-fold for f_{m-1}, is $(q-1)$-fold for this cone $A = 0$. The curve C is the complete intersection of the cone f_n and the monoid L, other than the generators (f_n, f_{m-1}), and is met by ϕ on the cone A. A q-fold generator of f_n arises from a q-fold chord of C, drawn from $(1, 0, 0, 0)$; and, in order that there may be q distinct points of C on this generator, f_{m-1} must, we know from the theory of plane curves, have this generator as $(q-1)$-fold at least. If the curve C has an actual q-fold point, this will give a q-fold generator of f_n, which need not be on f_{m-1} (or f_m), but will equally be $(q-1)$-fold on the cone A.

We thus see that the series on the curve C obtained by the surfaces S, described, as explained, through the residual intersection of two arbitrary surfaces F_1, F_2 which are put simply through C, is the series which projects, from an arbitrary general point, into the series obtained, on the projection of C, by *adjoint* curves of this projection (see Chap. IV, preceding, p. 60), and this, by the theory of the plane curve, is a complete series. In particular, the series by surfaces S of order $N_1 + N_2 - 4$, is the series projecting into that obtained on the plane curve by adjoint curves of order $n-3$; as this latter is the canonical series, the series on C by the surfaces S of order $N_1 + N_2 - 4$ is equally the canonical series. The series by surfaces S which are of order greater than $N_1 + N_2 - 4$ will project into series given by plane curves adjoint to the projection of C, of appropriately higher order. If there are surfaces of order less than $N_1 + N_2 - 4$ which contain the residual curves C_j (or when C is a complete intersection), then there exist adjoint curves of order less than $n-3$ for the projection. Conversely, any complete series on the plane curve corresponds to a series on the curve C obtainable by the rule; for we may take, for the surfaces F_1, F_2, the cone f_n and the monoid $x_0 f_{m-1} + f_m$. This completes the proof.

A very suggestive proof that if the surface S, in the general theorem, give a complete series on the curve C when the surfaces S are of sufficiently high order, then the same is true for existing surfaces S of lower order, is given by Severi, *Rend. Lincei*, XII, 1903, *Sulla deficienza della serie...*No. 2.

Ex. 1. Prove that the $\frac{1}{2}mM(m-1)(M-1)$ chords from an arbitrary general point to the curve of intersection of two surfaces of orders m and M, lie, with mM coplanar points of this curve, upon $4 + \frac{1}{2}mM(m + M - 6)$ linearly independent cones of order $mM - 3$, whose vertex is at the point.

Ex. 2. Prove that if the mnk intersections of three surfaces $f = 0$, $\phi = 0$, $\psi = 0$, of orders m, n, k, are all simple, then any surface through these points has an equation of the form $uf + v\phi + w\psi = 0$, where u, v, w are polynomials. Further, that the postulation of these points, for a surface of order N, is the coefficient of t^N in the expansion, in ascending powers of t, of the function $(1 - t^m)(1 - t^n)(1 - t^k)(1 - t)^{-4}$, and* is mnk only if $N > m + n + k - 4$.

Another proof of the determination of the canonical series on a curve in space. It has been proved that when two curves,

---

* If $\sigma = \overset{s}{\underset{k=1}{\Sigma}} (n_k - 1)$, $\varpi = \overset{s}{\underset{k=1}{\Pi}} n_k$, it may be proved that $(1 - t^{n_1})...(1 - t^{n_s})(1 - t)^{-s-1}$ is of the form $u + \varpi t^\sigma (1 - t)^{-1}$, where u is an integral polynomial, with positive integer coefficients, of order $\sigma - 1$.

C_1, of order n_1, and C_2, of order n_2, are, together, the complete intersection of two non-singular surfaces, of orders m and M, being simple on both these, and the only double points of C_1 or C_2 are simple contacts of the two surfaces, then the canonical series on C_1 is given by the surfaces of order $m+M-4$ through C_2 which contain the double points of C_1. We give now another proof of this result, depending on the Riemann-Roch theorem for the curve C_1, whose validity is a consequence of the $(1, 1)$ correspondence between this curve and a plane curve. Like the foregoing proof it may be assigned to Noether, whose great paper, of 1882, employs this $(1, 1)$ correspondence throughout.

After the detailed considerations of Ex. 11 (p. 215) preceding, it will be unnecessary to suppose that the curve C_2 is irreducible; but we assume, as in Ex. 11, that every component of C_2 has actual intersections with C_1.

Denoting the surfaces through C_1 which intersect further in C_2, by $F=0$, $\Phi=0$, we have proved (p. 219) that all surfaces of order $m+M-4$, containing both C_1 and C_2, have an equation of the form $AF+B\Phi=0$; the number of such surfaces (with $N=m+M-4$) is thus $(N-m+3, 3)+(N-M+3, 3)$, which is the same as

$$(N+3, 3)+(N-m-M+3, 3)-[NmM-\tfrac{1}{2}mM(m+M-4)],$$

or, for $N=m+M-4$, is $(m+M-1, 3)-1-\tfrac{1}{2}mM(m+M-4)$.

Now consider the surfaces of order $m+M-4$ containing C_2, which also contain the multiple points of the curve C_1. The surfaces of this order, conditioned only by containing C_2, are proved in Ex. 11 to be, in number, $(m+M-1, 3)-t-p_2+1-\delta_2-\kappa_2+\sigma$, where $\sigma \geqslant 0$. From the equation

$$2p_1-2=n_1(m+M-4)-2\delta_1-2\kappa_1-t$$

it follows, when the further condition of containing the multiple points of C_1 is imposed, that there arises on C_1 a linear series of $2p_1-2$ points in each set. We prove that the freedom of this series is $p_1-1-\sigma_1$, with $\sigma_1 \geqslant 0$; first, if the series be not special, and be complete, it will have a freedom p_1-2, and will have a less freedom if incomplete; and, if the series be special, it will have a freedom p_1-1, if complete (a set of $2p_1-2$ points not belonging to two canonical sets), but a less freedom if incomplete. Thus, the surfaces of order $m+M-4$, which contain the curve C_2 and the multiple points of C_1, will contain the curve C_1 entirely if put through other $p_1-\sigma_1$ points of this curve, of general position. If the number of conditions, for surfaces of order $m+M-4$ containing C_2, that they should pass through the multiple points of C_1, be denoted by $\delta_1+\kappa_1-\mu$, with $\mu \geqslant 0$, it follows then, that the number of surfaces of order $m+M-4$ containing both curves C_1 and C_2, is

$$[(m+M-1, 3)-t-p_2+1-\delta_2-\kappa_2+\sigma]-[p_1-\sigma_1]-[\delta_1+\kappa_1-\mu].$$

We found, however, that

$$p_1 + p_2 + t - 1 + \delta_1 + \kappa_1 + \delta_2 + \kappa_2 = 1 + \tfrac{1}{2}mM(m + M - 4);$$

thus, the number of surfaces of order $m + M - 4$ containing both curves C_1, C_2, is $(m + M - 1, 3) - 1 - \tfrac{1}{2}mM(m + M - 4) + \sigma + \sigma_1 + \mu$; comparing this with the number of such curves which we remarked above, based on the expression $AF + B\Phi$, we infer, therefore, that $\sigma = \sigma_1 = \mu = 0$.

The argument itself proves that the surfaces of order $m + M - 4$ through the curve C_2, even when put through the multiple points of C_1, do not necessarily all contain the curve C_1, for general values of p_1.

From $\sigma_1 = 0$ it follows that the surfaces of order $m + M - 4$, through C_2 and the multiple points of C_1, cut, on this curve C_1, a complete special series, of sets of $2p_1 - 2$ points, with freedom $p_1 - 1$, the canonical series. This is the main theorem we set out to prove.

It is also clear that surfaces of order $m + M - 4$, conditioned only by passing through the multiple points of C_1, and the t intersections of this curve with C_2, cut on C_1 a series of the same grade $(2p_1 - 2)$ as if they entirely contained the curve C_2, with freedom certainly not less than in that hypothesis (that is not less than $p_1 - 1$). Hence these surfaces do in fact give on C_1 the same series as that obtained by surfaces wholly containing C_2. For, if this series were not special its freedom would at most be $p_1 - 2$; and, being thence special, its freedom is at most $p_1 - 1$, so that it coincides with the canonical series obtained by surfaces wholly containing C_2.

From $\sigma = 0$, it follows that the numbers ζ_i used in the argument of Ex. 11 above (p. 215) are all zero. Thus, surfaces of order $m + M - 4$, drawn through the multiple points of all the components of the curve C_2, when this is composite, and through the mutual intersections of these components (but not their intersections with C_1), cut a complete series on each component, this freedom, in the notation of that example, being $p_{2i} - 2 + t_i$. In particular, when C_2 is irreducible, surfaces of order $m + M - 4$, through the multiple points of C_2, cut on C_2 a complete series of freedom $p_2 - 2 + t$. From this result, proved when C_1 is irreducible, we infer, by parity of reasoning, that surfaces of order $m + M - 4$, through the multiple points of C_1, cut on C_1 a series of freedom $p_1 - 2 + t - \lambda$, this series being complete, and λ zero, when C_2 is irreducible. From this we can prove that, for surfaces of order $m + M - 4$ through the multiple points of C_1, the t intersections with C_2 furnish only $t - 1 - \lambda$ independent conditions, for surfaces of this description to contain them. For denote the number of these conditions by ξ; we have shewn that surfaces of order $m + M - 4$, through the multiple points of C_1, cut on this curve a series of freedom $p_1-2+t-\lambda$; those through the multiple points and the t intersections thus cut a series of freedom $p_1-2+t-\lambda-\xi$; but we have shewn that the freedom of this series is p_1-1. Wherefore $\xi=t-1-\lambda$, where λ is zero when C_2 is irreducible, but otherwise is an integer $\geqslant 0$.

But, supposing C_2 is composite, we can go further. Consider the linear aggregate, Ω, of surfaces of order $m+M-4$, which pass through the multiple points of C_1 and through the multiple points of all the components of C_2, which pass also through all the intersections of the components of C_2 with C_1 and with one another. These surfaces cut on C_1 a linear series, of the same grade, $2p_1-2$, as if they contained the curves C_2 entirely, and of freedom not less than the freedom, p_1-1, which would then exist. This series on C_1 is thus special (not being of freedom equal to or less than p_1-2), and is also complete (or its freedom would be less than p_1-1). Thus these surfaces Ω cut the complete canonical series on C_1. Hence, by parity of argument, these surfaces cut the complete canonical series on every component of C_2.

Further, of the linear aggregate of surfaces Ω, those which do not contain the curve C_1 must be p_1 in number (in order to cut thereon a series of freedom p_1-1); and this is in fact the same number as of surfaces Ω, not containing C_1, which contain all the curves C_2, as was shewn. While, similar remarks may be made in regard to any one component of C_2 as are here made in regard to C_1. Form then, from Ω, the aggregate of the surfaces which contain C_2 but do not contain C_1; let this be Ω_1. Form also the aggregate of those which contain C_1 and also every component of C_2 except the curve (n_{2i}); denote this by Ω_{2i}, the number of surfaces in Ω_{2i} being (we have seen) p_{2i}. By the definition, no surface of Ω_{2i} is the same as any surface of Ω_1, or the same as any surface of Ω_{2j}. Finally, form the aggregate Ω_0 of all surfaces Ω which contain both C_1 and C_2; the surfaces of Ω_0, we have seen, are of number, say P, given by

$$P=(m+M-1,\,3)-1-\tfrac{1}{2}mM(m+M-4).$$

We have then $p_1+\Sigma p_{2i}+P$ linearly independent surfaces of the aggregate Ω. These are in fact the whole aggregate. For, if there were in Ω, for instance, a surface not containing either the curve (n_{2i}), or the curve (n_{2j}), which is not a linear aggregate of the surfaces already described, such surface would give on the curve (n_{2i})—as also on (n_{2j})—a set belonging to the canonical series on (n_{2i}), in addition to those already obtained, contrary to the fact proved that these define the complete canonical series on (n_{2i}).

Recurring however to Ex. 11 (p. 215), of which we have continued the notation, we have

$$p_1+\Sigma p_{2i}+P=(m+M-1,\,3)-[\delta_1+\kappa_1+\Sigma(\delta_{2i}+\kappa_{2i})+\Sigma t_i+\Sigma t_{ij}-s],$$

where s denotes the number of component curves in C_2. As the first member, we have proved, is the number of linearly independent surfaces of the aggregate Ω, *we infer, that the number of independent conditions for a surface of order $m + M - 4$ to belong to the aggregate Ω, that is, to contain all the multiple points of C_1 and C_2, and to pass through all the intersections of the $s + 1$ curves involved in C_1 and C_2, is* $\delta_1 + \kappa_1 + \Sigma(\delta_{2i} + \kappa_{2i}) + \Sigma t_i + \Sigma t_{ij} - s$. From this it follows that, if the conditions of passing through all the multiple points are independent, then the conditions, for the surfaces of order $m + M - 4$, passing through these multiple points, to pass through the $\Sigma t_i + \Sigma t_{ij}$ intersections, are not independent, having a defect equal to or less than the total number of curves into which the complete intersection of the two surfaces (F), (Φ) breaks up. We have already remarked that this is so when $s = 1$.

Ex. 1. In illustration of this theory when C_2 is irreducible, we consider two quintic curves, of genus 2, forming together the complete intersection of a quadric surface and a quintic surface.

Take a quadric surface and put through one generator of this a cubic surface. The remaining intersection is a quintic curve, C_1, for which $n_1 = 5$, and $p_1 = 2$, the quintic curve being met in two points by a variable plane through the generator. Through C_1 put a quintic surface, giving, by its remaining intersection with the quadric surface, another quintic curve, C_2, for which also, we easily find, $p_2 = 2$, having with C_1 a number, t, of intersections given by $t = 13$. The curves C_1, C_2 meet the generators of the two systems, of the quadric surface, in, respectively, 2 and 3, and in 3 and 2 points.

Now take 12 of the intersections of the two curves C_1, C_2; let O denote one of these; from O take the generator of the quadric surface which meets the curve C_1 in two points A_1, B_1 beside O; similarly let the other generator through O meet C_2 in A_2 and B_2, beside O. By what we have proved (the curves C_1, C_2 being supposed without multiple points), cubic surfaces (of order $m + M - 4$ for $m = 2$, $M = 5$) cut on C_1 a linear series of sets of 15 points with freedom $p_1 - 2 + t$, or 13; so that a cubic surface drawn through 14 independent points of the curve C_1 entirely contains the curve. Consider then cubic surfaces through the 12 chosen intersections, and through A_1 and B_1; assuming these conditions to be independent, these cubic surfaces will contain the curve C_1; and they will contain $20 - 14$, or 6, undetermined homogeneous coefficients. They will thus be given by an equation $(a_1 x + b_1 y + c_1 z + d_1 t) Q + \lambda_1 V_1 + \mu_1 W_1 = 0$, where x, y, z, t are the coordinates, $Q = 0$ is the quadric surface on which the curves C_1, C_2 lie, $V_1 = 0$, $W_1 = 0$ are definite cubic surfaces through the curve C_1, and a_1, b_1, c_1, d_1, λ_1, μ_1 are undetermined constants. In the same way, the cubic surfaces through the 12 chosen intersections of the two curves C_1, C_2, and through the points A_2, B_2, will be given by an equation $(a_2 x + b_2 y + c_2 z + d_2 t) Q + \lambda_2 V_2 + \mu_2 W_2 = 0$, where $V_2 = 0$, $W_2 = 0$ are definite cubic surfaces containing the curve C_2.

Whence, the general cubic surface through the 12 chosen intersections, which is of 8 members, has an equation of the form

$$(lx + my + nz + kt) Q + \lambda_1 V_1 + \mu_1 W_1 + \lambda_2 V_2 + \mu_2 W_2 = 0,$$

where every member belongs to a surface containing one, or both, of the two curves. All these cubic surfaces contain the 13th intersection of C_1 and C_2.

Ex. 2. Consider next the case in which the curve C_2 is composed of three skew lines; and the curve C_1 is the sextic intersection of two cubic surfaces through the lines. For the conventional genus of C_2 (see Ex. 11, p. 215), the formula $p_2 - 1 = \Sigma(p_{2i} - 1) + \Sigma t_{ij}$ gives $p_2 = -2$, while, from

$$t = n_2(m + M - n_2 - 1) + 2h_2, \quad = 3(3 + 3 - 4) + 6, \quad = 12,$$

and $\qquad t = n_1(m + M - 4) - (2p_1 - 2),$

we have $p_1 = 1$. Thus the sextic curve is of genus 1, and meets each of the three lines in 4 points, as is also easily seen by considering sections on an arbitrary plane through one of the three lines.

For quadric surfaces (of order $m + M - 4$), required to pass through the 12 intersections of the four curves which are the complete intersection of the two cubic surfaces, these 12 points are equivalent only to 9 conditions; and these quadric surfaces (one in number) cut the canonical series (of zero grade) on each of the four curves.

In this case, general quadric surfaces cut on the sextic curve, C_1, a series of sets of 12 points, of freedom 9 (there being no quadric surface through this curve). The complete series of sets of 12 points on C_1 has freedom 11, and is determinable by quartic surfaces through the three lines, as we have proved; there are $35 - 15$, or 20 such quartic surfaces; of these there are 8, with equation of the form $AF + B\Phi = 0$, where (F), (Φ) are the two cubic surfaces and (A), (B) are planes, which contain the sextic curve; the remaining 12 give the complete series in question. Among such quartic surfaces are the 10 which decompose into the quadric surface containing the three lines, together with another quadric surface.

Ex. 3. Next, let C_2 consist of a conic and a line not meeting this, while C_1 is the remaining sextic intersection of two cubic surfaces drawn through these. Then we find C_1 to be of genus 2, being met by the conic in 6 points, and by the line in 4 points.

By the result of the text, the 10 intersections furnish 8 independent conditions for quadric surfaces (of order $m + M - 4$) required to pass through them; and this is directly obvious. These quadric surfaces contain both the conic and the line, and break up into the plane of the conic, and a variable plane through the line, which determines the canonical series on the sextic curve.

Part V. Connexion with the theory of algebraic functions.

Consider, as before, an irreducible curve C_1, simple on each of two surfaces of orders m and M, whose further intersection is a curve C_2, not necessarily irreducible, there being no multiple points for either curve, other than contacts of the two surfaces. We have proved that surfaces of given order which pass through C_2, and through the multiple points of C_1, give a complete linear series on C_1.

Such a series, we know, from the theory of linear series on a plane curve, is determined by one of its sets. Hence, we see that, if any number of simple points, say (D), be taken arbitrarily on the curve C_1, and a surface (δ), of any sufficient order, be put through these points, through the multiple points of C_1, and through the curve C_2, and, if the set of remaining intersections of this surface (δ) with C_1 be denoted by (B), then, the most general surface (ν), of the same order, put through (B), the multiple points of C_1, and through the curve C_2, will give, as its remaining intersection with C_1, the most general set (N) of the linear series determined by the set (D) on C_1. And, the equations of these surfaces being $(\delta)=0$, $(\nu)=0$, where (δ), (ν) are homogeneous polynomials in the coordinates, the most general rational function, on the curve C_1, which has the set (D) as simple poles, will be given by the fraction $(\nu)/(\delta)$, where (ν) will involve a number, given by the Riemann-Roch theorem, of linearly entering arbitrary coefficients.

We also have the geometrical theorem that, if two particular surfaces $(\nu)=0$, $(\delta)=0$, of the same order, through the multiple points of C_1 and through the curve C_2, have, for simple inter- sections with C_1, respectively the sets (D), (B) and (N), (B), of which (B) is a common part; and, if any other surface (δ') be drawn, containing the curve C_2 and the multiple points of C_1, through the set (N), and if this meet the curve C_1 further in a set of points (B'), then there is a surface (ν'), through C_2 and the multiple points of C_1, whose further intersection with the curve C_1 consists of the sets (N) and (B'). This geometrical theorem is equivalent to saying that, on the curve C_1, the rational function $(\nu)/(\delta)$ is expressible also in the form $(\nu')/(\delta')$. Hence the polynomial $(\nu)(\delta')-(\nu')(\delta)$ vanishes on the curve C_1, that is, as a consequence of the equations $u=0$, $U=0$ of the two surfaces which contain C_1. By hypothesis, all the four polynomials (ν), ... vanish on the curve C_2.

We now apply these functional ideas to form an algebraic integral, on the curve C_1, which is everywhere finite thereon. This gives an interpretation of the formula we have found for the genus of this curve.

Let $\lambda=a_1x+b_1y+c_1z+d_1t$, $\mu=a_2x+...+d_2t$ be two arbitrary linear functions, and (u, U, λ, μ), as before, be the Jacobian of these and of the two surfaces $u=0$, $U=0$ which meet in the curves C_1 and C_2. Also let $\Psi=0$ be the equation of any surface of order $m+M-4$, drawn through the multiple points of the curves C_1 and C_2, and through the intersections of the component curves of C_2 with C_1 and with one another. We have proved that the aggregate of these surfaces cuts the canonical series on C_1, and on each component of C_2. Now consider the integral

$$\int \frac{\Psi(\lambda d\mu - \mu d\lambda)}{(u,\ U,\ \lambda,\ \mu)},$$

whose definition, as we have given it, is symmetrical in regard to the (say $s+1$) curves forming the complete intersection of $u=0$, $U=0$. We consider this integral on C_1.

As (u, U, λ, μ) is of order $m+M-2$, the integral is functional, that is, homogeneous of zero order in the current coordinates x, y, z, t. The form (u, U, λ, μ), we have seen, vanishes, at a node of the curve C_1, once on each branch, and three times at a cusp;

moreover, it vanishes at all points of the curve whereat the tangent line meets the line (λ, μ). Using θ as a parameter for the expression of the coordinates on the curve, we shall shew that $\lambda \, d\mu/d\theta - \mu \, d\lambda/d\theta$ vanishes once at each cusp, and also vanishes at the points of the curve whereat the tangent meets the line (λ, μ). The polynomial Ψ vanishes once on each branch of the curve C_1 at a node, and twice at each cusp. This shews that the integral is everywhere finite on the curve.

The zeros of (u, U, λ, μ) at non-singular points of the curve C_1 are what are known as the Jacobian points of the linear series determined on the curve by variable planes through the line (λ, μ); and what we have stated is thus in agreement with the known fact that the Jacobian set, of a linear series of freedom 1, on a curve, is equivalent (coresidual) with a set of the canonical series together with two sets of the linear series.

The verification that $\lambda \, d\mu/d\theta - \mu \, d\lambda/d\theta$ vanishes at the cusps of the curve, and also at the points whereat the tangent line meets the line (λ, μ) is simple. This function is a sum of terms such as $(a_1 b_2 - a_2 b_1)(x \, dy/d\theta - y \, dx/d\theta)$, $(c_1 d_2 - c_2 d_1)(z \, dt/d\theta - t \, dz/d\theta)$; and it can be proved at once that the integral is unaltered in form by a linear transformation of coordinates. We can thus suppose $t = 1$, and take, for the neighbourhood of a cusp, $x = \theta^2$, $y = b\theta^3 + \ldots$, $z = c\theta^4 + \ldots$; then the binary determinants in

$$\left\| \begin{array}{cccc} x & , & y & , & z & , & t \\ dx/d\theta, & dy/d\theta, & dz/d\theta, & dt/d\theta \end{array} \right\|,$$

or

$$\left\| \begin{array}{cccc} \theta^2, & b\theta^3 + \ldots , & c\theta^4 + \ldots , & 1 \\ 2\theta, & 3b\theta^2 + \ldots, & 4c\theta^3 + \ldots, & 0 \end{array} \right\|,$$

all vanish for $\theta = 0$.

Further, the integral is independent of the particular line (λ, μ) chosen. For, in virtue of the equations $u = 0$, $U = 0$, we have, at any point of the curve, the five equalities

$$\frac{x \, dy/d\theta - y \, dx/d\theta}{u_3 U_4 - u_4 U_3} = \ldots = \frac{z \, dt/d\theta - t \, dz/d\theta}{u_1 U_2 - u_2 U_1},$$

where $u_1 = \partial u/\partial x$, etc. Thus the integral may be replaced by

$$\int \frac{\Psi(x \, dy/d\theta - y \, dx/d\theta)}{u_3 U_4 - u_4 U_3} \, d\theta,$$

and, at a point of the curve whereat the tangent line meets the line $x = 0$, $y = 0$, the expression $x \, dy/d\theta - y \, dx/d\theta$ vanishes.

By the argument, this same integral gives the everywhere finite integrals on every component curve of the intersection of $u = 0$, $U = 0$. The method, moreover, as we shall illustrate below, is applicable to curves in space of more than three dimensions.

Part VI. The greatest genus possible for a curve of given order, in space of any number of dimensions. In the previous discussion of the linear series on a curve, the curve has been defined by the surfaces which pass through it. We may however seek to formulate results for a curve of which only the order and genus are given; and, if the curve be in space of dimension greater than 3, this is often more convenient. Castelnuovo has obtained some remarkable theorems from this point of view, which we proceed to state in part, with proof of some of these, referring the reader to the original paper for further detail (*Rend. Palermo*, VII, 1893: "Sui multipli di una serie lineare...una curva algebrica").

We consider an algebraical curve of order n, and genus p, lying in a space of r (and not fewer) dimensions; in particular, r may be 3. We denote by σ the greatest integer contained in the fraction $(n-2)/(r-1)$, namely

$$\sigma = E\left(\frac{n-2}{r-1}\right), \quad = \frac{n-2-\epsilon}{r-1}, \quad \epsilon = 0, 1, \ldots, (r-2);$$

then we define π by

$$\pi = \sigma[n - r - \tfrac{1}{2}(\sigma - 1)(r - 1)], \quad = \frac{1}{2}\left(\frac{n-2-\epsilon}{r-1}\right)(n - r + 1 + \epsilon),$$

where ϵ is such that $(n - 2 - \epsilon)/(r - 1)$ is an integer. One of the principal results obtained is that $p \leqslant \pi$.

Considering the linear series cut on the curve by unconditioned general primals of order m, it is proved that, for $m \geqslant \sigma$, this series is not special (namely its sets are not sets of $2p - 2$ points belonging to the canonical series, nor contained in such sets). For $m < \sigma$, no statement is made; it may be that a special series is not obtainable by unconditioned primals of any order. Thus, for $m \geqslant \sigma$, if the series be complete, its freedom will be $mn - p$. When the series is not complete, the incompleteness does not increase with m; but, after m has reached a certain value (depending on the multiple points which the curve may have), the incompleteness remains constant as m increases; in fact, for $m > \sigma + \pi - p$, the freedom is $mn - p - d$, where d does not depend on m. This number d is such that $\Sigma(s_i - 1) \leqslant d \leqslant \pi - p$, where s_i is the multiplicity of a general multiple point of the curve.

A corollary is that, when p reaches its extreme, or $p = \pi$, the curve is without multiple points, and primals of order σ, or more, cut a complete non-special series upon the curve.

The extreme π is reached by p for curves of three kinds:

(1) When $n = r + p < 2r$, in which case $\sigma = 1$. Then complete non-special series are cut upon the curve by primes of the space (and by primals of all orders).

(2) When $n = 2r = 2p - 2$, in which case $\sigma = 2$, and the curve is

canonical (see Chap. IV, p. 81). Then quadrics, and primals of higher order, give complete non-special series on the curve; but the primes of the space give complete special series.

(3) The curves without multiple points for which $n > 2r$, which lie on a rational ruled* surface of order $r-1$, and cut every generator of this in $\sigma_1 + 1$ points, where σ_1 is the greatest integer in $(n-1)/(r-1)$; thus $\sigma_1 = \sigma$ unless $(n-1)/(r-1)$ is an integer, but then $\sigma_1 = \sigma + 1$.

For curves in space of three dimensions, the value of π is $\frac{1}{4}(n-2)^2$ or $\frac{1}{4}(n-1)(n-3)$, according as n is even or odd; on such a curve, surfaces of order one less than the greatest integer in $\frac{1}{2}n$ give a non-special series, not necessarily complete; but surfaces of order $\geqslant E(\frac{1}{2}n) + \pi - p$ give a series whose incompleteness, which does not exceed $\pi - p$, is independent of the order of the surface.

Castelnuovo also proves that, for curves without multiple points, in space of any dimensions, the primals of order $n-2$, or more, cut a complete non-special series on the curve; and further, if there be multiple points of which a general multiplicity is s_i, primals (of order $\geqslant n-2$) passing $s_i - 1$ times through every such multiple point, have the same property. When the curve is not rational this remains true with $n-3$ instead of $n-2$. The proof is by projection on to a plane; and the condition for the primals determining the series, in more precise terms, is that they must meet each branch of the given curve, at a multiple point, as does an adjoint curve of the plane curve. For primals of order $k(\geqslant n-2)$, so passing through the multiple points, the freedom of the series which they determine on the curve is therefore $nk - \Sigma s_i(s_i - 1) - p$. Wherefore, the freedom of the series cut on the curve, by primals of order k not passing through the multiple points, is $nk - \Sigma s_i(s_i - 1) - p + \nu_k$, where ν_k is the number of independent conditions for the primals to pass through the multiple points.

For instance, consider a curve in ordinary space which has triple points; and first suppose that the three tangents at a particular triple point are not coplanar. For a surface to meet each of the three branches of the curve at this point in two points, it is necessary for the surface to have a double point, which is 4 conditions for the surface; the contribution to the number $\Sigma s_i(s_i - 1) - \nu_k$ from such a triple point is therefore 2. When however we consider a triple point of which the three tangents are in a plane, two intersections of the surface with each branch are secured by supposing the surface to touch the plane, which is only 3 conditions; and the contribution to $\Sigma s_i(s_i - 1) - \nu_k$ is then 3. Thus, the freedom of the series on the

---

* It was proved by Segre that rational ruled surfaces of order $r-1$ necessarily lie in space of r or lower dimensions, but in the latter case are obtainable by projection from such surfaces in space of r dimensions. Such surfaces are therefore said to be normal in a space $[r]$.

curve, cut by unrestricted surfaces of order k, is $nk - p - 2t_1 - 3t_2$, where t_1 is the number of general triple points, and t_2 of coplanar triple points; and the number of surfaces of order k containing the curve is $(k+3, 3) - (kn - p - 2t_1 - 3t_2 + 1)$. This number, in the form $(k+3, 3) - \nu_k - (kn - \mu - p + 1)$, where $\mu = \Sigma s_i(s_i - 1)$, we may also reach by considering surfaces through the multiple points, in number $(k+3, 3) - \nu_k$, which have μ intersections in all at these multiple points,

Consider now the proof of the extreme π which has been stated for the genus of a curve of order n. A lemma derived from the theory of special series is required: If a linear series, $g_R{}^N$, of sets of N points, of freedom R, not necessarily a complete series, be *special*, and we take a particular set of another series $g_r{}^n$, in which $n \leqslant R$, then a set of $g_R{}^N$ which contains $n - r$ of the points of the chosen set of $g_r{}^n$, contains all the n points of this set; in other words, a set of $g_r{}^n$ furnishes at most $n - r$ conditions for a set of $g_R{}^N$ which is to contain it. For, first, from $n \leqslant R$, there are sets of $g_R{}^N$ containing any given set of $g_r{}^n$, so that, as $g_R{}^N$ is a special series, so is $g_r{}^n$. If this latter be contained in the complete special series $g_\rho{}^n$, we have $\rho = n - (p - i)$, where i is the specialness of a set of $g_\rho{}^n$, or $g_r{}^n$, namely the number of unrestricted canonical sets (of $2p - 2$ points) of which this set of $g_r{}^n$ forms part; as there are p independent canonical sets, the set of $g_r{}^n$ furnishes $p - i$, or $n - \rho$, conditions for a canonical set which is to contain it; if $r = \rho - \delta$, the number $n - \rho$ is $n - r - \delta$, which is at most $n - r$. Now the sets of $g_R{}^N$, which is a special series, are presumably more restricted than the sets of the complete canonical series; and the necessary conditions for the more restricted sets to contain a set of $g_r{}^n$ are evidently not more, but may well be less, than for the general canonical sets. Thus, the number of conditions for a set of the special series $g_R{}^N$ to contain a particular set of $g_r{}^n$ is at most $n - r$. This is the lemma in question. The argument may be stated in terms of the canonical curve, in space $[p-1]$, as in Chap. IV preceding: The linear space which determines a set of $g_R{}^N$, contains the linear space which determines a set of $g_r{}^n$, if it contains points of the latter space sufficient to determine this.

Consider now the sets of n points, of the curve under consideration, determined by primes of the space $[r]$ in which the curve lies; these are sets of a linear series $g_r{}^n$. And consider the linear series, of sets of kn points, cut upon the curve by unrestricted primals of order k; these sets belong to a linear series $g_R{}^{kn}$. Let ν_k be the number, taken as small as possible, of points of a prime section of the series $g_r{}^n$, such that any set, of the series cut on the curve by the primals of order k, which contains any chosen ν_k points of a prime section, contains all the n points of this section. It is possible then to find $\nu_k - 1$ points, of the prime set, such that a primal of order k containing these, does not necessarily contain all the n points of this set. This number ν_k is such that $\nu_k \geqslant k(r-1)+1$, if the number on the right is less than n, namely if $k < (n-1)/(r-1)$. For, if $k(r-1)+1 < n$, we can choose, from the n points of a prime set, a number k, of sets of each $r-1$ points, still leaving at least 2 points of the set unchosen; and the k primes, each determined by one of these sets of $r-1$ points, form together a degenerate primal of order k which does not contain the whole set of n points, in general. Thus, at least $k(r-1)+1$ points of the prime set are necessary in order that a primal, of order k, passing through them, should entirely contain the set. Next, let r_k be the freedom of the series cut on the curve by primals of order k (not necessarily a complete linear series). Then the series cut on the curve by primals of order k which contain the points of a given prime section, will have freedom $r_k - \nu_k$; such primals may break up into the prime containing the base points, taken with a general primal of order $k-1$, or may have greater freedom than in that case; hence $r_k - \nu_k \geqslant r_{k-1}$, or $r_k - r_{k-1} \geqslant \nu_k$, which, if $k < (n-1)/(r-1)$, involves, we have seen, $r_k - r_{k-1} \geqslant k(r-1)+1$. We may put down this inequality for constantly diminishing values of k, until we come to

$$r_2 - r_1 \geqslant 2(r-1)+1, \text{ and } r_1, = r, = (r-1)+1;$$

hence, by addition,

$$r_k \geqslant \tfrac{1}{2}(r-1)k(k+1)+k, \quad k < (n-1)/(r-1).$$

The greatest integer less than $(n-1)/(r-1)$ is evidently

$$\sigma = E\left(\frac{n-2}{r-1}\right), \quad = \frac{n-2-\epsilon}{r-1}, \quad \epsilon = 0, 1, \dots, (r-2),$$

and as this is $\geqslant (n-r)/(r-1)$, we have $\nu_\sigma \geqslant \sigma(r-1)+1$, $\geqslant n-r+1$. This number is $> n-r$; hence, recurring to the lemma with which we began, we infer that the sets of the series $g_{r_\sigma}{}^{\sigma n}$ are not special. Wherefore $r_\sigma = \sigma n - p - d$, where d is the incompleteness of this series, which is $\geqslant 0$. This shews then that $p \leqslant \sigma n - r_\sigma$. We found however that $r_\sigma \geqslant \tfrac{1}{2}(r-1)\sigma(\sigma+1)+\sigma$. Therefore $p \leqslant \pi$, where

$$\pi = \sigma n - [\tfrac{1}{2}(r-1)\sigma(\sigma+1)+\sigma], \quad = \sigma[n-r-\tfrac{1}{2}(r-1)(\sigma-1)].$$

This is the result we desired to prove. For the other results we have quoted we refer to Castelnuovo's paper.

Ex. 1. The formula given above in Ex. 12, p. 216, leads, for a curve of order n, on a rational ruled surface of order $r-1$, which cuts each generator in k points, to the genus $\pi = (k-1)[n-r-\tfrac{1}{2}(r-1)(k-2)]$; this has the same value when k is replaced by h, given by

$$h+k-1 = 2(n-1)/(r-1).$$

Ex. 2. For a curve ${}^p c^{r+p}[r]$, of order n, equal to $r+p$, of genus p, lying in space of r dimensions, prove that, if $p < \tfrac{1}{2}r(r-1)$, the curve lies on

$\frac{1}{2}r(r-1)-p$ quadrics at least. And if the curve lies on exactly $\frac{1}{2}r(r-1)-p$ quadrics, then a complete series is determined on the curve by the quadrics of the space $[r]$.

On the sextic curve of genus 3 in space of three dimensions, obtained as the remaining intersection of a quadric surface with a quartic surface through 2 skew generators, the planes of the space, as also the cubic surfaces, cut a complete series; but quadric surfaces cut a series of incompleteness 1.

Ex. 3. For the canonical curve, of order $2p-2$ and genus p in space of $p-1$ dimensions, say the curve $^pc^{2p-2}[p-1]$, we have σ, equal to the greatest integer in $(2p-4)/(p-2)$, equal to 2, and

$$\pi = \sigma[n-r-\tfrac{1}{2}(r-1)(\sigma-1)] = 2[p-1-\tfrac{1}{2}(p-2)] = p.$$

On this curve quadrics cut a non-special complete series, of freedom $2(2p-2)-p$, or $3p-4$. The number of quadrics containing the curve is therefore $\frac{1}{2}p(p+1)-(3p-3)$, or $\frac{1}{2}(p-2)(p-3)$. The number of primals of order μ containing the curve is $(\mu>1)$, $(p-1+\mu,\ \mu)-(2\mu-1)(p-1)$ (Noether, *Math. Annal.* XVII, 1880, p. 263, and *Math. Annal.* XXVI, 1886, p. 143. Expounded in the writer's *Abel's Theorem*, 1897, p. 154).

The curve is the complete intersection of the $\frac{1}{2}(p-2)(p-3)$ quadrics, for general values of p, unless the curve contain a special linear series $g_1{}^3$; but for $p=6$, if the curve contain a series $g_2{}^5$, there is also a residual intersection. See pp. 96, 178, above (Enriques-Chisini, III, p. 106, or *Rend....Bologna*, XXIII, 1919). When the curve has a series $g_1{}^3$, the quadrics have, as residual intersection, a rational ruled surface of order $p-2$; when (for $p=6$) there is a $g_2{}^5$, all the quadrics contain a Veronese surface in addition to the curve.

Ex. 4. For quintic curves in ordinary space, $n=5$, $r=3$, the integer $E[(n-2)/(r-1)]$ is 1, and unconditioned surfaces of every order cut a non-special series upon the curve.

Ex. 5. In space $[4]$, the value of the extreme genus π, for curves of order n, is $\frac{1}{6}(n-2)(n-3)$, or $\frac{1}{6}(n-1)(n-4)$, according to the form of n (*mod.* 3). For example, when $n=6$, the greatest possible genus is 2. But in space $[5]$, for $n=6$, the greatest genus is 1.

Ex. 6. In our derivation of the canonical series upon a curve C_1, lying on the surfaces $u=0$, $U=0$, by means of surfaces, Ψ, of order $m+M-4$, passing through the residual intersection C_2 of $u=0$, $U=0$, or through the t intersections of C_2 with C_1, we have not imposed the natural condition that $u=0$, $U=0$ should be the surfaces of least order containing the curve; nor have we explicitly supposed that the surfaces Ψ are irreducible.

Consider, for instance, the sextic curve of genus 4 which is the complete intersection of a quadric surface $u=0$, and a cubic surface $U=0$. There are 5 linearly independent cubic surfaces through this curve, with equation of the form $(ax+by+cz+dt)u+U=0$. Take two of these $u'=0$, $U'=0$, where $u'=xu+U$, $U'=yu+U$. These meet further in a plane elliptic cubic curve, C_2, with equations $x-y=0$, $U=0$, which meets the sextic curve, C_1, in six points lying on the conic $x-y=0$, $u=0$. The surfaces of order $3+3-4$, that is the quadric surfaces through the curve C_2, break up into the plane $x-y=0$ and an arbitrary plane determining the canonical series on C_1. The quadric surfaces through the six points (C_1, C_2) are given by an equation $\Psi'=0$, where, with arbitrary parameters p, q, r, s, k, $\Psi'=(x-y)(px+qy+rz+st)+ku$, and the surfaces $\Psi'=0$ likewise determine the canonical series on the curve C_1. Putting $\Psi'=(x-y)\Psi+ku$, we can easily verify that, on the curve C_1, we have

$\Psi'/(u', U', \lambda, \mu) = \Psi/(u, U, \lambda, \mu)$, so that the everywhere finite integrals on C_1 can be determined by use of $u' = 0$, $U' = 0$.

Ex. 7. The everywhere finite integrals on the curve $^5c^8[4]$, of order 8 and genus 5, which is the complete intersection of three quadrics in space of four dimensions, say $u = 0$, $U = 0$, $V = 0$, are given by

$$\int \Psi (\lambda d\mu - \mu d\lambda)/(u, U, V, \lambda, \mu),$$

where λ, μ are arbitrary linear forms, and the denominator is the Jacobian determinant of 5 rows and columns.

Ex. 8. Consider the rational ruled cubic surface in space of four dimensions, which is obtainable by joining corresponding points of a conic and a line not meeting the conic. The curve which is the intersection of this surface with a general quadric is a sextic, of genus 2, meeting any conic of the ruled surface in 4 points. Thus primes through this conic meet the curve in the pairs of points of the canonical series on this curve $^2c^6$; these pairs are therefore the intersections of the curve with the generators. If a general quadric be drawn through one of these pairs, to meet the curve in 10 other points, the general quadric through these 10 points will also determine the canonical series on the curve. The curve in fact lies on 4 quadrics, and the 10 points furnish $15 - 4 - 2$, or 9, independent conditions for quadrics through them.

Now consider the canonical curve of order 10, and genus 6, in space of 5 dimensions, say $^6c^{10}[5]$. It is the intersection of a general quadric with a Del Pezzo surface $^2\psi^5$ (Bath, *Proc. Camb. Phil. Soc.* xxiv, 1928, p. 208). It can be shewn that three general quadrics (of the 5, linearly independent, which contain the surface), drawn through the Del Pezzo surface, meet again in a surface lying wholly in a space [4], namely in a ruled cubic surface therein; we have only to consider the section of the figure by a prime, or linear [4] space. Hence, four quadrics through the $^6c^{10}[5]$ meet again in a $^2c^6[4]$, meeting the former curve in 10 coprimal points. And, the canonical series on the $^6c^{10}[5]$, which (as it is a canonical curve) is given by the primes of the space [5], may also be given by the primals of order $(2 + 2 + 2 + 2 - 6)$, that is, by the quadrics, passing through these 10 intersections. The $^6c^{10}[5]$ lies on 6 quadrics, and the 10 points are equivalent only to $21 - 6 - 6$, or 9, conditions for quadrics passing through them.

If $\Psi = 0$ be the general quadric through these 10 points, the everywhere finite integrals on the $^6c^{10}[5]$ are given by

$$\int \Psi (\lambda d\mu - \mu d\lambda)/(u, U, V, W, \lambda, \mu),$$

where the denominator denotes the Jacobian of two arbitrary primes, λ, μ, and the four quadrics from which the 10 base points of Ψ are determined. This integral also gives the everywhere finite integrals on the residual curve $^2c^6[4]$.

Ex. 9. A general theorem (Severi, *Rend. Palermo*, xvii, 1903, § 3) is that, if, through an irreducible curve C_1, in space $[r]$, of order n_1 and genus p_1, without multiple points, there be drawn $(r-1)$ primals, of orders $m_1, m_2, \ldots, m_{r-1}$, meeting again in an irreducible non-singular curve C_2, then the canonical series $g_{p_1-1}{}^{2p_1-2}$ on C_1 is determined by primals through C_2, of order $k - 2$, where k is the sum of the $r - 1$ numbers $m_i - 1$; the whole number of linearly independent such primals is thus the sum of p_1 and the number of such primals which also contain C_1. Thus primals of order $> k - 2$, through C_2, cut a non-special series on C_1.

A slightly more general theorem would however seem to be true; namely, that primals of order $k - 2$, through the points of intersection of the curves C_1 and C_2, determine the canonical series on both C_1 and C_2;

and, further, if $\Psi = 0$ be the general primal of this character, while $u_1 = 0, \ldots, u_{r-1} = 0$ are the primals whose complete intersection consists of the curves C_1 and C_2, then the everywhere finite integrals of both curves are given by the formula $\int\Psi(\lambda d\mu - \mu d\lambda)/(u_1, u_2, \ldots, u_{r-1}, \lambda, \mu)$, where the notation is as before.

As in the case of ordinary space, the t intersections of C_1 and C_2 are equivalent only to $t - 1$ conditions for the primals Ψ, of order $k - 2$, required to pass through them. This follows from the fact that the postulation, $\chi(k-2)$, of the complete intersection $(C_1 + C_2)$ of $u_1 = 0, \ldots, u_{r-1} = 0$, for primals of order $k - 2$, is such that $\chi(k-2) = p_1 + p_2 + t - 1$, where p_2 is the genus of C_2. If n_2 be the order of C_2, we have, in fact,

$$t = n_1 k - (2n_1 + 2p_1 - 2) = n_2 k - (2n_2 + 2p_2 - 2),$$

and

$$\chi(k-2) = 1 + \tfrac{1}{2}m_1 \ldots m_{r-1}(k-2).$$

The number $\chi(k-2)$ is the coefficient of x^{k-2} in the ascending expansion of $(1 - x^{m_1})\ldots(1 - x^{m_{r-1}})(1 - x)^{-r-1}$. More generally, the postulation of the complete intersection of h primals, of orders $m_1, \ldots, m_h$, this intersection being supposed to be without multiple parts, for primals of order l, is the coefficient of x^l in the ascending expansion of

$$(1 - x^{m_1})\ldots(1 - x^{m_h})(1 - x)^{-r-1}.$$

It appears probable that the formula given above for the everywhere finite integrals remains valid, as in ordinary space, when the curve C_2 is reducible, provided $\Psi = 0$ be put through the mutual intersections of its components, as well as through their intersections with C_1.

Ex. 10. In particular, for the canonical curve $^p c^{2p-2}[p-1]$, it appears that the everywhere finite integrals are given by

$$\int\Psi(\lambda d\mu - \mu d\lambda)/(u_1, \ldots, u_{p-2}, \lambda, \mu),$$

where $u_1 = 0, \ldots, u_{p-2} = 0$ are $p - 2$ of the $\tfrac{1}{2}(p-2)(p-3)$ quadrics containing the curve $(p > 4)$, and $\Psi = 0$ is a primal of order $p - 4$ passing through the $2(p-1)(p-5)$ intersections of the canonical curve with the residual intersection of $u_1 = 0, \ldots, u_{p-2} = 0$. In fact, by a result quoted above (Ex. 3) there are $(2p-9)(p-1)$ primals of order $p - 4$ which do not contain the curve. (Cf. Klein, *Math. Annal.* xxxvi, 1889, § 3.)

The theorem assumes that the canonical curve can be defined as the complete intersection of quadrics (cf. Ex. 3).

Ex. 11. When, for $p = 6$, there exists, on the canonical curve, a g_2^5, this curve is the intersection of a Veronese surface with a cubic primal drawn through one of the conics of the surface. Obtain an expression for the everywhere finite integrals in this case.

Ex. 12. Prove Noether's theorem that, on any surface in ordinary space, the greatest genus, for curves of given order, arises for curves which form a complete section of the surface when taken with the whole, or part, of a plane section of the surface (Noether, "Grundzüge, u.s.w.", *Berlin. Abh.* 1882, § 6).

Ex. 13. The order of an irreducible algebraic curve which lies in a space of r dimensions (and not in space of lower dimensions) must be greater than $r - 1$. If it have the order r, it is a rational curve, without multiple points. If it have the order $r + 1$, it is elliptic or rational; in the latter case, the series cut upon the curve by primes of the space is incomplete, being contained in a series given by primals with equation of the form $u(\lambda_0 x_0 + \lambda_1 x_1 + \ldots + \lambda_r x_r) + U = 0$, where $x_0, x_1, \ldots, x_r$ are the coordinates, u and U are polynomials in these, and $\lambda_0, \ldots, \lambda_r$ are variable parameters. The curve is then obtainable by projection from a curve of order $r+1$ in space $[r+1]$, for which the coordinates of a point are given by

$$\frac{\chi_0}{ux_0} = \frac{\chi_1}{ux_1} = \dots = \frac{\chi_r}{ux_r} = \frac{\chi_{r+1}}{U}.$$

A curve of order n, in any space, which is not a projection of a curve of order n in higher space, is said to be *normal* in its own space.

Prove that a plane curve, of order n, with double points, is normal in the plane when the number of its double points is less than $n-2$, or its genus exceeds $\frac{1}{2}(n-2)(n-3)$; but is not normal when the number of its double points exceeds $\frac{1}{2}(n-2)(n-3)$, or its genus is less than $n-2$. Consider the case when the genus is $n-2$.

Ex. 14. For an irreducible algebraic manifold of dimension k, lying in space $[r]$, but not in lower space, we may speak of $r-k$ as the *complementary dimension of the manifold*. The order, n, of the manifold is the number of its intersections with a linear space $[r-k]$. It can be shewn that the order of the manifold must exceed the complementary dimension of the manifold. For any n points lie in a space of dimension $< n$; and a space $[r-k]$ thus meets the manifold in points lying in a space of dimension ρ with $\rho < n$ and $\rho \leqslant r-k$. If we assume that there is a space for which $\rho = r-k$, we thus have $r-k < n$. Thus, in particular, the order of a surface must be at least $r-1$; it can be shewn that, when the order of the surface is $r-1$, the surface is a rational ruled surface (normal in the space), there being one exception to this, however, namely when $r=5$, and the surface is the Veronese surface. In general, a manifold of dimension k and order $r-k+1$ can be regarded as the aggregate of ∞^1 linear spaces each of dimension $k-1$, there being again an exception to this, namely when $k=r-3$ and the manifold is a cone projecting the Veronese surface from a linear space $[r-6]$ (or the Veronese surface itself). For these results, cf. Clifford, *Collected Papers*, p. 305; Del Pezzo, *Rend. Palermo*, I, 1887, p. 241 (proving that a surface of order r in space $[r]$ is rational); and Bertini, *Geom. d. iperspazi*, 1907, p. 193.

INDEX

The references are to the pages

CAMBRIDGE: PRINTED BY WALTER LEWIS, M.A., AT THE UNIVERSITY PRESS

Printed in the United States
By Bookmasters